Grundlehren der mathematischen Wissenschaften 272

A Series of Comprehensive Studies in Mathematics

Editors

M. Artin   S. S. Chern   J. M. Fröhlich   E. Heinz
H. Hironaka   F. Hirzebruch   L. Hörmander   S. Mac Lane
W. Magnus   C. C. Moore   J. K. Moser   M. Nagata
W. Schmidt   D. S. Scott   Ya. G. Sinai   J. Tits
B. L. van der Waerden   M. Waldschmidt   S. Watanabe

Managing Editors

M. Berger   B. Eckmann   S. R. S. Varadhan

Grundlehren der mathematischen Wissenschaften

A Series of Comprehensive Studies in Mathematics

A Selection

180. Landkof: Foundations of Modern Potential Theory
181. Lions/Magenes: Non-Homogeneous Boundary Value Problems and Applications I
182. Lions/Magenes: Non-Homogeneous Boundary Value Problems and Applications II
183. Lions/Magenes: Non-Homogeneous Boundary Value Problems and Applications III
184. Rosenblatt: Markov Processes, Structure and Asymptotic Behavior
185. Rubinowicz: Sommerfeldsche Polynommethode
186. Handbook for Automatic Computation. Vol. 2. Wilkinson/Reinsch: Linear Algebra
187. Siegel/Moser: Lectures on Celestial Mechanics
188. Warner: Harmonic Analysis on Semi-Simple Lie Groups I
189. Warner: Harmonic Analysis on Semi-Simple Lie Groups II
190. Faith: Algebra: Rings, Modules, and Categories I
191. Faith: Algebra II, Ring Theory
192. Mallcev: Algebraic Systems
193. Pólya/Szegö: Problems and Theorems in Analysis I
194. Igusa: Theta Functions
195. Berberian: Baer*-Rings
196. Athreya/Ney: Branching Processes
197. Benz: Vorlesungen über Geometric der Algebren
198. Gaal: Linear Analysis and Representation Theory
199. Nitsche: Vorlesungen über Minimalflächen
200. Dold: Lectures on Algebraic Topology
201. Beck: Continuous Flows in the Plane
202. Schmetterer: Introduction to Mathematical Statistics
203. Schoeneberg: Elliptic Modular Functions
204. Popov: Hyperstability of Control Systems
205. Nikollskii: Approximation of Functions of Several Variables and Imbedding Theorems
206. André: Homologie des Algébres Commutatives
207. Donoghue: Monotone Matrix Functions and Analytic Continuation
208. Lacey: The Isometric Theory of Classical Banach Spaces
209. Ringel: Map Color Theorem
210. Gihman/Skorohod: The Theory of Stochastic Processes I
211. Comfort/Negrepontis: The Theory of Ultrafilters
212. Switzer: Algebraic Topology—Homotopy and Homology
213. Shafarevich: Basic Algebraic Geometry
214. van der Waerden: Group Theory and Quantum Mechanics
215. Schaefer: Banach Lattices and Positive Operators
216. Pólya/Szegö: Problems and Theorems in Analysis II
217. Stenström: Rings of Quotients
218. Gihman/Skorohod: The Theory of Stochastic Process II
219. Duvant/Lions: Inequalities in Mechanics and Physics
220. Kirillov: Elements of the Theory of Representations
221. Mumford: Algebraic Geometry I: Complex Projective Varieties
222. Lang: Introduction to Modular Forms
223. Bergh/Löfström: Interpolation Spaces. An Introduction
224. Gilbarg/Trudinger: Elliptic Partial Differential Equations of Second Order

Continued after Index

P. D. T. A. Elliott

Arithmetic Functions and Integer Products

Springer-Verlag
New York Berlin Heidelberg Tokyo

P. D. T. A. Elliott
Department of Mathematics
University of Colorado
Boulder, CO 80309
U.S.A.

AMS Subject Classifications: 10HXX, 10KXX, 10LXX, 20KXX

Library of Congress Cataloging in Publication Data
Elliott, P. D. T. A. (Peter D. T. A.)
  Arithmetic functions and integer products.
  (Grundlehren der mathematischen Wissenschaften ; 272)
  Bibliography: p.
  Includes index.
  1. Arithmetic functions. 2. Numbers, Natural.
I. Title. II. Series.
QA246.E55   1984   512'.72   84-13911

With 2 Illustrations.

© 1985 by Springer-Verlag New York Inc.
All rights reserved. No part of this book may be translated or reproduced in any form without written permission from Springer-Verlag, 175 Fifth Avenue, New York, New York 10010, U.S.A.

Typeset by J. W. Arrowsmith Ltd., Bristol, England.
Printed and bound by R. R. Donnelley & Sons, Harrisonburg, Virginia.
Printed in the United States of America.

9 8 7 6 5 4 3 2 1

ISBN 0-387-96094-5 Springer-Verlag New York Berlin Heidelberg Tokyo
ISBN 3-540-96094-5 Springer-Verlag Berlin Heidelberg New York Tokyo

Preface

Every positive integer m has a product representation of the form

$$m^v = \prod_{i=1}^{k} \left(\frac{3n_i+1}{5n_i+2}\right)^{\varepsilon_i},$$

where v, k and the n_i are positive integers, and each $\varepsilon_i = \pm 1$. A value can be given for v which is uniform in the m. A representation can be computed so that no n_i exceeds a certain fixed power of $2m$, and the number k of terms needed does not exceed a fixed power of $\log 2m$.

Consider next the collection of finite probability spaces whose associated measures assume only rational values. Let $h(x)$ be a real-valued function which measures the information in an event, depending only upon the probability x with which that event occurs. Assuming $h(x)$ to be non-negative, and to satisfy certain standard properties, it must have the form

$$-A(x \log x + (1-x) \log(1-x)).$$

Except for a renormalization this is the well-known function of Shannon.

What do these results have in common? They both apply the theory of arithmetic functions.

The two widest classes of arithmetic functions are the real-valued additive and the complex-valued multiplicative functions. Beginning in the thirties of this century, the work of Erdős, Kac, Kubilius, Turán and others gave a discipline to the study of the general value distribution of arithmetic functions by the introduction of ideas, methods and results from the theory of Probability. I gave an account of the resulting extensive and still developing branch of Number Theory in volumes 239/240 of this series, under the title *Probabilistic Number Theory*.

In the present volume my aim is to introduce a new and complementary aesthetic into the study of arithmetic functions by systematically applying them to certain problems in algebra. Singled out for attention is the problem of representing a given integer as a product of rationals of a prescribed type. This is not only of number-theoretic interest, but also touches upon the theory of denumerably infinite abelian groups.

A detailed study is made of the behavior of real-valued additive functions on the sequences of the form $(an+b)/(An+B)$ with varying positive integers n. The methods employed, rooted in elementary functional analysis, represent the simplest first steps in a general distributional treatment of the ring-theoretic properties of the integers. They lead to results which are at present otherwise unobtainable.

Acknowledgments

I would like to thank the John Simon Guggenheim Memorial Foundation whose generous support in 1980 enabled me to complete the work upon which much of this book is founded.

I would like to thank the Mathematics Department of Imperial College, London, whose hospitality I enjoyed in the Spring of 1980.

I would like to thank the National Science Foundation for past support on contracts MCS78-04374 and MCS82-00610.

My thanks to friends and colleagues who sent me corrections for my earlier book on Probabilistic Number Theory. Particular thanks to Charles Ryavec, and to F. Spilker and the Mathematics Seminar at Freiburg, Germany.

My thanks to the referee whose careful reading of it enabled me to correct a number of oversights in the typescript of the present volume.

My thanks to Colleen D'Oremus and especially to Liz Stimmel, who expertly typed the manuscript to completion.

An especial thanks to my wife Jean who, for a second time, saw the piles of yellow pages appear.

Contents

Notation	xiii
Introduction	1
Duality and the Differences of Additive Functions	7

First Motive   19

CHAPTER 1
Variants of Well-Known Arithmetic Inequalities   23
    Multiplicative Functions   23
    Generalized Turán–Kubilius Inequalities   27
    Selberg's Sieve Method   33
    Kloosterman Sums   34

CHAPTER 2
A Diophantine Equation   37

CHAPTER 3
A First Upper Bound   53
    The First Inductive Proof   64
    The Second Inductive Proof   71
    Concluding Remarks   76

CHAPTER 4
Intermezzo: The Group Q^*/Γ 78

CHAPTER 5
Some Duality   81
    Duality in Finite Spaces   81
    Self-adjoint Maps   83
    Duality in Hilbert Space   92
    Duality in General   93

Second Motive 97

CHAPTER 6
Lemmas Involving Prime Numbers 101
    The Large Sieve and Prime Number Sums 101
    The Method of Vinogradov in Vaughan's Form 114
    Dirichlet L-Series 119

CHAPTER 7
Additive Functions on Arithmetic Progressions with Large Moduli 121
    Additive Functions on Arithmetic Progressions 121
    Algebraicanalytic Inequalities 149

CHAPTER 8
The Loop 155

Third Motive 177

CHAPTER 9
The Approximate Functional Equation 183

CHAPTER 10
Additive Arithmetic Functions on Differences 204
    The Basic Inequality 204
    The Decomposition of the Mean 232
    Concluding Remarks 239

CHAPTER 11
Some Historical Remarks 244

CHAPTER 12
From L^2 to L^∞ 250

CHAPTER 13
A Problem of Kátai 259

CHAPTER 14
Inequalities in L^∞ 264

CHAPTER 15
Integers as Products 277
    More Duality; Additive Functions as Characters 277
    Divisible Groups and Modules 278
    Sets of Uniqueness 281
    Algorithms 287

Contents   xi

CHAPTER 16
The Second Intermezzo   291

CHAPTER 17
Product Representations by Values of Rational Functions   297
    A Ring of Operators   297
    Practical Measures   307

CHAPTER 18
Simultaneous Product Representations by Values of
Rational Functions   309
    Linear Recurrences in Modules   310
    Elliptic Power Sums   318
    Concluding Remarks   328

CHAPTER 19
Simultaneous Product Representations with $a_i x + b_i$ 329

CHAPTER 20
Information and Arithmetic   343
    Transition to Arithmetic   346
    Information as an Algebraic Object   353

CHAPTER 21
Central Limit Theorem for Differences   356

CHAPTER 22
Density Theorems   372
    Groups of Bounded Order   379
    Measures on Dual Groups   380
    Arithmic Groups   392
    Concluding Remarks   393

CHAPTER 23
Problems   394
    Exercises   394
    Unsolved Problems   417

SUPPLEMENT
Progress in Probabilistic Number Theory   423
    Analogues of the Turán–Kubilius Inequality   435

References   449

Subject Index   459

Notation

I list here some of the more important symbols/definitions which occur in this book.

$\mathbb{Z}$	the ring of rational integers.
Q	the field of rational numbers.
Q^*	the multiplicative group of the positive rationals.
$\mathbb{R}$	the real numbers.
$\mathbb{C}$	the complex numbers.
$s = \sigma + i\tau$	complex variable.
n	will generally denote a positive (natural) integer.
p	will generally denote a positive (natural) prime.
$[a, b]$	the least common multiple of the integers a and b. It also denotes the closed interval of real numbers x, $a \leq x \leq b$. In Chapter 5 and the Supplement it will denote an inner product on a vector space.
(a, b)	the highest common factor of the integers a and b. It also denotes the open interval of real numbers x, $a < x < b$.
$(a, b]$	the interval of real numbers x, $a < x \leq b$.
$\|y\|$	denotes the distance of the real number y from a nearest integer. In Chapter 5 and the Supplement it will denote the norm on a vector space induced by the inner product $[\ ,\ ]$.
$K[x]$	the ring of polynomials in the variable x with coefficients in the ring K. A typical such polynomial will be denoted by $F(x)$.
$K(x)$	for commutative rings K without divisors of zero, is the quotient field of $K[x]$, the field of rational functions in x with coefficients in K.
$v(\alpha)$	denotes a pre-valuation on a $\mathbb{Z}$-module; Chapter 18.
E	denotes the shift operator, as in Chapter 17; or the expectation with respect to a probability measure, as in Chapter 22.
An arithmetic function	is a function which is defined on the positive natural integers.

An additive function	which will generally be denoted by $f(n)$, is an arithmetic function which satisfies $f(ab) = f(a) + f(b)$ whenever a and b are coprime integers. Classically assumed to be real-valued.
A strongly additive function	is an additive function which also satisfies $f(p^m) = f(p)$ for every prime-power p^m, $m \geq 1$.
A completely additive function	satisfies $f(ab) = f(a) + f(b)$ for every pair of (positive) integers a and b.
A multiplicative function	which will generally be denoted by $g(n)$, is an arithmetic function which satisfies $g(ab) = g(a)g(b)$ whenever a and b are coprime integers. Classically assumed to be complex-valued.
A strongly multiplicative function	is a multiplicative function which also satisfies $g(p^m) = g(p)$ for every prime-power p^m, $m \geq 1$.
A completely multiplicative function	satisfies $g(ab) = g(a)g(b)$ for every pair of positive integers a and b.
$\pi(x)$	the number of primes not exceeding x.
$\Lambda(n)$	von Mangoldt's function; $= \log p$ if n is a power of a prime p, $= 0$ otherwise.
$\omega(n)$	the number of distinct prime divisors of the integer n.
$\Omega(n)$	the number of prime divisors of the integer n, counted with multiplicity.
$[x]$	the largest integer not exceeding x. Thus $[\tfrac{3}{2}] = 1$ and $[-\tfrac{3}{2}] = -2$.
$\tau(n)$ $d_2(n)$ $d(n)$	the number of positive integers which divide the integer n.
$p^k \| n$	the prime-power p^k divides n, p^{k+1} does not.
$\pi(x, D, l)$	the number of primes not exceeding x which satisfy $p \equiv l \pmod{D}$.
$\chi(n)$	denotes a Dirichlet character, as in lemma (19.3).
$\mu(n)$	Möbius' function; zero if n is not squarefree, $(-1)^{\omega(n)}$ if it is.
$A(x)$	if $A: a_1 < a_2 < \cdots$ is a sequence of positive integers, $A(x)$ will often be used to denote the number of these which do not exceed x.
Asymptotic density	a sequence of positive integers A is said to have an asymptotic density d if $$d = \lim_{x \to \infty} x^{-1} A(x)$$ exists.
$\underline{d}(A)$	lower asymptotic density. For a sequence of positive integers A $$\underline{d}(A) = \liminf_{x \to \infty} x^{-1} A(x).$$

Notation

$\bar{d}(A)$ upper asymptotic density. For a sequence of positive integers A
$$\bar{d}(A) = \limsup_{x \to \infty} x^{-1} A(x).$$

$\sigma(A)$ Schnirelmann density. For a sequence of positive integers A
$$\sigma(A) = \inf_{n \geq 1} n^{-1} A(n).$$

$\nu_x(n; \ldots)$ let $N_x(n; \ldots)$ denote the number of positive integers not exceeding x which have the property $\ldots$; then
$$\nu_x(n; \ldots) = \text{the frequency } [x]^{-1} N_x(n; \ldots).$$

For a random variable X, with associated distribution function $F(z)$,

$\bar{X}$, mean
$$= \int_{-\infty}^{\infty} z \, dF(z),$$

σ^2, D^2, variance
$$= \int_{-\infty}^{\infty} (z - \bar{X})^2 \, dF(z).$$

$\rho(F, G)$, Lévy metric   for any two distribution functions $F (=F(z))$, $G (=G(z))$, we define $\rho(F, G)$ to be the greatest lower bound of those numbers h which have the property that the inequality
$$G(z - h) - h \leq F(z) \leq G(z + h) + h$$
holds for all real values of z.

$\mathscr{F}(x)$ denotes the information in an event which occurs with probability x, Chapter 20.

$c_1, c_2, \ldots$ will denote constants. These are renumbered from the beginning of each chapter, and on occasion, when no confusion thereby arises, from the beginning of a new section.

$f(x) = O(g(x))$ for a range of x-values, means that there is a constant A so the inequality
$$|f(x)| \leq A g(x)$$
holds over the range.

$f(x) = o(g(x))$ as $x \to \infty$, means
$$\lim_{x \to \infty} \frac{f(x)}{g(x)} = 0.$$

$f(x) \sim g(x)$ as $x \to \infty$, means
$$\lim_{x \to \infty} \frac{f(x)}{g(x)} = 1.$$

These last two will only be used with functions $g(x)$ which do not vanish when x is sufficiently large. The same meaning is attached to these symbols when $x \to \infty$ is formally replaced by $x \to \alpha$, for any fixed α.

Introduction

A function is said to be *arithmetic* if it is defined on the positive integers.
Those arithmetic functions which are real-valued and satisfy

(1) $$f(ab) = f(a) + f(b)$$

for every pair of integers with $(a, b) = 1$ are called *additive*.

If an arithmetic function has complex values and satisfies

(2) $$g(ab) = g(a)g(b)$$

when $(a, b) = 1$, it is called *multiplicative*.

These are the two widest classes of arithmetic functions, and contain many well-known functions. Examples of additive functions are $\log n$, and $\omega(n)$ which counts the number of distinct prime divisors of n. The fixed powers n^α, Euler's function $\varphi(n)$, Dirichlet's divisor function, and Ramanujan's function $\tau(n)$ from modular arithmetic are all mutliplicative functions.

The positive rational numbers, with multiplication as law of combination, form a group. We shall denote it by Q^*. Since every positive rational can be uniquely expressed as a ratio of integers r/s with $(r, s) = 1$, we can extend the definition of an additive function to the whole of Q^* by

$$f(r/s) = f(r) - f(s).$$

Likewise, if a multiplicative function does not assume the value zero we may extend it by

$$g(r/s) = g(r)/g(s).$$

The requirement that relations (1) and (2) only hold when $(a, b) = 1$ reflects that Q^* is a free group with the prime numbers as generators.

A *completely additive* function satisfies the relation (1) for all choices of the integers a and b. When extended it becomes a group homomorphism of Q^* into the additive reals $(\mathbb{R}, +)$.

A non-vanishing *completely multiplicative* function, which satisfies (2) for all positive a, b, is a homomorphism of Q^* into the multiplicative group of non-zero complex numbers. Since each complex $z \neq 0$ has a polar representation $r\exp(2\pi i\theta)$, $r > 0$, the map $z \mapsto (\log r \oplus \theta)$ shows that this last group is isomorphic to the direct sum of the additive reals, and the additive reals modulo the integers. One may thus regard a multiplicative function as being a pair of additive functions, one with the traditional real values, and one with values in the group $(\mathbb{R}/\mathbb{Z}, +)$.

In this volume I view (with a slight abuse of meaning) additive and multiplicative arithmetic functions to be group homomorphisms, and treat them as characters on Q^*.

Given any group D the homomorphisms of Q^* into D themselves form a group, $\text{Hom}(Q^*, D)$. Beginning in Chapter 15 I systematically study such collections of arithmetic functions. By choosing various image groups D one may construct in this manner analogues of the dual group of a finite abelian group. In order to motivate choices for D, consider the following problem:

Let $a_1, a_2, \ldots$ be a sequence of positive integers, repetition being allowed. Which positive integers n have a representation

(3) $$n = a_{j_1}^{\varepsilon_1} \cdots a_{j_k}^{\varepsilon_k},$$

where each ε_i has a value ± 1?

There is a well-known conjecture that there are infinitely many pairs of primes p, $p+2$. This amounts to solving the equation

$$x - y + 2 = 0$$

in primes x and y. Similarly, if $n > 0$ is a fixed integer then there are probably infinitely many prime solutions x, y to the equation

$$nx - y + n - 1 = 0.$$

Each such solution would give a representation

$$n = \frac{p+1}{q+1}.$$

If we weaken the requirement to

$$n = \prod_{i=1}^{k} (p_i + 1)^{\varepsilon_i},$$

with primes p_i and each $\varepsilon_i = \pm 1$, then we have an example of (3).

Introduction   3

It would be interesting to obtain representations of the type (3) with k independent of n. If we allow k to vary with n then we can reformulate our question. Let Γ be the subgroup of Q^* which is generated by the a_j, and define the quotient group $G = Q^*/\Gamma$. Then every integer $n > 0$ has a representation of the form (3), with a value of k possibly depending upon n, if and only if G is trivial.

This suggests that the groups D in $\text{Hom}(Q^*, D)$ should have enough structure to distinguish between interesting subgroups of Q^*, yet be familiar and advantageous to work in. In order to localize the a_{j_i} it is convenient to require D to be injective.

In Chapter 15 I give an account of the necessary algebra, in terms of modules rather than groups. *Divisible* and *extra-divisible* modules are introduced as candidates for D. The general procedure suggested to obtain the triviality of G consists of three steps, proving in turn that

(4)
  (i) G is finitely generated,
  (ii) G is finite (or of finite order),
  (iii) G is trivial.

The first two of these steps are to be established by considering the ($\mathbb{Z}$-module) homomorphisms of G into the modules $Q/\mathbb{Z}$ and the additive reals, respectively. For the third step let p be a prime, and let pG denote the set of pth-powers of the elements of G. Then the finite field F_p of p elements acts upon the quotient group G/pG and we study the homomorphisms of G/pG into F_p, as F_p-modules. Such modules are in fact vector spaces.

In this way the existence of representations of the form (3) is reduced to the study of those arithmetic functions, derived from homomorphisms of Q^*, which vanish on the sequence of integers a_n. Since I feel addition to be more familiar than multiplication, I shall generally consider these arithmetic functions to be additive.

The characterization of those additive functions, with values in a suitable group, which vanish on a prescribed integer or (rational) sequence $a_1, a_2, \ldots$, is a large program. Moreover, it is desirable to give this program a quantitative aspect by, as far as possible, explicitly relating the value of a given $f(n)$ to the appropriate values amongst the $f(a_j)$.

In this generality such a program amounts to the study an arbitrary denumerable abelian group. One can arithmetize such a group G by regarding it as a homomorphic image of a free group F with generators g_j, $j = 1, 2, \ldots$, using relations $a_1 = e$, $a_2 = e$, $\ldots$. Here e denotes the identity of F, and each a_i will be a product of finitely many of the generators g_j and of the g_j^{-1}. Identifying g_i with the ith (positive) rational prime number now presents G in the form Q^*/Γ considered above.

It follows from the theories of logic and abelian groups that there can be no (general) algorithm for determining whether a given member of Q^*/Γ is trivial. However, we shall be interested in groups Γ given by integers a_n

which have recognizable arithmetic properties, usually of a *ring-theoretic* nature. By considering its homomorphisms into the real or complex numbers there is introduced the possibility of studying Q^*/Γ with methods from both algebra and analysis. This leads to problems which seem interesting and difficult.

To illustrate the difficulties involved let $P(x)$ be a polynomial in x, with rational integer coefficients, which assumes positive values for $x = 1, 2, \ldots$. Let $g(\ )$ be a (complex-valued) multiplicative function which satisfies $|g(n)| \leq 1$ for each integer $n > 0$. What is the necessary and sufficient condition, in terms of the values of $g(q)$ for prime-powers q, in order that

$$\lim_{x \to \infty} x^{-1} \sum_{n \leq x} g(P(n))$$

exist? Only in the linear case $P(x) = ax + b$ can we presently give a satisfactory answer to this question. As soon as the degree of $P(x)$ is at least as large as two, we are stuck. Thus we cannot presently prove that the Möbius function $\mu(n)$ satisfies

$$\lim_{x \to \infty} x^{-1} \sum_{n \leq x} \mu(n(n+1)) = 0,$$

although this seems very likely.

For a detailed discussion of the case $P(x) = x$, see my book, Elliott [11], $P(x) = ax + b$ may be dealt with by introducing Dirichlet characters mod $a(a, b)^{-1}$.

Let

$$R(x) = \prod_{i=1}^{k} (x + a_i)^{b_i}$$

be a rational function, all of whose zeros and poles are rational integers. For simplicity of exposition only I shall assume that these are not positive, so that every $a_i \geq 0$. I shall also assume that the multiplicities b_i, $i = 1, \ldots, k$ have highest common factor 1, so that $R(x)$ is not identically a power.

As an example in the application of the procedure (4) I prove in Chapter 17 that every positive integer n has a representation

$$n = \prod_{i=1}^{j} R(m_i)^{\varepsilon_i},$$

with $\varepsilon_i = \pm 1$, and where the positive integers m_i do not exceed $c_0 n$ for an effective constant, c_0. This bound on the m_i holds uniformly in n. Other size restrictions could also be put on the m_i, if desired.

To obtain this result I study arithmetic functions f, in particular with values in a $\mathbb{Z}$-module, which satisfy $f(R(n)) = 0$ for an interval of positive

Introduction

integers n. Since f is a homomorphism, we can write this relation in the form

$$\sum_{i=1}^{k} b_i f(n+a_i) = 0.$$

The method is algebraic, and employs a ring of polynomial operators.

The reduction of representation problems (3) to a study of arithmetic functions can equally be applied to the study of simultaneous representations. In Chapter 18 I prove that for any positive integers n_1, n_2 and t, there are infinitely many simultaneous representations

$$n_1 = \prod_{i=1}^{j} R(m_i)^{\varepsilon_i},$$

$$n_2 = \prod_{i=1}^{j} R(m_i + t)^{\varepsilon_i}.$$

Here I do not localize the m_i, although with some extra effort it could be done. Note that the rational functions $R(x)$ and $R(x+t)$ may well have roots/poles in common.

In this case we are reduced to the study of pairs of additive functions $f_1(\ )$, $f_2(\ )$ which satisfy

$$f_1(R(n)) + f_2(R(n+t)) = 0$$

for all sufficiently large integers n. The algebraic method of the previous chapter does not apply and new ideas are required. An investigation is made of linear recurrences in a $\mathbb{Z}$-module, that is to say of the orders of elements in an abelian group which satisfy certain relations. Besides this I apply an asymptotic estimate for elliptic power sums

$$\sum_{j=1}^{k} \alpha_j z_j^{n^2},$$

where the α_j, z_j are fixed, and $n \to \infty$. These arguments reduce the proof to a study of the differences

$$f(An-1) - f(n),$$

where $A \geq 1$ is fixed.

Let $a > 0$, b, $A > 0$, B be integers which satisfy $aB - Ab \neq 0$. In Chapter 19 I establish the existence of an integer $v > 0$ so that every pair of integers n_1, n_2 which satisfy $(n_1, a) = 1$, $(n_2, A) = 1$, respectively, has a (simultaneous) representation

(5)
$$n_1^v = \prod_{i=1}^{k} (am_i + b)^{\varepsilon_i},$$

$$n_2^v = \prod_{i=1}^{k} (Am_i + B)^{\varepsilon_i},$$

where, as usual, $\varepsilon_i = \pm 1$. If required a bound for the m_i in terms of n_1 and n_2 and the constants a, b, A, B could also be obtained. In particular, we are reduced to the study of the more general differences

(6) $$f_1(am+b) - f_2(Am+B)$$

for additive functions f_1 and f_2.

Congruence conditions show that the integer v which appears in the representation (5) cannot always be given the value 1. I am not presently able to specify its best value in every case. As an example of what is to be expected I prove in the same chapter that if $(n_1, 3) = 1 = (n_2, 3)$ then a representation

$$n_1 = \prod_{i=1}^{k} (3m_i - 17)^{\varepsilon_i},$$

$$n_2 = \prod_{i=1}^{k} (3m_i + 19)^{\varepsilon_i},$$

exists if and only if $n_1 \equiv n_2 \equiv 1$, 4 or 7 (mod 9). In particular, the smallest uniform value for v is in this case 6.

These later results depend upon a study of those arithmetic functions $f(\ )$ for which

$$f\left(\frac{an+b}{An+B}\right)$$

is prescribed.

The first ten chapters of the present volume are devoted to the exposition of a (new) method for quantitatively characterizing real-valued additive functions in terms of their differences

$$f(an+b) - f(An+B).$$

With this formulation the condition of coprimality at (1) may be taken into account.

In particular the method given enables us to carry out steps (i) and (ii) of the procedure (4) for the sequence $(an+b)/(An+B)$. In order to complete step (iii) in this case, and to determine theoretically the exponent v at (5), an analogue of this method is required for additive functions which take values in the finite fields of p elements. Here I have at present only partial results.

Duality and the Differences of Additive Functions

The notion of duality has occurred implicitly or explicitly in number theory many times. For a fixed complex s the map $n \mapsto n^{-s}$ may be regarded as a character on Q^*. A Dirichlet series is practically a Fourier transform with respect to a frequency measure on the integers. Dirichlet's celebrated proof of the infinitude of primes in an arithmetic progression exploys the mixed character $\chi(n)n^{-s}$, which for some positive integer m is defined essentially on the direct product of the multiplicative reduced residue classes (mod m) and the multiplicative group of rationals having no prime factor in common with m. It is therefore a kind of (double) Fourier analysis (see Davenport [1]).

More recently, H. Weyl's classic criterion [1] for a sequence of real numbers to be uniformly distributed (mod 1) is precisely a study of the appropriate Fourier (–Stieltjes) transform.

In both of these cases the original problem was studied in terms of an appropriate (dual) group of characters. One may enhance this approach by systematically applying it to finite spaces as well.

Let $\alpha > 1$, $\beta > 1$ be given. Let m and n be positive integers. We regard the space $\mathbb{C}^m$ of complex m-tuples as normed by

$$\|\mathbf{x}\|_m = \left(\sum_{i=1}^{m} |x_i|^\alpha \right)^{1/\alpha},$$

where as usual $x_1, \ldots, x_m$ denote the coordinates of the vector $\mathbf{x}$. Similarly we give a norm to $\mathbb{C}^n$ by

$$\|\mathbf{y}\|_n = \left(\sum_{j=1}^{n} |y_j|^\beta \right)^{1/\beta}.$$

If now c_{ij}, $i=1,\ldots,m$, $j=1,\ldots,n$, are complex numbers then we can define a linear map $T: \mathbb{C}^n \to \mathbb{C}^m$ by

$$\mathbf{x} = \mathbf{C}\mathbf{y},$$

where $\mathbf{C}$ denotes the matrix with typical entry c_{ij}. As is usual, one defines a norm for T by

$$\|T\| = \sup_{y \neq 0} \|Ty\|_m / \|y\|_n.$$

There is then a dual map T^*, between the space of linear functionals on $\mathbb{C}^m$ and the space of linear functionals on $\mathbb{C}^n$. This dual map is represented by the matrix (c_{ji}), $j = 1, \ldots, n$, $i = 1, \ldots, m$, and has a norm which satisfies

$$\|T^*\| = \|T\|.$$

This relation enables us to switch between T and its dual, as is convenient. Whilst this is a suggestive medium, it is not yet a method.

We give direction to the proceedings by adopting the philosophy

$$\text{operator} \to \text{sufficiency},$$

$$\text{dual of operator} \to \text{necessity}.$$

This diagram represents the (metamathematical) notion that there is a correspondence: If an operator (argument) can be applied to prove that certain conditions force a result to be true, then the dual of that operator (argument) can be used to prove that those same conditions are also necessary.

This procedure is exemplified in Projective Geometry, where the duality is between point and line. Thus the dual of Desargues' theorem is, in fact, already its converse. Generally, however, one expects to supplement the dual argument with further results. In fact, by studying the scheme

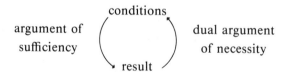

one may be able to determine precisely what form the conditions should take.

We can illustrate this procedure with an example taken from Probabilistic Number Theory.

For each $x \geq 1$ let

(7) $$\nu_x(n; \ldots)$$

denote the ratio of the number of positive integers $n \leq x$ which have the property $\ldots$, to the total number of positive integers not exceeding x.

We shall outline a proof of the theorem, due to Erdös [1], and Erdös and Wintner [1], that for a real-valued additive function $f(\ )$ the frequencies

$$\nu_x(n;f(n)\leq z) \qquad (x\to\infty)$$

converge weakly to some distribution function if and only if the three series

(8) $$\sum_{|f(p)|>1}\frac{1}{p}, \qquad \sum_{|f(p)|\leq 1}\frac{f^2(p)}{p}, \qquad \sum_{|f(p)|\leq 1}\frac{f(p)}{p},$$

taken over the prime numbers, converge.

To simplify the exposition we shall assume that $f(\ )$ is completely additive.

A form of the Turán–Kubilius inequality (see Elliott [11]) states that every real-valued completely-additive arithmetic function satisfies the inequality

(9) $$\sum_{n\leq x}|f(n)-E(x)|^2 \leq c_1 x D(x)^2, \qquad x\geq 1,$$

where c_1 is an absolute constant,

$$E(x)=\sum_{p\leq x}\frac{f(p)}{p}, \qquad D(x)=\left(\sum_{p\leq x}\frac{f(p)^2}{p}\right)^{1/2}\geq 0.$$

For a probabilist this reflects the fact that whether an integer is divisible by one prime is essentially independent from it being divisible by another. Here the approximate independence is implicitly with respect to the frequency measure (7). For an algebraist the Turán–Kubilius inequality expresses the freedom of the group Q^*.

Suppose now that the three series at (8) converge.
For $r\geq 1$ define

$$h_r(n)=\sum_{\substack{p^k\|n\\p\leq r}} kf(p),$$

where $p^k\|n$ denotes that p^k divides n but p^{k+1} does not. Then by the Turán–Kubilius inequality

$$\sum_{n\leq x}\left|f(n)-h_r(n)-\sum_{r<p\leq x}\frac{f(p)}{p}\right|^2 \leq c_1 \sum_{r<p\leq x}\frac{|f(p)|^2}{p}.$$

We deduce from our hypothesis (8) that for each fixed $\varepsilon>0$

$$\nu_x(n;|f(n)-h_r(n)|>\varepsilon)\leq \delta(\varepsilon,r), \qquad x\geq 2,$$

where $\delta(\varepsilon, r) \to 0$ as $r \to \infty$. This is an argument of Tchebyshev type from the theory of probability.

To study the limiting behavior of the frequencies

$$\nu_x(n; f(n) \leq z)$$

as $x \to \infty$ it is therefore enough to study the frequencies

$$\nu_x(n; h_r(n) \leq z)$$

since these frequencies are at a Lévy distance (see Notation) of $O(\varepsilon + \delta(\varepsilon, r))$ from each other.

But $h_r(n)$ involves only the finitely many primes not exceeding r, and a straightforward application of the sieve of Eratosthenes (or of the Chinese Remainder Theorem for congruences between integers) shows that

$$\text{(weak)} \lim_{x \to \infty} \nu_x(n; h_r(n) \leq z) = F_r(z)$$

exists; moreover, so does

$$F(z) = \lim_{r \to \infty} F_r(z).$$

This last assertion may be justified by again applying the Turán–Kubilius inequality together the Tchebyshev's argument to obtain, for any fixed $\varepsilon > 0$,

$$\nu_x(n; |h_r(n) - h_s(n)| > \varepsilon) \leq \lambda(r, s), \qquad x \geq 2,$$

where $\lambda(r, s) \to 0$ if $r \geq s \to \infty$. In obtaining this estimate it is helpful to consider separately the contribution towards the additive functions $h_r(n)$ and $h_s(n)$ which arises from the $f(p)$ which satisfy $|f(p)| \leq 1$, $|f(p)| > 1$, respectively. Thus the $F_r(z)$ form a Cauchy sequence.

It is then not difficult to conclude that

$$\nu_x(n; f(n) \leq z) \Rightarrow F(z), \qquad x \to \infty$$

in the (weak) sense of probability.

To fit this into our above philosophy we define

$$c(p, n) = \begin{cases} p^{1/2} - p^{-1/2} & \text{if } p \mid n, \\ -p^{-1/2} & \text{if } p \nmid n. \end{cases}$$

Replacing $f(p)$ in the Turán-Kubilius inequality (9) by $f(p)p^{1/2}$ we see that it can be represented in the form

$$\sum_{n \leq x} |c(p,n)f(p)|^2 \leq c_1 x \sum_{p \leq x} |f(p)|^2.$$

We can regard this as an L^2-norm inequality for an operator T between $\mathbb{C}^{\pi(x)}$ where $\pi(x)$ denotes the number of primes not exceeding x, and $\mathbb{C}^{[x]}$ where $[x]$ is the number of positive integers not exceeding x. It asserts that the norm of T does not exceed $(c_1 x)^{1/2}$.

According to our philosophy we dualize T, and after some rearrangement obtain the inequality

(10) $$\sum_{p \leq x} p \left| \sum_{\substack{n \leq x \\ n \equiv 0 (\bmod p)}} a_n - p^{-1} \sum_{n \leq x} a_n \right|^2 \leq c_1 x \sum_{n \leq x} |a_n|^2,$$

with the same constant c_1.

We now dualize as far as possible our previous argument.

Let

(11) $$\nu_x(n; f(n) \leq z) \Rightarrow F(z), \qquad x \to \infty$$

in the usual (weak) sense of probability.

The Fourier–Stieltjes transform

$$\phi_x(t) = \int_{-\infty}^{\infty} e^{itz} \, d\nu_x(n; f(n) \leq z)$$

of the frequency is readily computed to be

$$[x]^{-1} \sum_{n \leq x} e^{itf(n)}.$$

As $x \to \infty$

$$\phi_x(t) \to \phi(t),$$

where $\phi(t)$ is the transform of $F(z)$.

We set $a_n = \exp(itf(n))$ in inequality (10), the dual of the Turán–Kubilius inequality. From the complete additivity of $f(\ )$

$$x^{-1} \sum_{\substack{n \leq x \\ n \equiv 0 (\bmod p)}} a_n = e^{itf(p)} x^{-1} \sum_{m \leq x/p} e^{itf(m)},$$

and as $x \to \infty$ this last expression converges to $\phi(t) \exp(itf(p)) p^{-1}$. It follows that

$$|\phi(t)|^2 \sum_p p^{-1} |e^{itf(p)} - 1|^2 \leq c_1.$$

Since $\phi(t)$ is a continuous function of t and has the value 1 when $t=0$, there is an interval of t-values around the origin on which $\phi(t)$ is bounded uniformly from below. This allows us to deduce the convergence of the series

$$\sum_{|f(p)|>1} \frac{1}{p}, \quad \sum_{|f(p)|\leq 1} \frac{f(p)^2}{p}.$$

These are the first two series which appear in the original condition (8).

Looking back at the early part of our argument we see that it will also establish the weak convergence of the frequencies

$$\nu_x(n; f(n) - E(x) \leq z)$$

assuming only the convergence of these same two series. But this together with (11) guarantees the finite existence of

$$\lim_{x \to \infty} E(x),$$

and hence the convergence of the third series at (8).

This completes our treatment of the example.

I discuss duality from a functional analysis point of view in Chapter 5. In this volume I shall confine myself to norms of type L^2 or L^∞. The dual of an L^2 operator is generally another L^2 operator, which gives the argument a uniform appearance. However, the method is not particularly bound up with this choice, and could probably be modified to work in terms of any L^α-norm with $\alpha > 1$.

As I mentioned before, duality from an algebraic point of view is considered in Chapter 15.

The following inequality, taken from Chapter 10, typifies what I can presently obtain for real-valued additive functions $f(\ )$.

Let the integers $a > 0$, b, $A > 0$, B satisfy $\Delta = aB - Ab \neq 0$. Then there are positive constants c_2 and c so that for all $x \geq 2$

$$(12) \sum_{\substack{q \leq x \\ (q, aA|\Delta|)=1}} \frac{|f(q) - F(x) \log q|^2}{q} \leq c_2 \sup_{x < w \leq x^c} \frac{1}{w} \sum_{x < n \leq w} |f(an+b) - f(An+B)|^2.$$

Here the variable q runs over prime-powers, and the function $F(x)$ can be explicitly defined in terms of the $f(q)$ with $q \leq x$. The function $f(\ )$ is understood to be additive in the usual weak sense (1).

To illustrate the strength of this inequality let $a = A = 3$, $b = 2$, $B = 1$, so that $\Delta = -3$. Suppose that

$$f(3n+2) - f(3n+1) = 0$$

for each integer n in the interval $N < n \leq (4N)^c$, where N is an integer greater than 1. Then

$$f(q) = F(4N) \log q$$

holds for all prime-powers $q \leq 4N$ which are not powers of 3. In particular

$$F(4N) \log\left(\frac{3N+2}{3N+1}\right) = f(3N+2) - f(3N+1) = 0,$$

showing that $F(4N) = 0$. Out temporary hypothesis leads to the conclusion that $f(n)$ vanishes on the interval $1 \leq n \leq N$, $(n, 3) = 1$.

The following diagram illustrates the logical construction of the proof of the main inequalities of Chapter 10, numbers referring to chapters.

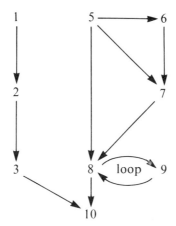

With few exceptions the necessary preliminary material is collected in Chapters 1 and 6. These exceptions are that in Chapters 8, 9 and 10 I apply the Prime Number Theorem in the form

$$\sum_{p \leq x} \frac{\log p}{p} = \log x + \text{constant} + O((\log x)^{-K})$$

with arbitrarily large (but fixed) values of K, and in Chapter 7 I employ the Siegel–Walfisz theorem concerning primes in arithmetic progression. Proofs of these last results may be found in Davenport [1] or Prachar [1]. For use in Chapter 2 I state in Chapter 1 an estimate for a Kloosterman exponential sum. For a prime modulus an elementary proof of this bound, using the method of Stepanov, may be found in the Springer notes of Schmidt [1].

In Chapter 7 I employ the simplest means of Dirichlet series, and apply Cauchy's integral theorem. I there establish inequalities which measure the distribution of additive functions in residue classes to large prime moduli.

Let d, r be integers, c, k positive integers, σ, y positive real numbers. Define

$$E(y,k,r) = \sum_{\substack{n \leq y \\ n \equiv d \pmod{c} \\ n \equiv r \pmod{k}}} \frac{f(n)}{n^\sigma}\left(1-\frac{n}{y}\right) - \frac{1}{\varphi(k)} \sum_{\substack{n \leq y \\ n \equiv d \pmod{c} \\ (n,k)=1}} \frac{f(n)}{n^\sigma}\left(1-\frac{n}{y}\right),$$

where $\varphi(k)$ denotes Euler's function. Then if $\frac{1}{2} < \sigma < 1$ the inequality

$$\sum_{\substack{\log x < p \leq Q \\ (p,c)=1}} (p-1) \max_{(r,p)=1} \max_{y \leq x} |E(y,p,r)|^2 \leq c_3 \left(\frac{x^{1-\sigma}}{\log x} + Q^\mu\right) \sum_{q \leq x} \frac{|f(q)|^2}{q^\sigma}$$

holds uniformly for all additive functions $f(n)$, real $x \geq 2$ and $Q \geq 1$. Here μ can be given explicitly in terms of σ. The constant c_3 depends at most upon c, d and σ.

An important feature of this inequality, which is of L^2-norm type, is that it is abstract, that is to say no *a priori* assumptions are made concerning the function $f(n)$ other than that it be additive.

During the proof of the inequality (12) and the complementary inequalities of Chapter 10 I employ an iterative procedure, labelled "loop" on the above logical diagram, which is constructed in Chapter 8. It leads to an approximate functional equation which is solved in Chapter 9.

One could say that the proof method of the main argument is largely "elementary". At least it is carried out "locally", near to some arbitrary parameter. However, it is quite complicated. In order to clarify the proceedings there are three un-numbered chapters, labelled First, Second and Third Motive, respectively, in which I discuss the general aim and drift of the following chapters.

Even with the accompanying motivation, the proofs in Chapters 2 and 3 may appear as a swamp of detail. It would perhaps be better to begin reading at Chapter 5, continuing through the chapters up to 10. As the logical design shows, the results of Chapter 3 are not applied until Chapter 10, and after seeing their place in the proof of the main inequalities, the interested reader may then feel the details of the earlier chapters to be worthwhile.

In the second half of the book, including Chapter 11 onwards, the results of Chapter 10 play an important rôle. It is not necessary to have read the first ten chapters to appreciate this section, since the *results* of theorems (10.1) and (10.2) are applied rather than the method of their proof.

In Chapters 13 and 14 I settle a question of Kátai [3], and sharpen and extend a result of Wirsing [3]. In particular, I characterize all real additive functions f for which

$$\lim_{n \to \infty} f(an+b) - f(An+B)$$

exists and is finite, $a > 0$, $A > 0$, $aB \neq Ab$.

The following chapters are concerned with the representation of integers by products of prescribed integers, as described in the earlier part of this introduction.

The book is rounded out with applications of the main inequalities of Chapter 10 to the Theory of Information, and to Probabilistic Number Theory.

Let $\beta(x) > 0$ be a measurable function which satisfies $\beta(x) \to \infty$ and

(13) $$\beta(x^y)/\beta(x) \to 1, \qquad x \to \infty$$

for each fixed $y > 0$.

In Chapter 21 I prove that in order that for a real-valued additive function $f(n)$ the frequencies

(14) $$\nu_x(n; f(n+1) - f(n) \leq z\beta(x))$$

possess a limiting distribution as $x \to \infty$, and that the mean and variance of these frequencies be uniformly bounded, it is both necessary and sufficient that there exist a constant A and a non-decreasing function $K(u)$ of bounded total variation, so that for prime-powers q the function $h(q) = f(q) - A \log q$ satisfies

(15) $$\sum_{\substack{q \leq x \\ h(q) \leq u\beta(x)}} \frac{1}{q}\left(\frac{h(q)}{\beta(x)}\right)^2 + \sum_{\substack{q \leq x \\ -h(q) \leq u\beta(x)}} \frac{1}{q}\left(\frac{h(q)}{\beta(x)}\right)^2 \Rightarrow K(u), \qquad x \to \infty.$$

on bounded intervals, and that these sums are bounded uniformly in u, x.

For additive functions $f(\ )$ which are not assumed to satisfy any *a priori* bounds on the size of the $f(q)$ this result is the first of its type. Doubtless a form of this result holds without the requirement (13) that $\beta(x)$ does not grow too rapidly. However, there is evidence that if this requirement is omitted then the condition (15) is no longer appropriate (see Elliott [11], Chapter 18).

If we define the complex-valued multiplicative function

$$g_t(n) = \exp\left(\frac{2\pi i t f(n)}{\beta(x)}\right)$$

then the Fourier–Stieltjes transform of the frequency (14) is

$$\phi_x(t) = [x]^{-1} \sum_{n \leq x} g_t(n+1)\overline{g_t(n)}.$$

To decide whether the frequencies (14) converge weakly is then the same as giving necessary and sufficient conditions that these $\phi_x(t)$ converge as $x \to \infty$, uniformly over every bounded interval of real t-values. The presence of the renormalizing function $\beta(x)$ complicates matters, but this is offset by the existence of the parameter t. I do not consider this correlation directly, but show that a satisfactory treatment can be given if $\partial^2 \phi_x(t)/\partial t^2$ at $t=0$ is bounded uniformly for $x \geq 2$.

In Chapter 20 quantitative characterizations of Shannon's entropy function are given for those finite probability spaces whose events occur with rational probabilities. These are the simplest (and perhaps oldest) examples of probability spaces. The surprising link between Shannon's entropy function and the differences of real-valued additive functions was first exhibited by Fadeev [1].

In Chapter 22 I give a quantitative measure of how well a dense sequence of integers will multiplicatively generate the positive rationals. The method is a synthesis of the theory of infinite abelian groups, and of the treatment of the mean-value of complex multiplicative functions in the manner of Halász [1]. Besides these I employ Ruzsa's [1] notion of random characters.

Chapter 23, the last chapter, contains as exercises discussions of topics related to the subject of the present volume, and a number of open problems.

The logical construction of this second half of the book is as follows:

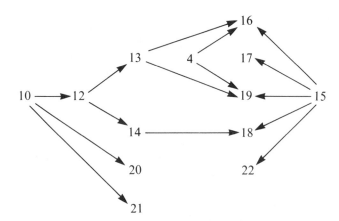

There is essentially no overlap between the present book and my two-volume work devoted to Probabilistic Number Theory. Only in Chapters 21 and 22 do I here apply results from that work, and these chapters may be omitted without affecting the others. In *Probabilistic Number Theory* additive and multiplicative arithmetic functions are considered as analytic (and in some sense "moving") objects. In the present volume the emphasis shifts towards regarding them as algebraic (and so "static") objects. The rôle of continuity in analysis is then taken over by the consideration of collections of arithmetic functions. There is of course not such a clear division between these points of view in practice, and experience in probabilistic number theory certainly motivates some of the argument in the proof of the results of Chapter 10.

I follow here, as I did in *Probabilistic Number Theory*, the practice of giving, as far as possible, the authorship of each result. If there is no credit of this type given then generally the result is either considered straightforward, or it is new. This last means that it is new to me and that I have constructed the proof given. For example, theorem (20.5) from Volume II of *Probabilistic Number Theory* was new with its publication. It will be seen that there are far fewer references in the present book. This reflects the fact that almost all of it is new.

The typescript of the present book was largely completed by the end of 1983. Shortly before Christmas of that year Walter Kaufmann-Bühler, editor, was kind enough to telephone me that it had been accepted by Springer-Verlag for publication. At the same time he suggested that I might add some remarks concerning progress in Probabilistic Number Theory since my two volume work of that title.

Accordingly, I have added a Supplement. In order not to burden this extra chapter with technical lemmas I have enlarged Chapter 5 to incorporate a more extensive treatment of self-adjoint maps than is needed for the rest of the volume.

Even restricting myself to topics related to those considered in my original book there is a wealth of material that can only be touched upon. I have included a treatment of advances in the study of the Turán–Kubilius inequality. More than ever it seems fruitful to regard it as a bound for an operator norm, and to study the underlying operator. In particular I discuss the results of Chapter 9 from this point of view.

First Motive

In the first three chapters my aim is to obtain an upper bound for an additive function $f(n)$ for which $f(an+b)-f(An+B)$ is bounded.

Consider the special case when

$$|f(n+1)-f(n)|\le c$$

holds with some constant c for all positive integers n. Assume, for simplicity, that $f(rs)=f(r)+f(s)$ holds for *every* pair of positive integers r and s. Following Wirsing [1] we argue that

$$f(n)=\begin{cases} f(n/2)+f(2) & \text{if } n \text{ is even,} \\ f(n-1)+\theta, |\theta|\le c & \text{if } n \text{ is odd.} \end{cases}$$

In either case there is an integer $n_1 \le n/2$ so that

$$|f(n)-f(n_1)|\le c_1$$

for a certain constant c_1.

Proceeding inductively

$$|f(n)-f(n_k)|\le kc_1, \qquad n_k \le 2^{-k}n$$

for $k=1,2,\ldots$, and after $O(\log n)$ steps we reach a bound

$$|f(n)| \ll \log n.$$

However, attempts to generalize this argument immediately run into difficulties. Consider the hypothesis

$$|f(3n+1)-f(n)|\le c, \qquad n\ge 1.$$

In this situation the analogous reduction argument becomes

$$|f(n)-f(T(n))|\le c_2,$$

where the map T is given by

$$T(n) = \begin{cases} n/2 & \text{if } n \text{ is even,} \\ 3n+1 & \text{if } n \text{ is odd.} \end{cases}$$

This is the celebrated Kakutani/Syracuse map, and it is an open question whether successive iterations of this map will take an arbitrary integer to 1. Indeed, Conway [1] has shown that for a wide class of procedures of this type it is impossible to decide whether successive iterations reach 1.

The above argument is therefore not promising.

In the general situation we, instead, argue as follows.

If $(n, a) = 1$ then we look for an integer k so that $kn \equiv b \pmod{a}$, say $kn = am + b$. Then

$$f(n) = f(kn) - f(k) = f(am + b) - f(k)$$

$$= f(Am + B) - f(k) + O(1).$$

We further restrict k so that $Am + B$ becomes a product xy of integers less than n. It proves possible to do this with (essentially) $k \ll n^{3/4}$, and $x, y \ll n^{7/8}$. Then for large values of n

$$f(n) = f(x) + f(y) - f(k) + O(1),$$

where x, y, k are all less than n. This procedure lends itself to induction, and leads to an upper bound for $|f(n)|$.

Speaking algebraically, let a, b, A, B be integers satisfying $a > 0$, $A > 0$, $aB - Ab \neq 0$. Let Q^* denote the group of positive rationals with multiplication as group law, and let Γ denote the subgroup generated by the positive fractions of the form $(an + b)/(An + B)$, where n is a positive integer. Then we prove that the quotient group Q^*/Γ is finitely generated. If $g_1, \ldots, g_k$ are integer generators, then a typical integer n has a representation

$$n \equiv g_1^{t_1} \ldots g_k^{t_k} \pmod{\Gamma}.$$

Assuming that $f(an + b) - f(An + B)$ is small, we expect that $f(n)$ will approximately have the value

$$\sum_{i=1}^{k} t_i f(g_i).$$

At least three circumstances lead to complications in our treatment:

(i) We need a bound on the size of the exponent t_i.
(ii) We work in L^2 (mean square) rather than L^∞ (absolute values).
(iii) $f(rs) = f(r) + f(s)$ holds only if $(r, s) = 1$.

However, these technical complications can be overcome. As an example, assuming that

$$|f(an+b) - f(An+B)| \leq c, \qquad n \geq n_0$$

for some positive n_0, we can deduce the bound

$$|f(n)| \leq c_3 (\log n)^3$$

for all integers $n \geq 2$ which are prime to (a, A).

Chapter 1

Variants of Well-Known Arithmetic Inequalities

Multiplicative Functions

In this chapter I derive new variants of some well-known inequalities.

An arithmetic function $g(n)$ is said to be *submultiplicative* if it satisfies the relation $g(ab) \leq g(a)g(b)$ for every pair of mutually prime integers a, b.

Lemma (1.1). *Let $g(n)$ be a non-negative submultiplicative arithmetic function. Then for $x \geq 2$*

$$\sum_{n \leq x} g(n) \leq \left(\frac{x}{\log x} + \frac{10x}{(\log x)^2} \right) \Delta \sum_{n \leq x} \frac{g(n)}{n},$$

with

$$\Delta = \sup_{1 \leq y \leq x} y^{-1} \sum_{p^\alpha \leq y} g(p^\alpha) \log p^\alpha.$$

Moreover

$$\frac{1}{\log x} \sum_{n \leq x} \frac{g(n)}{n} \leq c_0 \exp\left(\sum_{p^\alpha \leq x} \frac{g(p^\alpha) - 1}{p^\alpha} \right)$$

for some absolute constant c_0.

Lemma (1.2). *With certain constants the estimates*

$$\pi(x) \ll \frac{x}{\log x},$$

$$\sum_{p \leq x} \frac{\log p}{p} = \log x + O(1),$$

$$\sum_{p \leq x} \frac{1}{p} = \log \log x + c_1 + O\left(\frac{1}{\log x} \right),$$

$$\prod_{p \leq x} \left(1 - \frac{1}{p} \right) = \frac{c_2}{\log x} \left(1 + O\left(\frac{1}{\log x} \right) \right)$$

hold for $x \geq 2$.

These results have mirror proofs.

Proof of Lemma (1.1). By means of the identity

$$\log m = \sum_{p^\alpha \| m} \log p^\alpha, \qquad m > 0,$$

we have

$$S(x) = \sum_{m \leq x} g(m) \log m = \sum_{p^\alpha \leq x} \log p^\alpha \sum_{t \leq xp^{-\alpha}, (t,p)=1} g(p^\alpha t).$$

Here a typical innersum does not exceed

$$g(p^\alpha) \sum_{t \leq xp^{-\alpha}, (t,p)=1} g(t) \leq g(p^\alpha) \sum_{t \leq xp^{-\alpha}} g(t)$$

so that

$$S(x) \leq \sum_{p^\alpha \leq x} \log p^\alpha g(p^\alpha) \sum_{t \leq xp^{-\alpha}} g(t)$$

$$= \sum_{t \leq x} g(t) \sum_{p^\alpha \leq xt^{-1}} g(p^\alpha) \log p^\alpha$$

$$\leq x\Delta \sum_{t \leq x} g(t) t^{-1}.$$

An integration by parts gives

$$\sum_{n \leq x} g(n) = g(1) + \int_{2-}^{x} \frac{1}{\log w} \, dS(w)$$

$$= \frac{S(x)}{\log x} + \int_{2}^{x} \frac{S(w)}{w(\log w)^2} \, dw + g(1) - g(2).$$

The desired result now follows by applying the upper bound

$$S(w) \leq w\Delta \sum_{t \leq x} g(t) t^{-1}, \qquad 2 \leq w \leq x$$

to estimate the last integral, noting that for $x \geq 2$

$$\int_{2}^{x} \frac{dy}{(\log y)^2} \leq \frac{10x}{(\log x)^2} - \frac{2}{\log 2}.$$

We postpone the proof of the final proposition of lemma (1.1) until after the proof of lemma (1.2).

Proof of lemma (1.2). We modify the proof of lemma (1.1), now using the representation

$$\log m = \sum_{p^\alpha | m} \log p,$$

with $g(n)$ identically one. This gives

$$T(x) = \sum_{p^\alpha \leq x} \log p \left[\frac{x}{p^\alpha} \right] = \sum_{m \leq x} \log m$$

$$= x \log x + O(x),$$

the last estimate being obtained with an integration by parts.

Noting that for real y

$$[2y] - 2[y] \begin{cases} =1 & \text{if } \tfrac{1}{2} \leq y < 1, \\ \geq 0 & \text{always}, \end{cases}$$

we form the expression $T(2x) - 2T(x)$ to obtain the inequalities

$$\sum_{x < p^\alpha \leq 2x} \log p \leq T(2x) - 2T(x) \ll x.$$

In particular

$$\pi(2x) - \pi(x) \leq \frac{c_3 x}{\log x}.$$

Replacing x by $x2^{-k}$, $k = 1, 2, \ldots$, and summing over k we obtain the first desired estimate of lemma (1.2).

Removing the brackets [ ] in the first of the above representations for $T(x)$ introduces an error which is, from what we have already proved, $O(x)$. Since

$$\sum_{p^\alpha \leq x, \alpha \geq 2} p^{-\alpha} \log p \leq \sum_{p \leq x} \frac{\log p}{p(p-1)},$$

which is bounded for all $x \geq 2$, we obtain

$$x \left\{ \sum_{p \leq x} \frac{\log p}{p} + O(1) \right\} = x \log x + O(x).$$

This gives the second estimate in the statement of lemma (1.2).

The third desired estimate now follows with an integration by parts. The fourth is essentially an exponentiated form of the third, since

$$\log \prod_{p \leq x} (1-p^{-1})^{-1} = \sum_{p \leq x} -\log(1-p^{-1})$$

$$= \sum_{p \leq x} p^{-1} + c_4 + O((\log x)^{-1}).$$

This completes the proof of lemma (1.2).

Since

$$\sum_{n \leq x} \frac{g(n)}{n} \leq \prod_{p \leq x} \left(1 + \sum_{\alpha \leq \log x / \log p} \frac{g(p^\alpha)}{p^\alpha}\right)$$

$$\leq \exp\left(\sum_{p^\alpha \leq x} \frac{g(p^\alpha)}{p^\alpha}\right)$$

the final assertion of lemma (1.1) is obtained from the third estimate of lemma (1.2).

This completes the proof of lemma (1.1).

Remarks. An elementary proof in Hardy and Wright [1] shows that the c_2 in lemma (1.2) has the value $e^{-\gamma}$ where

$$\gamma = -\int_0^\infty e^{-x} \log x \, dx$$

is Euler's constant.

The general design of lemma (1.1) is to estimate $g(n)$ in terms of its values on the prime-powers. In the proof of lemma (1.2) we use the same initial ideas as in the proof of lemma (1.1) but deduce the behaviour of $\log p$ from the sharper known behaviour of $\log n$.

As an example in the application of lemma (1.1), consider the multiplicative function $d(n)^\lambda$, where $d(n)$ denotes the number of divisors of the integer n, and λ is a non-negative real. When n is a prime-power p^α, $d(n)^\lambda$ has the value $(\alpha+1)^\lambda$. Applying lemma (1.2) we have

$$\sum_{p^\alpha \leq y} (\alpha+1)^\lambda \log p^\alpha \ll \sum_{p \leq y} \log p + (\log y)^{\lambda+1}\{\pi(y^{1/2}) + \pi(y^{1/3}) + \cdots\}$$

$$\ll y.$$

In the notation of lemma (1.1), Δ is bounded. Applying lemma (1.2) again gives

$$\sum_{n \leq x} d(n)^{\lambda} \ll x(\log x)^{2^{\lambda}-1}, \qquad x \geq 2.$$

An asymptotic estimate of the same order can readily be obtained by the method of Dirichlet series.

Generalized Turán–Kubilius Inequalities

Let $f(n)$ be a real additive function. Define

$$E(x) = \sum_{p^{\alpha} \leq x} \frac{f(p^{\alpha})}{p^{\alpha}}\left(1 - \frac{1}{p}\right),$$

$$D(x) = \left(\sum_{p^{\alpha} \leq x} \frac{|f(p^{\alpha})|^2}{p^{\alpha}}\right)^{1/2} \geq 0.$$

Lemma (1.3). *Let $\rho(n)$ be a non-negative arithmetic function which satisfies*

$$\sum_{\substack{m \leq x \\ m \equiv 0 (\bmod L)}} \rho^2(m) \leq c_0 x L^{-1}$$

uniformly for L the product of at most two prime-powers, $L \leq x$, $x \geq 2$. Then there is further constant c_1, depending at most upon c_0, so that

$$\sum_{n \leq x} \rho(n)|f(n) - E(x)|^2 \leq c_1 x D(x)^2$$

holds for all additive functions $f(n)$, for all $x \geq 2$.

It is convenient to establish a preliminary result. We introduce the modified additive function

$$h(n) = \begin{cases} f(p^{\alpha}) & \text{if } |f(p^{\alpha})| \leq D(x), \\ 0 & \text{otherwise,} \end{cases}$$

and the corresponding

$$H = \sum_{p^{\alpha} \leq x} \frac{h(p^{\alpha})}{p^{\alpha}}.$$

For this next lemma only we shall denote $D(x)$ by D.

Lemma (1.4). *For each integer $k \geq 0$*

$$\sum_{n \leq x} |h(n) - H|^k \ll D^k.$$

Proof of lemma (1.4). It will suffice to obtain this inequality for even positive integers k. For odd integers it may then be deduced from an application of the Cauchy–Schwarz inequality.

By considering real and imaginary parts it will be enough to treat the case when $f(n)$ is real-valued.

We shall assume that $D > 0$, otherwise there will be nothing to prove. For real r define the multiplicative function

$$g(n) = \exp\left(\frac{rh(n)}{D}\right).$$

If $|r| \leq 1$ then in the notation of lemma (1.1) Δ is bounded, so that

$$\sum_{n \leq x} g(n) \leq x \exp\left(\sum_{p^\alpha \leq x} \frac{g(p^\alpha) - 1}{p^\alpha}\right).$$

Using the inequality

$$e^w - 1 \leq w + w^2 e^{|w|}, \qquad w \text{ real},$$

many times, we see that the argument of the exponential function in the preceding upper bound does not exceed

$$rD^{-1}H + r^2 eD^{-2} \sum_{p^\alpha \leq x} \frac{|h(p^\alpha)|^2}{p^\alpha} \leq rD^{-1}H + e.$$

Hence

$$x^{-1} \sum_{n \leq x} g(n) \exp(-rD^{-1}H) \ll 1, \qquad |r| \leq 1.$$

If z is a complex number, then the function defined by

$$\psi(z) = x^{-1} \sum_{n \leq x} \exp\left(z\left\{\frac{h(n) - H}{D}\right\}\right)$$

is analytic in the whole complex z-plane and satisfies the inequality

$$|\psi(z)| \leq \psi(\operatorname{Re}(z)),$$

a form of the ridge property for characteristic functions in the theory of probability. In view of the above remarks $\psi(z)$ is (absolutely) bounded in the disc $|z| \leq 1$.

For each integer $k \geq 0$ we apply Cauchy's integral formula:

$$x^{-1} \sum_{n \leq x} \left| \frac{h(n) - H}{D} \right|^{2k} = \frac{\psi^{2k}(0)}{(2k)!} = \frac{1}{2\pi i} \int_{|z|=1} z^{-2k-1} \psi(z) \, dz \ll 1.$$

Lemma (1.4) is proved.

Proof of lemma (1.3). In the notation of lemma (1.4), we apply the Cauchy–Schwarz inequality:

$$\sum_{n \leq x} \rho(n) |h(n) - H|^2 \ll \left\{ \sum_{n \leq x} \rho(n)^2 \right\}^{1/2} \{xD(x)^4\}^{1/2}$$

$$\ll xD(x)^2.$$

Those prime-powers $l = p^\alpha$ for which $|f(p^\alpha)| > D(x)$ are few in number, in fact

$$\sum_{l \leq x} \frac{1}{l} \leq D(x)^{-2} \sum_{p^\alpha \leq x} \frac{|f(p^\alpha)|^2}{p^\alpha} = 1.$$

With these we argue directly:

$$\sum_{n \leq x} \rho(n) |f(n) - h(n)|^2 \leq \sum_{n \leq x} \rho(n) \left(\sum_{l \| n} |f(l)| \right)^2$$

$$\leq \sum_{l \leq x} |f(l)|^2 \sum_{\substack{n \leq x \\ l \| n}} \rho(n) + \sum_{\substack{l_1, l_2 \leq x \\ l_1 \neq l_2}} |f(l_1)| |f(l_2)| \sum_{\substack{n \leq x \\ l_j \| n, j=1,2}} \rho(n).$$

In view of our hypothesis concerning $\rho(n)$ these sums are

$$\ll x \sum_{l \leq x} |f(l)|^2 l^{-1} + x \left(\sum_{l \leq x} |f(l)| l^{-1} \right)^2 \ll xD(x)^2,$$

the last step by an application of the Cauchy–Schwarz inequality.

Similarly

$$|H - E(x)|^2 \leq \left(\sum_{l \leq x} |f(l)| l^{-1} + \sum_{\substack{p^\alpha \leq x \\ |f(p^\alpha)| \leq D(x)}} |f(p^\alpha)| p^{-\alpha-1} \right)^2 \ll D(x)^2.$$

Lemma (1.3) now follows readily.

Remarks. The particular case $\rho(n) = 1$ for all n gives the standard Turán–Kubilius inequality

$$\sum_{n \leq x} |f(n) - E(x)|^2 \leq c_1 x D(x)^2.$$

For a discussion of this and some other forms of the inequality we refer to Volume I, Chapter 4 of the author's book [11].

We shall also need the following variant.

Lemma (1.5). *If $0 \leq \sigma < 1$ then*

$$\sum_{n \leq x} n^{-\sigma} |f(n) - E(x)|^2 \ll x^{1-\sigma} D(x)^2$$

uniformly for all additive functions $f(n)$, and real $x \geq 2$. The implied constant depends only upon σ, and can be made uniform over any interval of the type $0 \leq \sigma \leq 1 - \delta(<1)$.

Proof. We essentially follow the classical method of Turán [1] and Kubilius [1]. During the proof we shall apply the following estimate(s)

$$\sum_{n \leq y} n^{-\sigma} = \int_1^y w^{-\sigma} \, dw + O(1) = \frac{y^{1-\sigma}}{1-\sigma} + O(1),$$

valid for $y \geq 1$. Once again $f(n)$ may be assumed real-valued.

If we define additive functions by

$$f^+(p^\alpha) = \begin{cases} f(p^\alpha) & \text{if } f(p^\alpha) > 0, \\ 0 & \text{otherwise,} \end{cases} \qquad f^-(p^\alpha) = \begin{cases} -f(p^\alpha) & \text{if } f(p^\alpha) \leq 0, \\ 0 & \text{otherwise,} \end{cases}$$

with the corresponding $E^+(x)$, $E^-(x)$, then

$$|f(n) - E(x)|^2 \leq 2|f^+(n) - E^+(x)|^2 + 2|f^-(n) - E^-(x)|^2.$$

We shall therefore lose no generality by assuming that $f(n)$ takes only non-negative values.

Define the functions

(1) $$w(n) = \sum_{p^\alpha \| n, \, p^\alpha \leq \sqrt{x}} f(p^\alpha),$$

$$W = \sum_{p^\alpha \leq \sqrt{x}} p^{-\alpha}(1 - p^{-1}) f(p^\alpha),$$

and consider the sum

$$\sum_{n\leq x} n^{-\sigma}|w(n)-W|^2 = \sum_{n\leq x} n^{-\sigma}|w(n)|^2 - 2W\sum_{n\leq x} n^{-\sigma}w(n) + W^2 \sum_{n\leq x} n^{-\sigma}$$

$$= S_1 - 2S_2 + S_3,$$

say.

Expanding $w(n)^2$ by means of the representation (1), and inverting the order of summation shows that

$$S_1 = \sum_{p^\alpha \leq \sqrt{x}} |f(p^\alpha)|^2 \sum_{\substack{n\leq x \\ p^\alpha \| n}} n^{-\sigma} + \sum_{\substack{p^\alpha q^\beta \leq x \\ p \neq q}} \sum f(p^\alpha)\overline{f(q^\beta)} \sum_{\substack{n\leq x \\ p^\alpha \| n, q^\beta \| n}} n^{-\sigma}.$$

The innersums here may be respectively estimated by

$$\frac{x^{1-\sigma}}{1-\sigma} \frac{1}{p^\alpha}\left(1-\frac{1}{p}\right) + O\left(\frac{1}{p^{\alpha\sigma}}\right),$$

and

$$\frac{x^{1-\sigma}}{1-\sigma} \frac{1}{p^\alpha}\left(1-\frac{1}{p}\right)\frac{1}{q^\beta}\left(1-\frac{1}{q}\right) + O\left(\frac{1}{(p^\alpha q^\beta)^\sigma}\right).$$

For example, the first innersum can be given the form

$$\sum_{m\leq xp^{-\alpha}} (p^\alpha m)^{-\sigma} - \sum_{m\leq xp^{-\alpha-1}} (p^{\alpha+1} m)^{-\sigma}.$$

Since, applying the Cauchy–Schwarz inequality,

$$\left(\sum_{p^\alpha\leq\sqrt{x}} |f(p^\alpha)|p^{-\alpha\sigma}\right)^2 = \left(\sum_{p^\alpha\leq\sqrt{x}} |f(p^\alpha)|p^{-\alpha/2}p^{\alpha(1/2-\sigma)}\right)^2$$

$$\leq D(x)^2 \sum_{p^\alpha\leq\sqrt{x}} p^{\alpha(1-2\sigma)} \ll \frac{x^{1-\sigma}}{\log x} D(x)^2,$$

we see that for some constant c

$$S_1 \leq \frac{x^{1-\sigma}}{1-\sigma}(W^2 + cD(x)^2).$$

Similar (lower and upper) estimates with W^2 in the leading term can be made for the sums S_2 and S_3. Putting these estimates together, the terms involving W^2 cancel and we obtain

$$\sum_{n \leq x} n^{-\sigma} |w(n) - W|^2 \ll x^{1-\sigma} D(x)^2.$$

Since each integer not exceeding x can have at most one exact prime-power divisor $p^\alpha > \sqrt{x}$,

$$\sum_{n \leq x} n^{-\sigma} |w(n) - f(n)|^2 = \sum_{\sqrt{x} < p^\alpha \leq x} |f(p^\alpha)|^2 \sum_{\substack{n \leq x \\ p^\alpha \| n}} n^{-\sigma}$$

$$\leq \sum_{p^\alpha \leq x} |f(p^\alpha)|^2 p^{-\alpha\sigma} \sum_{m \leq xp^{-\alpha}} m^{-\sigma} \ll x^{1-\sigma} D(x)^2.$$

Moreover

$$|E(x) - W| \leq \sum_{\sqrt{x} < p^\alpha \leq x} |f(p^\alpha)| p^{-\alpha}$$

$$\leq D(x) \left\{ \sum_{\sqrt{x} < p^\alpha \leq x} p^{-\alpha} \right\}^{1/2} \ll D(x).$$

Combining these results we complete the proof of lemma (1.5).

Remarks. If $\sigma = 1$ the above proof can be modified to give

(2) $$\sum_{n \leq x} n^{-1} |f(n) - E(x)|^2 \ll \log x \cdot D(x)^2.$$

For $\sigma > 1$ (cf. Elliott [9], Lemma 5) one obtains the analogue

$$\sum_{n=1}^{\infty} n^{-\sigma} \left| f(n) - \sum_{p^\alpha} p^{-\alpha\sigma} (1 - p^{-\sigma}) f(p^\alpha) \right|^2$$

$$= \zeta(\sigma) \sum_{p^\alpha} p^{-\alpha\sigma} (1 - p^{-\sigma}) |f(p^\alpha)|^2 - \zeta(\sigma) \sum_{p} \left(\sum_\alpha p^{-\alpha\sigma} (1 - p^{-\sigma}) f(p^\alpha) \right)^2$$

$$\leq \zeta(\sigma) \sum_{p^\alpha} p^{-\alpha\sigma} (1 - p^{-\sigma}) |f(p^\alpha)|^2,$$

where

$$\zeta(\sigma) = \sum_{n=1}^{\infty} n^{-\sigma}$$

is the well-known Riemann zeta function. Setting $\sigma = 1 + (\log x)^{-1}$ and $f(p^\alpha) = 0$ for $p^\alpha > x$ we can again obtain the inequality (2).

If $f(n)$ is strongly additive, $f(p^m) = f(p)$ for every prime-power p^m, $m \geq 1$, then an application of the Cauchy–Schwarz inequality enables one to easily deduce from lemma (1.5) the further inequality

$$(3) \qquad \sum_{n \leq x} n^{-\sigma} \left| f(n) - \sum_{p \leq x} \frac{f(p)}{p} \right|^2 \ll x^{1-\sigma} \sum_{p \leq x} \frac{|f(p)|^2}{p}.$$

When $\sigma = 0$ this inequality is also sometimes called the Turán–Kubilius inequality.

An analogue of the Turán–Kubilius inequality which pertains to the differences of additive functions may be obtained from

$$|f(n) - f(n-1)|^2 \leq 2|f(n) - E(x)|^2 + 2|E(x) - f(n-1)|^2,$$

giving

$$(4) \qquad \sum_{n \leq x} |f(n) - f(n-1)|^2 \ll xD(x)^2.$$

We shall use this simple argument many times during the following chapters.

Selberg's Sieve Method

We shall need both upper and lower sieve estimates. The following result, of a type generally known as *the fundamental lemma of Kubilius*, will suffice.

Lemma (1.6). *Let $k(n)$ be a real-valued non-negative arithmetic function. Let a_n, $n = 1, \ldots, N$, be a sequence of integers. Let r be a positive real number, and let $p_1 < p_2 < \cdots < p_s \leq r$ be primes. Set $Q = p_1 \cdots p_s$. If $d | Q$ then let*

$$\sum_{\substack{n=1 \\ a_n \equiv 0 (\mathrm{mod}\ d)}}^{N} k(n) = \eta(d) X + R(N, d)$$

where X, $R(N, d)$ are real numbers, $X \geq 0$, and $\eta(d_1 d_2) = \eta(d_1) \eta(d_2)$ whenever d_1 and d_2 are coprime divisors of Q.

Assume that for each prime p, $0 \leq \eta(p) < 1$.

Let $I(N, Q)$ denote the sum

$$\sum_{\substack{n=1 \\ (a_n, Q) = 1}}^{N} k(n).$$

Then the estimate

$$I(N, Q) = \{1 + 2\theta_1 H\} X \prod_{p|Q} (1 - \eta(p)) + 2\theta_2 \sum_{\substack{d|Q \\ d \leq z^3}} 3^{\omega}(d) |R(N, d)|$$

holds uniformly for $r \geq 2$, $\max(\log r, s) \leq \frac{1}{8} \log z$, where $|\theta_1| \leq 1$, $|\theta_2| \leq 1$, and

$$H = \exp\left(-\frac{\log z}{\log r}\left\{\log\left(\frac{\log z}{S}\right) - \log\log\left(\frac{\log z}{S}\right) - \frac{2S}{\log z}\right\}\right),$$

$$S = \sum_{p|Q} \frac{\eta(p)}{1 - \eta(p)} \log p.$$

When these conditions are satisfied there is a positive absolute constant c so that $2H \leq c < 1$.

Remarks. If only an upper bound for $I(N, Q)$ is desired then one may replace the condition $d \leq z^3$ by $d \leq z^2$, and change the definition of S to

$$\sum_{p|Q} \eta(p) \log p.$$

This then allows the possibility that $\eta(p) = 1$.

Here $\omega(d)$ denotes the number of distinct prime divisors of the integer d.

A permissible value for c is $\exp(-0.006)$.

In the sequence a_n repetition is allowed.

Proof. This result is proved in Volume I, Chapter 2 of the author's book [11].

Kloosterman Sums

The following basic result concerning exponential sums will be useful several times.

Lemma (1.7). *If $k \geq 1$ and a are integers, then*

$$\frac{1}{k} \sum_{r=1}^{k} \exp\left(\frac{2\pi i a r}{k}\right) = \begin{cases} 1 & \text{if } a \equiv 0 \pmod{k}, \\ 0 & \text{otherwise.} \end{cases}$$

Proof. If k divides a the result is trivial. Otherwise, with $\rho = \exp(2\pi i a k^{-1})$, so that $\rho^k = 1$, the sum in question is a geometric progression, with the value

$$\frac{\rho}{k}\left(\frac{1-\rho^k}{1-\rho}\right) = 0.$$

Remark. This last argument shows that if k does not divide a, then uniformly in all integers m we have

$$\left|\sum_{r=1}^{m} \exp\left(\frac{2\pi i a r}{k}\right)\right| \leq \frac{2}{|1-\rho|} \leq \left(2\left\|\frac{a}{k}\right\|\right)^{-1},$$

where $\|y\|$ denotes the distance from y to the nearest integer. For

$$|1-\rho| \geq 2\left|\sin\frac{\pi a}{k}\right| = 2\left|\sin\pi\left\|\frac{a}{k}\right\|\right|,$$

and $\sin\theta \geq 2\theta/\pi$ holds for $0 \leq \theta \leq \pi/2$.

For integers $k > 0$, a, b we define a *Kloosterman Sum* by

$$S(a, b; k) = \sum_{\substack{h=1 \\ (h,k)=1}}^{k} \exp\left(\frac{2\pi i}{k}(ah + b\bar{h})\right),$$

where $h\bar{h} \equiv 1 \pmod{k}$.

Lemma (1.8). *The estimate*

$$|S(a, b; k)| \leq k^{1/2} d(k)(b, k)^{1/2}$$

holds, where (b, k) denotes the highest (positive) common factor of b and k.

Proof. For k a prime this result was first established by Weil. An elementary proof using the approach of Stepanov may be found in Schmidt [1].

When k is a prime-power p^α with $\alpha \geq 2$, Salié [1] showed by elementary means that

$$|S(a, b; p^\alpha)| \leq 3p^{\alpha/2}, \quad (b, p) = 1.$$

These results together give the desired inequality for every prime-power k. The general result then follows from the multiplicative property of Kloosterman sums (see Hardy and Wright [1], Theorem 68, pp. 56–57).

We shall need the following estimate for a (Kloosterman) sum over a shorter range than the length of the modulus.

Lemma (1.9). *The estimate*

$$\sum_{\substack{h=1 \\ (h,k)=1}}^{t} \exp\left(\frac{2\pi i b \bar{h}}{k}\right) \leq k^{1/2}(b,k)^{1/2} d(k) \log(4k)$$

holds uniformly for $1 \leq t \leq k$.

Proof. In view of lemma (1.7) the sum in question may be represented by

$$\sum_{\substack{h=1 \\ (h,k)=1}}^{k} \exp\left(\frac{2\pi i b \bar{h}}{k}\right) \sum_{w=1}^{t} \frac{1}{k} \sum_{r=1}^{k} \exp\left(\frac{2\pi i (h-w) r}{k}\right)$$

$$= \frac{1}{k} \sum_{r=1}^{k} S(r, b; k) \sum_{w=1}^{t} \exp\left(\frac{-2\pi i w r}{k}\right).$$

For $r = k$ there is a contribution of

$$\frac{t}{k} S(0, b; k).$$

Otherwise, from lemma (1.8) and the remark following lemma (1.7) the remaining terms are in absolute value not more than

$$k^{1/2}(b,k)^{1/2} d(k) k^{-1} \sum_{r \leq k/2} k r^{-1}.$$

Since

$$\sum_{r \leq y} r^{-1} \leq 1 + \int_{1}^{y} \frac{dx}{x} = 1 + \log y, \quad y \geq 1,$$

these together give the asserted bound.

Remark. In our applications of lemma (1.9) the factor $d(k) \log(4k)$ is unimportant. It is essential that k occurs to a power less than one. With $k^{3/4}$ in place of $k^{1/2}$ bounds analogous to those in these last two lemmas can be obtained using the short elementary answer given by Vinogradov to Exercise 15(a) in Chapter 6 of his book [2]. This would qualitatively suffice for the purposes of the present volume.

Lemma (1.10) (Pólya–Vinogradov). *The inequality*

$$\sum_{n \leq y} \chi(n) \ll k^{1/2} \log k$$

holds uniformly for all y, for all non-principal characters mod k, $k \geq 2$.

Proof. This result is established in Chapter 23 of Davenport [1].

Chapter 2

A Diophantine Equation

In accordance with the first motive I study the diophantine equation

$$\alpha k + \beta = xy.$$

Here α, β are (large) integers of comparable size, and we look for solution triplets $\{k, x, y\}$ in integers of smaller size.

Lemma (2.1). *Let $a_i > 0$, b_i, $i = 1, 2, 3$ be integers which satisfy $\lambda = a_1 b_2 - a_2 b_1 \neq 0$, $(a_3, b_3) = 1$. Let j be a further integer so that*

$$\rho = a_3(a_2 b_3 - a_3 b_2) - j(a_1 b_3 - a_3 b_1) \neq 0.$$

Let m be a positive integer and set $\alpha = a_1 m + b_1$, $\beta = a_2 m + b_2$, $\gamma = a_3 m + b_3$.
For each τ, $\frac{3}{4} < \tau < 1$, there are positive integers m_0, p_0 ($>a_1|\lambda|$), t_0, and positive constants c_1, c_2, c_3 and c_4 so that whenever $m > m_0$ the number $\mathfrak{N}(m)$, of triplets $\{k, x, y\}$ which satisfy the equation

(1) $$\alpha k + \beta = xy$$

and the conditions:

(i) $m^\tau/2 < x \le m^\tau$, $(x, \alpha) = 1$, $m^\tau/2 < y \le m^\tau$, $c_1 m^{2\tau-1} < k \le c_2 m^{2\tau-1}$,
(ii) $(a_3 k + j, \gamma(\gamma, a_3 \lambda)^{-1}) = 1$,
(iii) *if* $p | a_3 \lambda$ *and* $p' \| (a_3 k + j)$ *then* $t \le t_0$,
(iv) *if* $p^2 | xy$ *then* $p \le p_0$,
(v) *if* $p \le p_0$ *and* $p' \| x$, *or* $p' \| y$, *then* $t \le t_0$,

is bounded above and below by

$$c_3 \frac{\varphi(\alpha)\varphi(\gamma)}{\alpha \gamma} m^{2\tau-1} \le \mathfrak{N}(m) \le c_4 \frac{\varphi(\alpha)\varphi(\gamma)}{\alpha \gamma} m^{2\tau-1}.$$

Furthermore, whilst the constants p_0, t_0, c_j, $j = 1, \ldots, 4$, may depend upon the a_i, b_i and upon j, they do not depend upon m.

Remarks (see Notation). Here $p^t \| n$ denotes that p^t divides n but p^{t+1} does not. $\varphi(n)$ denotes Euler's totient function.

We shall also need the following variant of this result.

Lemma (2.2). *Let the notation of lemma (2.1) be in force save that we assume that γ is a prime-power, and that α has the form $D\gamma$ for some fixed integer D.*

Let $\tilde{\mathfrak{N}}(m)$ denote the number of solutions $\{k, x, y\}$ to the equation (1) which satisfy the conditions (i)–(v) of lemma (2.1) and which have the additional property that every prime p which divides xy must satisfy either $p \leq p_0$ or $p > m^\delta$.

Then for a suitably chosen (fixed) value of δ, $0 < \delta < 1$,

$$c_5 \frac{m^{2\tau-1}}{(\log m)^2} \leq \tilde{\mathfrak{N}}(m) \leq c_6 \frac{m^{2\tau-1}}{(\log m)^2}$$

holds with some positive constants c_5, c_6 and all sufficiently large values of m.

Remarks. The factor $\varphi(\alpha)\alpha^{-1}\varphi(\gamma)\gamma^{-1}$ does not appear in this estimate for the number of solutions since if γ is the power of a prime p then $\varphi(\gamma)\gamma^{-1} = 1 - p^{-1}$, which is bounded above and below for all p; similarly for $\varphi(\alpha)\alpha^{-1}$.

In these solutions x and y have only a bounded number of prime factors. In a slight abuse of classical notation (cf. Linnik [1]) we shall call them *quasiprimes*.

Broadly speaking, lemma (2.2) shows that we can find quasiprimes P_i, P_j so that

$$P_i P_j \equiv \beta \pmod{\alpha},$$

and each quasiprime lies in an interval of the type

$$\alpha^\tau/2 < P_i \leq \alpha^\tau,$$

with $0 < \tau < 1$. Some simplification of the argument of the next chapter would be possible if we could obtain a result of this type where P_i, P_j were (distinct) primes. At the moment such a result seems out of reach, even in an average sense.

In the proofs of both of these results we combine the method of exponential sums with Selberg's sieve.

Proof of lemma (2.1). We begin by obtaining solutions to the equation (1) which are bounded according to condition (i). Note that if we ensure that x and y both lie in the interval $(m^\tau/2, m^\tau]$ then the constraints of the equation guarantee that

$$m^{2\tau-1} \ll k \ll m^{2\tau-1}.$$

It is convenient to formulate a slightly more general problem.

Let u, v be integers, for the moment restricted only by $u > 0$, $(u, v) = 1$. Let τ be a real number, $\frac{3}{4} < \tau < 1$, and let $2M_1$, $2M_2$ be positive real numbers not exceeding u. Let $E(u, v)$ denote the number of solutions to the equation

(2) $$uk + v = xy,$$

in integers x, y, with $(x, u) = 1$, which are restricted to lie in the intervals $M_1 < x \leq 2M_1$ and $M_2 < y \leq 2M_2$, and in integers k.

For each solution $\{k, x, y\}$ to the equation (2) we have $xy \equiv v \pmod{u}$. If $\bar{x}$ is defined by $x\bar{x} \equiv 1 \pmod{u}$ then $v\bar{x} \equiv y \pmod{u}$. Retracing our steps we see that a solution to this congruence with x, y in the specified ranges gives a solution, for some integer k, to the original equation (2). Our (more general) problem is equivalent to the estimation of the number of solutions to the congruence (in two variables x, y)

$$v\bar{x} \equiv y \pmod{u},$$

where $(x, u) = 1$, $x\bar{x} \equiv 1 \pmod{u}$ and x, y are restricted to lie in the intervals

$$M_1 < x \leq 2M_1,$$

$$M_2 < y \leq 2M_2.$$

In view of lemma (1.7) there is a representation

$$E(u, v) = \frac{1}{u} \sum_{h=0}^{u-1} \sum_{\substack{M_1 < x \leq 2M_1 \\ (x,u)=1}} \sum_{M_2 < y \leq 2M_2} \exp\left(\frac{2\pi i h}{u}(v\bar{x} - y)\right),$$

where $x\bar{x} \equiv 1 \pmod{u}$.

Those terms involving $u = 0$ contribute

$$u^{-1} M_2 \sum_{\substack{M_1 < x \leq 2M_1 \\ (x,u)=1}} 1.$$

The remaining terms have a sum whose absolute value does not exceed

$$\frac{1}{u} \sum_{h=1}^{u-1} \left| \sum_{\substack{M_1 < x \leq 2M_1 \\ (x,u)=1}} \exp\left(\frac{2\pi i h v\bar{x}}{u}\right) \right| \left| \sum_{M_2 < y \leq 2M_2} \exp\left(-\frac{2\pi i h y}{u}\right) \right|.$$

Estimating the partial Kloosterman sum by lemma (1.9), and applying the remark following lemma (1.7) shows that this amount is

$$(3) \qquad \ll u^{-1} \sum_{h=1}^{u-1} u^{(1/2)+\varepsilon} (hv, u)^{1/2} \left\| \frac{h}{u} \right\|^{-1}$$

for every fixed $\varepsilon > 0$, the implied constant depending only upon ε. Since $(hv, u) \leq (h, u)(v, u)$, those terms with $h \leq u/2$ contribute

$$\ll u^{(1/2)+\varepsilon}(v, u)^{1/2} \sum_{h \leq u/2} (h, u)^{1/2} h^{-1} \ll u^{(1/2)+2\varepsilon}(v, u)^{1/2}.$$

Here we have made use of the simple bounds

$$\sum_{h \leq u/2} (h, u)^{1/2} h^{-1} \leq \sum_{d | u} d^{1/2} \sum_{t \leq u/2d} d^{-1} t^{-1} \ll \log u \sum_{d | u} d^{-1/2} \ll u^{\varepsilon}.$$

In the sum at (3) the terms with $h > u/2$ may be treated similarly if one notes that $(u - h, u) = (h, u)$ and $\|hu^{-1}\| = (u - h)u^{-1}$.

Hence

$$(4) \qquad E(u, v) = u^{-1} M_2 \sum_{\substack{M_1 < x \leq 2M_1 \\ (x, u) = 1}} 1 + O(u^{(1/2)+2\varepsilon}(v, u)^{1/2}).$$

To apply this estimate to the equation (1) with $u = \alpha$, $v = \beta$, we note that if $d | \alpha$, $d | \beta$ then d divides $a_2 \alpha - a_1 \beta = a_2 b_1 - a_1 b_2$. Thus (α, β) is bounded in terms of the a_i, b_i, $1 \leq i \leq 2$. In this particular case

$$2M_1 = 2M_2 = m^{\tau},$$

and for each $\varepsilon > 0$

$$\sum_{\substack{M_1 < x \leq 2M_1 \\ (x, \alpha) = 1}} 1 = \frac{\varphi(\alpha)}{\alpha} m^{\tau} + O(m^{\varepsilon})$$

(cf. Vinogradov [2], Chapter 2, Exercise 19), so that

$$E(\alpha, \beta) = \frac{\varphi(\alpha)}{4a_1 \alpha} m^{2\tau - 1} + O(m^{(1/2)+\varepsilon}).$$

In particular $E(\alpha, \beta) > 0$ if ε is sufficiently small in terms of $\tau - \frac{3}{4}$, and m is sufficiently large in terms of τ and ε.

We have established that part of lemma (2.1) which involves the condition (i).

Let us write g for $\gamma(\gamma, a_3\lambda)^{-1}$.

In order to embody the condition $(a_3k+j, g) = 1$ we apply the Selberg sieve method, and so need an estimate for the number of solutions to the equation (1) when $a_3k+j \equiv 0 \pmod{d}$, and d is a squarefree divisor of g, g assumed to be non-zero. Note that since $(a_3, b_3) = 1$, the condition $(d, a_3) = 1$ is automatically satisfied. Thus the condition $a_3k+j \equiv 0 \pmod{d}$ is equivalent to $k \equiv k_0 \pmod{d}$ for a certain k_0, $0 \le k_0 < d$. We seek an estimate for the number of solutions $\{k', x, y\}$ to

(5) $$\alpha(dk'+k_0)+\beta = xy,$$

where x, y satisfy the same conditions

$$m^\tau/2 < x, y \le m^\tau; \quad (x, \alpha) = 1,$$

as at the previous stage. We reduce this problem to another which allows us to apply our estimate for $E(u, v)$.

Let $\Lambda = a_3\beta - \alpha j \; (= a_3(\alpha k_0 + \beta) - \alpha(a_3 k_0 + j))$.

If $l | (d, x)$ in (5) then $l | (\alpha k_0 + \beta)$. But by definition $l | d$, and d divides $a_3 k_0 + j$, so that l divides Λ. Retracing our steps: If l divides both Λ and d then l divides $a_3(\alpha k_0 + \beta)$. Since $(a_3, d) = 1$, $l | (\alpha k_0 + \beta)$ and so the left-hand side of the equation (5).

We classify the solutions to equation (5) according to the value l of (x, d). Then the number of solutions to that equation is

(6) $$Y(d) = \sum_{l | (d, \Lambda)} E(\alpha dl^{-1}, (\alpha k_0 + \beta)l^{-1}),$$

where it is to be understood that in the summand corresponding to a typical value of l, the (implicit) x-variable is restricted by

$$m^\tau/(2l) < x \le m^\tau/l \quad \text{and} \quad (x, \alpha dl^{-1}) = 1,$$

and the y-variable (as before) by

$$m^\tau/2 < y \le m^\tau.$$

From (4) a typical summand is then estimated by

(7) $$\frac{m^\tau}{4\alpha dl^{-1}} \cdot \frac{m^\tau}{l} \cdot \frac{\varphi(\alpha dl^{-1})}{\alpha dl^{-1}} + O\left(\left(\frac{\alpha d}{l}\right)^{(1/2)+2\varepsilon} d^{1/2}\right).$$

Note that

$$(\alpha dl^{-1}, (\alpha k_0 + \beta)l^{-1}) \le (\alpha d, \alpha k_0 + \beta) \le (\alpha, \beta)d \ll d.$$

Indeed, if $\Lambda \neq 0$ then $(d, \alpha k_0 + \beta)|\Lambda$, so that the extra factor of $d^{1/2}$ in the error term of the estimate (7) may be removed. We shall not need such a refinement.

From (6) and (7), with a little manipulation, we obtain

$$Y(d) = \frac{m^{2\tau-1}}{4a_1} \cdot \frac{\varphi(\alpha)}{\alpha} \cdot \eta(d) + O(m^{1/2} d(md)^{3\varepsilon}),$$

where

$$\eta(d) = \frac{1}{d^2} \cdot \frac{\varphi(\alpha d)}{\varphi(\alpha)} \sum_{l|(d,\Lambda)} \frac{l\varphi(\alpha)}{\varphi(\alpha l)}.$$

For fixed values of α and Λ, the functions $\varphi(\alpha l)/\varphi(\alpha)$ and (d, Λ) are multiplicative in l and d, respectively. Therefore $\eta(d)$ is multiplicative in d. Note that for a prime p

$$\eta(p) = \begin{cases} (1/p)(1-1/p) & \text{if } p\nmid\Lambda \text{ and } p\nmid\alpha, \\ 1/p & \text{if } p\nmid\Lambda \text{ but } p|\alpha, \\ (1/p)(2-1/p) & \text{if } p|\Lambda \text{ but } p\nmid\alpha, \\ 2/p & \text{if } p|\Lambda \text{ and } p|\alpha. \end{cases}$$

If $p|(\Lambda, d)$ then from the definition of Λ, p divides $(a_3\beta, \alpha)$, and so p divides $a_3\lambda$ (see the remark following (4)). For each prime divisor p of g we therefore have $0 < \eta(p) < 1$.

We are now in a position to apply a lower bound Selberg sieve.

Let Q denote the product of those prime divisors p of g which do not exceed m^μ. Here μ is a positive number, $0 < \mu < 1$, to be chosen presently. Then according to lemma (1.6) with

$$r = m^\mu, \qquad z = m^{\mu^{1/2}},$$

the number of solutions to the equation (1) which have the property that $(a_3k+j, Q) = 1$ in addition to the property (i) is

$$\{1 + O(\exp(-K\mu^{1/2}))\} \cdot \frac{\varphi(\alpha)}{4a_1\alpha} m^{2\tau-1} \cdot \prod_{p|Q} (1 - \eta(p))$$

$$+ O\left(\sum_{d \leq m^{3\mu^{1/2}}} m^{1/2} d(md)^{4\varepsilon}\right)$$

for some positive constant K. This estimate will be at least as large as

(8) $$\frac{\varphi(\alpha) m^{2\tau-1}}{8a_1\alpha} \prod_{\substack{p|g \\ p \leq m^\mu}} \left(1 - \frac{1}{p}\right) \prod_{\substack{p|(g,\Lambda) \\ p \leq m^\mu}} \left(1 - \frac{1}{p}\right)$$

A Diophantine Equation

provided that μ, ε are fixed at sufficiently small values, and m is large enough in terms of μ, τ, ε and the various a_i, b_i.

If $p|(g, \Lambda)$ then $p|(\gamma, \Lambda)$. From our definitions

$$\gamma = a_3 m + b_3,$$

$$\Lambda = (a_3 a_2 - a_1 j) m + a_3 b_2 - b_1 j,$$

so that

$$(a_3 a_2 - a_1 j)\gamma - a_3 \Lambda = \rho,$$

and p must divide ρ. Thus

$$\prod_{p|(g,\Lambda)} \left(1 - \frac{1}{p}\right) \geq 16 a_1 c > 0$$

for some positive constant c, and the lower bound (8) is at least as large as

$$\frac{\varphi(\alpha)\varphi(\gamma)}{\alpha\gamma} \cdot cm^{2\tau - 1}.$$

Moreover, the number of solutions to the equation (1) with $a_3 k + j$ divisible by some prime q, which itself divides g and exceeds m^μ, is not more than

$$\sum_{\substack{q|g \\ q > m^\mu}} \sum_{r \leq L}{}' d(r),$$

where $L = \max(xy) = m^{2\tau}$, and $'$ indicates that the summation is over those r which satisfy the congruence condition

$$a_3 r \equiv a_3 \beta - j\alpha \mod(q\alpha).$$

This may be seen by collecting together those solutions to (1) for which xy has the value r, and noting the identity

$$(a_3 k + j)\alpha + a_3 \beta - j\alpha = a_3(k\alpha + \beta).$$

Since (a_3, α) is bounded independently of m, and $q \nmid a_3$ when m is sufficiently large, the sum over r is

$$O\left(m^{\varepsilon_0}\left(\frac{m^{2\tau}}{q\alpha} + 1\right)\right)$$

for every fixed $\varepsilon_0 > 0$. However, γ and so g can have only boundedly many distinct prime factors q of the above type, therefore the double sum over q and r is

$$O(m^{\varepsilon_0}(m^{2\tau-1-\mu}+1)).$$

We have now shown that there are at least

$$\frac{\varphi(\alpha)\varphi(\gamma)}{\alpha\gamma} \cdot \tfrac{1}{2}cm^{2\tau-1}$$

solutions to the equation (1) which satisfy conditions (i) and (ii) in the statement of lemma (2.1).

Clearly one can obtain an upper bound for the number of such solutions which is of the same size, apart from the numerical constant c. This gives the desired upper bound for $\mathfrak{N}(m)$.

Continuing with our lower bound, suppose now that p is a prime not exceeding a given p_0, and that t is an integer so that $p^t \leq m^{\varepsilon_1}$ for some $\varepsilon_1 > 0$. In what follows we regard p_0, ε_1 as being (temporarily) fixed. We estimate the number of solutions to the equation

$$\alpha k + \beta = xy,$$

with

$$m^\tau/2 < x, y \leq m^\tau, \qquad (x, \alpha) = 1, \qquad (a_3 k + j, g) = 1,$$

and $a_3 k + j$ divisible by p^t.

Here $a_3 k + j$ cannot be divisible by p^t at all unless the highest power of p which divides a_3 also divides j. Assuming this to be the case our condition becomes (say)

$$k \equiv k_1 \pmod{p^{t-s}}, \qquad 0 \leq k_1 < p^{t-s},$$

where $s \leq (\log a_3)/\log 2$.

We are reduced to looking for solutions $\{k'', x, y\}$ to the equation

$$\alpha p^{t-s} k'' + (\alpha k_1 + \beta) = xy,$$

with x and y subject to the above restrictions. Our earlier method shows that if ε_1 is fixed at a sufficiently small value then there are at most

$$c_7 p^{-t} m^{2\tau-1} \cdot \frac{\varphi(\alpha)\varphi(\gamma)}{\alpha\gamma}$$

A Diophantine Equation

such solutions. If t_0 is fixed at a large enough value

$$c_7 \sum_{p|a_3\lambda} p^{-t_0} \leq c_72^{-t_0+2} \sum_p p^{-2} < c_72^{-t_0+2} < \frac{c}{4}.$$

We can thus obtain at least

(9) $$\frac{\varphi(\alpha)\varphi(\gamma)}{\alpha\gamma} \cdot \frac{cm^{2\tau-1}}{4}$$

solutions to equation (1) with the first three conditions of lemma (2.1) satisfied.

Our next task is to estimate how many of the solutions have the property that xy is divisible by the square of some prime.

Let ε_2 be a positive real number, $2\varepsilon_2 < 2\tau - 1$. Consider first those solutions to equation (1) with

$$m^\tau/2 < x, y \leq m^\tau \quad \text{and} \quad xy \equiv 0 \pmod{p^2},$$

where p is a prime in the range

$$m^{2\varepsilon_2} < p \leq m^{2\tau-1-2\varepsilon_2}.$$

For any fixed k, there are $O(m^{\varepsilon_2})$ pairs (x, y) for which $\alpha k + \beta = xy$. Hence the total number of solutions of this type is

$$\ll m^{\varepsilon_2} \sum_{m^{2\varepsilon_2} < p \leq m^{2\tau-1-2\varepsilon_2}} \left(\frac{m^{2\tau-1}}{p^2} + 1\right) \ll m^{2\tau-1-\varepsilon_2},$$

a number which is negligible in comparison with the estimate at (9). Note that from the condition $xy \equiv 0 \pmod{p^2}$ we obtain $\alpha k \equiv -\beta \pmod{p^2}$. Here there will be no solution in k unless (p^2, α) divides β. This condition will be satisfied unless $p|\alpha$, when $p|\beta$ will have to hold. But (α, β) is bounded independently of m and the condition $p > m^{2\varepsilon_2}$ with m large rules out the possibility that $p|(\alpha, \beta)$.

The solutions to (1) with $\alpha k \equiv -\beta \pmod{p^2}$ and $p \leq m^{2\varepsilon_2}$ must be treated with a little more care. If we choose p_0 sufficiently large (in terms of the a_i, b_i) and restrict p by $p > p_0$, then once again $(p^2, \alpha) = 1$ will hold.

There are three (not necessarily distinct) ways in which p^2 can divide xy. We can have $p^2|x$, $p^2|y$ or $p|x$ together with $p|y$.

We shall suppose that $p|x$ and $p|y$, the other cases being similar in outcome. We count the number of solutions to the equation

$$\alpha k + \beta = p^2 x_1 y_1,$$

where

$$m^\tau/(2p) < x_1, y_1 \le m^\tau/p; \quad (px_1, \alpha) = 1, \quad (a_3k+j, g) = 1.$$

The method used in the consideration of condition (ii) leads here to an upper bound

$$O\left(\frac{\varphi(\alpha)}{\alpha} \frac{m^{2\tau-1}}{p^2} \prod_{\substack{p|g \\ p \le m^\mu}} \left(1 - \frac{1}{p}\right) + m^{(1/2)+3\mu^{1/2}+2\varepsilon}\right).$$

Moreover,

$$0 \le \prod_{\substack{p|g \\ p \le m^\mu}} \left(1 - \frac{1}{p}\right) - \frac{\varphi(g)}{g} \ll \sum_{\substack{p|g \\ p > m^\mu}} \frac{1}{p} \ll m^{-\mu},$$

so that the number of solutions with $x \equiv 0 \pmod{p}$, $y \equiv 0 \pmod{p}$, $p_0 < p \le m^{2\varepsilon_2}$, and for which condition (ii) holds, is

$$\ll \frac{\varphi(\alpha)\varphi(\gamma)}{\alpha\gamma} \cdot m^{2\tau-1} \sum_{p > p_0} \frac{1}{p^2} + m^{(1/2)+3\mu^{1/2}+2\varepsilon+2\varepsilon_2}.$$

If $\varepsilon_2 \le \varepsilon$, ε, μ are fixed at suitably small values, and p_0 is sufficiently large (but fixed), the number of these solutions does not exceed one quarter of the amount given at (9).

Similar arguments may be applied if $p^2|x$, and so on, and we see that if the constant 4 at (9) is replaced by 16, then we may safely assume that whenever p^2 divides xy we must have

$$p \le p_0 \quad \text{or} \quad p > m^{2\tau-1-2\varepsilon_2}.$$

Consider the situation when $p > m^{2\tau-1-2\varepsilon_2}$. Then we shall have

(10) $$\alpha k + \beta = p^2 h,$$

where the integer h satisfies

$$h \le p^{-2} \max(xy) \ll m^{2(1-\tau+2\varepsilon_2)}.$$

Let us consider h to be (temporarily) fixed. Let the congruence $\alpha k + \beta \equiv 0 \pmod{h}$ be equivalent to $k \equiv k_2 \pmod{h_1}$, where $h_1 = h(\alpha, h)^{-1}$ and $0 \le k_2 < h_1$. Note that (h, α) divides β and so (α, β), and thus $(h, \alpha) \le c_8$ for some constant c_8 which depends upon the a_i, b_i but not upon m. Let

ns
A Diophantine Equation

$k = k_2 + h_1 w$. Then equation (10) becomes

$$\alpha(k_2 + h_1 w) + \beta = p^2 h,$$

that is

(11) $$\alpha_1 w + \sigma = p^2,$$

where $\alpha_1 = \alpha(\alpha, h)^{-1}$, $\sigma = (\alpha k_2 + \beta) h^{-1}$. Here σ may be as large as m in size, but w is contained in an interval of integers of length at most

$$c_9 h_1^{-1} m^{2\tau - 1}.$$

Moreover

$$m^{2\tau - 1} h_1^{-1} \geq c_{10} m^{\theta},$$

with

$$\theta = 2\tau - 1 - 2(1 - \tau + 2\varepsilon_2) = 4\tau - 3 - 4\varepsilon_2 > 0$$

if ε_2 is sufficiently small.

We estimate (in terms of w) the number of solutions to the equation (11), making use of the fact that the left-hand side of the equation is to be a square.

Let r be a positive real number, to be chosen presently.

Let Q be the product of those odd primes q, not exceeding r, which do not divide α_1. Set

$$X = (c_9 h_1^{-1} m^{2\tau - 1}) + 1.$$

For each prime q which divides Q, $\alpha_1 w + \sigma$ is a square (mod q), so that w does not represent a certain set of $(q-1)/2$ residue classes (mod q). Let $G_q(x)$ be a (monic) polynomial which has as its roots (mod q) precisely these classes. Let

$$\gamma(q) \equiv \begin{cases} 1 & (\text{mod } q), \\ 0 & (\text{mod } Qq^{-1}), \end{cases}$$

and define a polynomial

$$G(x) = \sum_{q \mid Q} \gamma(q) G_q(x).$$

The number of possible values w which satisfy equation (11), does not exceed the number J of integers n, in the range $w_0 \le n \le w_0 + X - 1$, for which $G(n)$ is prime to Q. We estimate this last number by means of Selberg's sieve.

If $d|Q$ the condition $G(n) \equiv 0 \pmod{d}$ is equivalent to the simultaneous conditions

$$G_q(n) \equiv 0 \pmod{d}, \quad q|d.$$

The number of positive integers in $[w_0, w_0 + X - 1]$ which satisfy these is

$$\frac{\varphi(d)}{2^{\omega(d)}d} \cdot X + \psi \frac{\varphi(d)}{2^{\omega(d)}}, \quad |\psi| \le 1.$$

The conditions of lemma (1.6) are satisfied with

$$\eta(d) = \frac{\varphi(d)}{2^{\omega(d)}d},$$

and an error term $R(N, d)$ which does not exceed $2^{-\omega(d)}\varphi(d)$ in absolute value. In the notation of that lemma

$$S = \sum_{p|Q} \eta(p) \log p \le \tfrac{1}{2} \sum_{p \le r} \log p \le c_{10} r$$

for some constant $c_{10} \ge 1$. We apply lemma (1.6) with $z = \exp(8c_{10}r)$ to obtain the bound

$$J \ll X \prod_{P|Q} (1 - \eta(p)) + \sum_{\substack{d|Q_2 \\ d \le z^2}} (\tfrac{3}{2})^{\omega(d)} \varphi(d).$$

The second of these terms is (arguing wastefully)

$$\le z^6 \sum_{d|Q} (\tfrac{3}{2})^{\omega(d)} \frac{\varphi(d)}{d^3} \le z^6 \prod \left(1 + \frac{3(p-1)}{2p^3}\right) \ll z^6 \ll \exp(48c_{10}r).$$

Therefore

$$J \ll \frac{m^{2\tau - 1}}{h_1} 2^{-\pi(r)} \log r + \exp(48 c_{10} r)$$

(12)
$$\ll h^{-1} m^{2\tau - 1} \exp(-\log m)^{1/2}),$$

provided r is chosen suitably.

Summing over the possible h-values, each of which can occur with a multiplicity of at most $3d(h)$, we obtain for the number of (triplet) solutions

A Diophantine Equation

to the equation (10) the upper bound

$$O(m^{2\tau-1} \exp(-(\log m)^{1/3})),$$

which is much less than the number at (9).

In this way we can guarantee that the condition (iv) of lemma (2.1) is also satisfied.

It is perhaps worthwhile to remark here that while the bound (12) is ample for our present purposes, a much sharper upper bound for J may be obtained by applying a result of Montgomery [1], Chapter 3.

It is now straightforward to satisfy the condition (v) of lemma (2.1). Suppose that p^t divides x, with $p \leq p_0$. Then so long as p^t does not exceed a certain power of m, we may argue as in the method used to satisfy the condition (iv), to obtain for the number of solutions to the equation

$$\alpha k + \beta = p^t x_1 y,$$

with

$$m^\tau (2p^t)^{-1} < x_1 \leq m^\tau p^{-t}, \qquad m^\tau/2 < y \leq m^\tau, \qquad (x_1, \alpha) = 1,$$

$$(a_3 k + j, g) = 1,$$

an upper bound

$$c_{11} p^{-t} \frac{\varphi(\alpha)\varphi(\gamma)}{\alpha \gamma} m^{2\tau-1}.$$

Since

$$c_{11} \sum_{p \leq p_0} p^{-t} < c_{11} 2^{-t+2} \sum_p p^{-2} = 4c_{11} 2^{-t},$$

and this last number can be made as small as desired by choosing t sufficiently large, we may safely remove from the solutions to the equation (1) which so far remain, all those which are divisible by a high power ($>$ some particular t_0) of a prime not exceeding p_0. We may clearly take the same value of t_0 in conditions (iii), (v).

We are left with at least

$$\frac{\varphi(\alpha)\varphi(\gamma)}{\alpha \gamma} \cdot \frac{cm^{2\tau-1}}{20}$$

suitable solutions to the equation (1).

This completes the proof of lemma (2.1).

Proof of lemma (2.2). The proof of this result is very similar to that of lemma (2.1). We give details only of the main change, which comes right at the beginning.

Let us consider the solutions to the equation

(1) $$\alpha k + \beta = xy$$

subject to the conditions

$$m^\tau/2 < x, y \leq m^\tau, \qquad (x, \alpha) = 1,$$

to be a *sequence* of triplets $\{k, x, y\}$. Corresponding to this sequence there is another sequence, of integers xy. In this sequence there can be repetition. Let Q be the product of primes p in an interval

$$p_1 < p \leq m^\delta,$$

where p_1, for the moment restricted only by $p_1 > |\lambda|$ (see the statement of lemma (2.1)), and δ are presently to be given suitable fixed values. We seek an estimate for N_1, the number of members of the sequence xy which have no factor in common with Q. In order to apply lemma (1.6) we need an estimate for $Z(d)$, the number of solutions to the above equation which satisfy $xy \equiv 0 \pmod{d}$ for a typical divisor d of Q.

Let $d|Q$ and let $d_1 d_2 = d$. Consider those solutions to the equation (1) for which $(y, d) = d_2$, so that $d_1|x$. The number of such solutions is the same as the number of solutions $\{k, x_1, y_1\}$ to the equation

(13) $$\alpha k + \beta = d_1 x_1 d_2 y_1,$$

where

$$m^\tau (2d_1)^{-1} < x_1 \leq m^\tau d_1^{-1}, \qquad m^\tau (2d_2)^{-1} < y_1 \leq m^\tau d_2^{-1},$$

and

$$(d_1 x_1, \alpha) = 1, \qquad (y_1, d) = 1.$$

If we set

$$M_1 = m^\tau (2d_1)^{-1}, \qquad M_2 = m^\tau (2d_2)^{-1}, \qquad u = \alpha,$$

the number of such solutions may be represented, in a manner similar to that used for $E(u, v)$ in the proof of lemma (2.1), by

(14) $$\frac{1}{u}\sum_{h=0}^{u-1}\sum_{\substack{M_1<x_1\leq 2M_1\\(x_1,u)=1}}\sum_{\substack{M_2<y_1\leq 2M_2\\(y_1,d)=1}}\exp\left(\frac{2\pi ih}{u}(\beta\overline{d_1x_1}-d_2y_1)\right).$$

Note that if $p > |\lambda|$ then p cannot divide (α, β). However, if in a solution to the equation (13) both d_1 and α are divisible by a prime p, then p divides β and so (α, β). Thus $(d_1, \alpha) = 1$, so that $\overline{d_1 x_1}$ is well defined (mod α), and has the alternative expression (in an obvious notation) $\overline{d_1} \cdot \overline{x_1}$.

In the sum at (14) the main term, arising from $h = 0$, is

$$u^{-1}\sum_{\substack{M_1<x_1\leq 2M_1\\(x_1,u)=1}}\sum_{\substack{M_2<y_1\leq 2M_2\\(y_1,d)=1}}1,$$

which in our present circumstances can be estimated by

$$\frac{m^{2\tau-1}}{4a_1}\cdot\frac{\varphi(\alpha)}{\alpha}\cdot\frac{\varphi(d)}{d^2}+O(m^{2\tau-1+\varepsilon}).$$

The application of the bound for incomplete Kloosterman sums goes as before.

The sum

$$\sum_{M_2<y\leq 2M_2}\exp\left(-\frac{2\pi ihy}{u}\right)$$

in the earlier analysis is now replaced by

(15) $$\sum_{\substack{M_2<y\leq 2M_2\\(y_1,d)=1}}\exp\left(-\frac{2\pi ihd_2y_1}{u}\right).$$

The condition that $(y_1, d) = 1$ can be represented by

$$\sum_{\substack{s|y_1\\s|d}}\mu(s)=\begin{cases}1 & \text{if }(y_1, d)=1,\\0 & \text{otherwise,}\end{cases}$$

where $\mu(\ )$ denotes the arithmetic function of Möbius. By means of this representation the sum at (15) can be written as

$$\sum_{s|d}\mu(s)\sum_{\substack{M_2<y_1\leq 2M_2\\y_1\equiv 0\,(\text{mod }s)}}\exp\left(-\frac{2\pi ihd_2y_1}{u}\right).$$

Here a typical innersum has the absolute value

$$\left| \sum_{M_2 s^{-1} < w \leq 2M_2 s^{-1}} \exp\left(-\frac{2\pi i h d_2 s w}{u}\right) \right| \leq \left\| \frac{h d_2 s}{u} \right\|^{-1}.$$

The main argument proceeds along the lines of that used in the consideration of $E(u, v)$ at the beginning of the proof of lemma (2.1), save that in the corresponding error term a factor $2^{\omega(d)}$ appears.

The equation (13) has

$$\frac{m^{2\tau-1}}{4a_1} \cdot \frac{\varphi(\alpha)}{\alpha} \cdot \frac{\varphi(d)}{d^2} + O(m^{(1/2)+\varepsilon} d^{\varepsilon})$$

solutions, and

$$Z(d) = \frac{m^{2\tau-1}}{4a_1} \cdot \frac{\varphi(\alpha)}{\alpha} \cdot \frac{2^{\omega(d)} \varphi(d)}{d^2} + O(m^{(1/2)+\varepsilon} d^{2\varepsilon}).$$

We can now apply lemma (1.6) (which was obtained by Selberg's sieve method) and for a small enough δ will obtain an estimate for $\widetilde{\mathfrak{N}}(m)$ with the dominating term

$$\frac{m^{2\tau-1}}{4a_1} \cdot \frac{\varphi(\alpha)}{\alpha} \prod_{p_1 < p \leq m^\delta} \left(1 - \frac{2}{p} + \frac{2}{p^2}\right).$$

By lemma (1.2) this has the form

$$c_0 m^{2\tau-1} \cdot \frac{\varphi(\alpha)}{\alpha} \cdot \frac{1}{(\log m)^2} \left\{1 + O\left(\frac{1}{\log m}\right)\right\}.$$

The proof of lemma (2.2) can be continued by modifying the proof of lemma (2.1), using this last estimate in the rôle previously assigned to $E(\alpha, \beta)$.

This completes our remarks concerning the proof of lemma (2.2).

Chapter 3

A First Upper Bound

Let $a > 0$, b, $A > 0$, B be integers for which

$$\Delta = \det\begin{pmatrix} a & b \\ A & B \end{pmatrix} \neq 0.$$

There is a natural limitation upon the information which can be deduced from a knowledge of the difference values

$$f(an+b) - f(An+B)$$

of an additive arithmetic function $f(n)$.

If p is a prime which does not divide $aA\Delta$ then it is easy to find an integer n so that one of $an+b$, $An+B$, as desired, is exactly divisible by a given integer p^m, whilst the other is coprime to p. For primes p which divide $aA\Delta$ some complications may occur.

Let $\mu = (a, b)$; $\nu = (A, B)$. Clearly $an+b$ is divisible by μ for every n. Let $a = \mu a_1$, $b = \mu b_1$; $A = \nu A_1$, $B = \nu B_1$. It will sometimes be convenient to arrange that the $a_1 n + b_1$ and $A_1 n + B_1$ have as few prime factors as possible in common with $aA\Delta$.

Suppose first that p is an odd prime. The two conditions

$$a_1 n + b_1 \not\equiv 0 \pmod{p}, \qquad A_1 n + B_1 \not\equiv 0 \pmod{p}$$

can be met simultaneously since there are at least three residue classes (mod p) available from which to choose n. If $p = 2$ these conditions can still be met except when $(2, a_1 A_1) = 1$, and b_1 and B_1 are distinct (mod 2). Note that if 2 divides a_1 then it does not divide b_1, since $(a_1, b_1) = 1$.

An application of the Chinese Remainder Theorem shows that if $aA\Delta$ is odd then there exists an n (and so infinitely many) for which

$$an + b = \mu(a_1 n + b_1), \qquad An + B = \nu(A_1 n + B_1),$$

where $a_1 n + b_1$, $A_1 n + B_1$ have no factors in common with $aA\Delta$.

When $aA\Delta$ is odd we call $(\mu; \nu)$ the *fixed-divisor pair* associated with the sequences $\{an+b; n=1,2,\ldots\}$, $\{An+B; n=1,2,\ldots\}$.

A similar definition is made in every case except when $2|aA\Delta$, $(2, a_1 A_1) = 1$ and $b_1 + B_1 \equiv 1 \pmod{2}$. We can then arrange an n such that 2μ divides $an+b$ and

$$((an+b)/2\mu, aA\Delta) = 1 \quad \text{with} \quad ((An+B)/\nu, aA\Delta) = 1;$$

and also an(other) n so that 2ν divides $An+B$ and

$$((an+b)/\mu, aA\Delta) = 1 \quad \text{with} \quad ((An+B)/2\nu, aA\Delta) = 1.$$

We then call $(2\mu; \nu)$, $(\mu; 2\nu)$ the *fixed-divisor pairs* associated with the sequences $\{an+b\}$ and $\{An+B\}$.

For example, let $a = 8$, $b = 8$, $A = 12$, $B = 0$. Then $aA\Delta = -2^{10} \cdot 3^2$. Here $\mu = 8$, $a_1 = 1 = b_1$; $\nu = 12$, $A_1 = 1$, $B_1 = 0$. The fixed-divisor pairs are therefore $(16; 12)$ and $(8; 24)$.

As another example let $a = 21$, $b = 7$, $A = 15$, $B = 10$. Then $aA\Delta = 3^3 \cdot 5 \cdot 7$. Here $\mu = 7$, $\nu = 5$, $a_1 = 3 = A_1$, $b_1 = 1$, $B_1 = 2$. There is a fixed-divisor pair $(7; 5)$.

In our treatment of the differences of additive functions it is convenient to bracket together the contributions corresponding to these fixed-divisor pairs. If, moreover, p divides $aA\Delta$ and for some i, j we have $p^i \| (an+b)$, $p^j \| (An+B)$, then the corresponding contribution $f(p^i) - f(p^j)$ towards the difference $f(an+b) - f(An+B)$ is conveniently bracketed with $-\{f(p^r) - f(p^s)\}$, where $p^r \| u$, $p^s \| v$ and $(u; v)$ is an appropriate *fixed-divisor pair*.

To illustrate this in the first of our examples, suppose that for some $i \geq 4$ we have $2^i \| (8n+8)$. Then n is odd, the appropriate fixed-divisor pair is $(16; 12)$, and we bracket

$$f(8n+8) - f(12n) = \sum_{\substack{p^i \| (8n+8) \\ (p,6)=1}} f(p^i) - \sum_{\substack{p^j \| 12n \\ (p,6)=1}} f(p^j)$$

$$+ (f(2^i) - f(2^2) - \{f(2^4) - f(2^2)\})$$

$$+ (f(3^{k_1}) - f(3^{k_2}) - \{0 - f(3)\})$$

$$+ (f(16) - f(12)),$$

where $3^{k_1} \| (8n+8)$ and $3^{k_2} \| 12n$.

If, on the other hand, $2^j \| 12n$ for some $j \geq 3$, then n is even, $2^3 \| (8n+8)$

A First Upper Bound

and the appropriate fixed-divisor pair is $(8; 24)$. We then bracket

$$f(8n+8) - f(12n) = \sum_{\substack{p^i \| (8n+8) \\ (p,6)=1}} f(p^i) - \sum_{\substack{p^m \| 12n \\ (p,6)=1}} f(p^m)$$

$$+ (f(2^3) - f(2^j) - \{f(2^3) - f(2^3)\})$$

$$+ (f(3^{t_1}) - f(3^{t_2}) - \{0 - f(3)\})$$

$$+ (f(8) - f(24)),$$

where $3^{t_1} \| (8n+8)$ and $3^{t_2} \| 12n$.

These differing representations reflect the fact that the conclusions of our study will only give information concerning the bracketed expressions. Thus, if we know that an additive function satisfies

$$f(8n+8) - f(12n) = 0$$

for all positive n, and is zero on the powers of all primes other than 2 or 3, we shall be able to deduce only that

$$f(2^i) - f(2^2) - \{f(2^4) - f(2^2)\} = 0, \qquad i \geq 4,$$

$$f(3^{k_1}) - f(3^{k_2}) - f(3) = 0, \qquad k_1 \geq 0, \ k_2 \geq 1,$$

$$f(16) - f(12) = 0,$$

$$f(2^3) - f(2^j) = 0, \qquad j \geq 3,$$

$$f(8) - f(24) = 0.$$

These relations in turn imply that $f(3^m) = 0$ for every $m \geq 1$, and that the values of $f(2^i)$ for $i \geq 2$ are the same, *not necessarily zero*. As inspection of the hypothesis shows, this result will be the best possible. In particular, we can say nothing concerning the value of $f(2)$.

Corresponding to the fixed-divisor pair (μ, ν) we define

$$L_1 = f(\mu) - f(\nu).$$

In the case when there are two fixed-divisor pairs we define

$$L_1 = f(2\mu) - f(\nu), \qquad L_2 = f(\mu) - f(2\nu).$$

It will be convenient to denote by $L(p^i, p^j)$ the expression

$$f(p^i) - f(p^j) - \{f(p^r) - f(p^s)\},$$

where p^r, p^s exactly divide the corresponding members of the fixed-divisor pair associated with L_1, unless this is not appropriate when that of L_2 is to be used.

It follows from the identity

$$a(An + B) - A(an + b) = \Delta$$

that if $p^i \| (an+b)$ and $p^j \| (An+B)$ then $\min(i,j) \leq t$, where $p^t \| \Delta$.

For $y > 0$ define

$$L(y) = \max_{\substack{p^k \leq y; (p, aA\Delta) = 1}} |f(p^k)| + \max_{p^{\max(i,j)} \leq y} |L(p^i, p^j)| + |L_1| + |L_2|,$$

where $|L_1|$, $|L_2|$ and the terms $|L(p^i, p^j)|$ are to be included only if they can occur. In what follows one may view $L(y)$ as small. It will not appear in the final inequalities (of Chapter 10).

We can now state the main result of the present chapter.

Theorem (3.1). *Let $f(n)$ be an additive function. Let $a > 0$, b, $A > 0$, B be integers for which*

$$\Delta = \det \begin{pmatrix} a & b \\ A & B \end{pmatrix} \neq 0.$$

Then for each $c > \frac{3}{2}$ there are constants c_1 and c_2, depending at most upon c, a, b, A and B, so that the inequality

$$\sum_{\substack{p^k \leq x \\ (p, aA\Delta) = 1}} p^{-k} |f(p^k)|^2 + \sum_{p | aA\Delta} \sum_{p^{\max(i,j)} \leq x} p^{-\max(i,j)} |L(p^i, p^j)|^2$$

$$\leq (\log x)^{c_1} \sum_{n \leq x^c} n^{-1} |f(an+b) - f(An+B)|^2 + (\log x)^{c_1} L(c_2)^2$$

holds for all $x \geq 2$.

As an example, when $a = 3$, $b = 1$, $A = 3$, $B = 2$ we have $\Delta = 3$, and theorem (3.1) gives

$$\sum_{\substack{p^k \leq x \\ (p, 3) = 1}} p^{-k} |f(p^k)|^2 \leq (\log x)^{c_1} \sum_{n \leq x^c} n^{-1} |f(3n+1) - f(3n+2)|^2$$

$$+ (\log x)^{c_1} \max_{\substack{p^k \leq c_2 \\ (p, 3) = 1}} |f(p^k)|^2,$$

since neither of $3n + 1$ or $3n + 2$ can be divisible by 3.

A First Upper Bound

It is convenient to begin our proof of the theorem by estimating from above the sum

$$\sum_{\substack{n \leq x \\ (n, aA\Delta) = 1}} |f(n)|.$$

For $x \geq 2$ define the function

$$h(x) = \left(\sup_{1 \leq w \leq x} w^{-1} \sum_{n \leq w} |f(an+b) - f(An+B)|^2 \right)^{1/2}.$$

This function is non-decreasing as x increases. Using an integration by parts we obtain the inequalities

$$\{h(x)\}^2 \leq \sum_{n \leq x} n^{-1} |f(an+b) - f(An+B)|^2 \leq \{h(x)\}^2 (1 + \log x),$$

valid uniformly for $x \geq 2$.

Lemma (3.2). *For each τ, $\tfrac{3}{4} < \tau < 1$, there are constants c_4, c_5 and c_6, so that*

$$\sum_{\substack{n \leq x \\ (n, aA\Delta) = 1}} |f(n)| \ll x (\log x)^{c_4} \{ h(c_5 x^{2\tau}) + L(c_6) \}$$

uniformly for $x \geq 2$.

Proof. Consider a typical integer n, prime to $aA\Delta$, say $n = am + c$ where $1 \leq c \leq a$. Let j be the unique integer in the range $1 \leq j \leq a$ for which $jc \equiv b \pmod{a}$. We look for integers k so that $ak + j$ is prime to $am + c$. Then since $f(\ )$ is additive

(1) $$f(n) = -f(ak+j) + f(an'+b),$$

where

$$n' = (ak+j)m + kc + a^{-1}(jc - b),$$

$$an' + b = (ak+j)(am+c) = (ak+j)n.$$

For integers t define

$$\theta(t) = |f(at+b) - f(At+B)|.$$

We further restrict the values of k so that

(2) $$An' + B = xy,$$

where x and y are small compared to m. To do this we apply lemma (2.1) with $j = j$ and

$$a_1 = aA, \qquad b_1 = cA,$$

$$a_2 = jA, \qquad b_2 = Aa^{-1}(jc - b) + B,$$

$$a_3 = a, \qquad b_3 = c.$$

With these choices (in the notation of lemma (2.1))

$$\lambda A^{-1} = \rho a^{-1} = aB - bA \neq 0,$$

(3)

$$\alpha = An, \qquad \beta = a^{-1}(jAn + aB - bA), \qquad \gamma = n.$$

If, in equation (2), x and y were mutually prime then

$$|f(n)| \leq |f(ak + j)| + \theta(n') + |f(x)| + |f(y)|$$

would follow. We can obtain essentially this result. Using the notation of lemma (2.1) let

$$R = \prod_{p \leq p_0} p^{t_0}.$$

Consider a solution triplet $\{k, x, y\}$ to the equation

(4) $$An' + B = \alpha k + \beta = xy,$$

which satisfies the conditions (i)–(v) of lemma (2.1). In the present circumstances $a_3\lambda = aA\Delta$, which is prime to n. Condition (ii) of lemma (2.1) thus ensures that $(ak + j, n) = 1$. Moreover, conditions (iv) and (v) of that lemma assert that x and y have representations

$$x = x_1 r, \qquad y = y_1 s,$$

where r and s divide R, x_1 and y_1 are prime to R, and $(x_1, y_1) = 1$. Let ρ_2 be that part of xy (and so of $An' + B$) which is made up of the powers of the primes p which divide $aA\Delta$. ρ_2 will divide R^2. Then

$$f(xy) - f(\rho_2) = f\left(\frac{xy}{\rho_2}\right)$$

$$= f\left(x_1 y_1 \cdot \frac{rs}{\rho_2}\right) = f(x_1) + f(y_1) + f\left(\frac{rs}{\rho_2}\right),$$

so that

$$|f(xy)-f(\rho_2)-f(x_1)-f(y_1)| \le \max_{\substack{d|R^2 \\ (d,aA\Delta)=1}} |f(d)|.$$

If ρ_1 denotes that part of $ak+j$ (and so $an'+b$) which is made up of the powers of p which divide $aA\Delta$ then from (1) we obtain

$$|f(n)+f(\rho_1)-f(xy)| \le \left|f\left(\frac{ak+j}{\rho_1}\right)\right| + \theta(n').$$

Here

$$f(\rho_1)-f(\rho_2) = \sum_{\substack{p^i\|(an'+b),p^j\|(An'+B) \\ i,j \le t_0, p|aA\Delta}} \sum L(p^i, p^j) + (L_1 \text{ or } L_2).$$

Hence

$$(5) \qquad |f(n)| \le \left|f\left(\frac{ak+j}{\rho_1}\right)\right| + \theta(n') + |f(x_1)| + |f(y_1)| + M,$$

with

$$M = \max_{\substack{d|R^2 \\ (d,aA\Delta)=1}} |f(d)| + L(R)$$

holds for each of the solutions $\{k, x, y\}$ to equation (4) which are counted by $\mathfrak{N}(m)$ in the statement of lemma (2.1).

We remark that the constants such as p_0, t_0 in the statement of lemma (2.1) depend upon the τ, a_i, b_i, $1 \le i \le 3$, and j of that lemma. In the present circumstances τ is considered fixed, and the various a_i, b_i depend upon the numbers a, b, A, B, j and c. Here j and c are restricted by $1 \le c \le a$, $(c, a) = 1$ and $1 \le j \le a$, $jc \equiv b \pmod{a}$. We can therefore arrange that the above argument holds with the same values of p_0, t_0 for each of the finitely many possible values of j and c.

We sum the inequalities (5) over all the solutions to equations (4) guaranteed by lemma (2.1), divide both sides by the lower bound

$$c_3 \frac{\varphi(\alpha)\varphi(\gamma)}{\alpha\gamma} m^{2\tau-1}$$

for $\mathfrak{N}(m)$ given in that lemma, and then sum over the integers $n \ (= am + c)$

which are prime to $aA\Delta$ and lie in the interval $N/2 < n \le N$. This gives

$$\sum_{\substack{N/2<n\le N \\ (n,aA\Delta)=1}} |f(n)| \ll N^{1-2\tau} \sum_{N/2<n\le N}^{\dagger} \psi(n) \sum{}' \left| f\left(\frac{ak+j}{\rho_1}\right) \right|$$

(6)
$$+ N^{1-2\tau} \sum_{N/2<n\le N}^{\dagger} \psi(n) \sum{}' (|f(x_1)| + |f(y_1)|)$$

$$+ N^{1-2\tau} \sum_{N/2<n\le N}^{\dagger} \psi(n) \sum{}' \theta(n') + N^{1-2\tau} M \sum_{\substack{N/2<n\le N \\ (n,aA\Delta)=1}}{}' \psi(n)$$

$$\ll N^{1-2\tau}(\Sigma_1 + \Sigma_2 + \Sigma_3 + \Sigma_4),$$

where

$$\psi(n) = \frac{\alpha\gamma}{\varphi(\alpha)\varphi(\gamma)},$$

the α, γ being defined in terms of n ($=am+c$) as at (3) above (in accordance with the notation of lemma (2.1)), ' denoting summation over the solutions to equation (3) guaranteed by lemma (2.1), and where † denotes that $(n, aA\Delta) = 1$ holds.

We consider the sums Σ_j, $j = 3, 4, 1, 2$ in turn.

It is convenient to introduce the multiplicative function $g(n) = n/\varphi(n)$. Note that $g(n) \ge 1$ always, $g(p^k) = g(p)$ for every prime-power p^k, $k \ge 1$, and

$$\max(g(r), g(s)) \le g(rs) \le g(r)g(s)$$

for every pair of positive integers r and s.

The sums Σ_3 and Σ_4. We begin (in the notation of (3)) with

$$|\Sigma_3| \le \sum_{\substack{k \ m \\ (n,aA\Delta)=1}} \psi(n)\theta(n') \sum_{xy=\alpha k+\beta} 1 \ll \sum_{k,m} \psi(n)\theta(n') d_2(An'+B),$$

where $d_2(t)$ denotes the number of divisors of the integer t. In the present circumstances $n' \le c_7 N^{2\tau}$. Recall that n' may be written in the form

$$n' = \{(ak+j)(am+c) - b\}a^{-1},$$

and the number of pairs (k, m) which will so generate n' does not exceed

A First Upper Bound

$d_2(an'+b)$. Hence, applying the Cauchy–Schwarz inequality,

$$|\Sigma_3|^2 \le \left(\sum_{k,m}' \theta(n') d_2(an'+b)^{-1/2} \cdot d_2(an'+b)^{1/2} \psi(n) d_2(An'+B) \right)^2$$

$$\le \sum_{n' \le c_7 N^{2\tau}} \theta(n')^2 \cdot \sum_{k,m}' \psi(n)^2 d_2(an'+b) d_2^2(An'+B).$$

From our definition of the function $h(x)$ the sum involving $\theta(n')^2$ is

$$\ll N^{2\tau} \{h(c_7 N^{2\tau})\}^2.$$

We estimate the sum over the divisor functions crudely, applying the Cauchy–Schwarz inequality several times. It does not exceed

$$\left\{ \sum_{n' \le c_7 N^{2\tau}} d_2^2(an'+b) d_2^4(An'+B) \cdot \sum_{k,m}' \psi(n)^4 \right\}^{1/2}.$$

The first of these sums is not more than

$$\left\{ \sum_{t \le c_8 N^{2\tau}} d_2^4(t) \cdot \sum_{t \le c_8 N^{2\tau}} d_2^8(t) \right\}^{1/2} \ll N^{2\tau} (\log N)^{135}.$$

Moreover,

$$\sum_{k,m}' \psi(n)^4 \ll N^{2\tau - 1} \sum_m g^8(am+c),$$

since $\gamma = n = am + c$, $\alpha = A\gamma$. This last sum does not exceed

$$\sum_{t \le N} g(t)^8$$

which we estimate by

$$\sum_{t \le N} g(t)^8 \ll \sum_{t \le N} \prod_{p|t} (1 + 8p^{-1}) = \sum_{t \le N} \sum_{d|t} \mu^2(d) 8^{\omega(d)} d^{-1}$$

$$= \sum_{d \le N} \mu^2(d) 8^{\omega(d)} d^{-1} \sum_{\substack{t \le N \\ t \equiv 0 \pmod{d}}}$$

$$\le N \sum_{d=1}^{\infty} \mu^2(d) 8^{\omega(d)} d^{-2}$$

$$\ll N.$$

Altogether we obtain

(7) $$\Sigma_3 \ll N^{2\tau} \cdot h(c_7 N^{2\tau}) \cdot (\log N)^{34}.$$

The argument given immediately above also shows that

(8) $$\Sigma_4 \ll N^{2\tau} M.$$

The sum Σ_1. For a fixed n ($\equiv c \pmod{a}$)

$$\sum{}' \left| f\!\left(\frac{ak+j}{\rho_1}\right) \right| \leq \sum_{\rho | P} \sum_{\substack{w \leq c_9 N^{2\tau-1} \\ \rho w \equiv j (\mathrm{mod}\, a)}}^{\dagger} |f(w)| \sum_{\alpha a^{-1}(\rho w - j) + \beta = xy}{}' 1,$$

where

$$\alpha = A(am+c), \qquad \beta = jAm + Aa^{-1}(jc-b) + B, \qquad P = \prod_{p | aA\Delta} p^{t_0},$$

and †, as usual, denotes that w has no prime factor in common with $aA\Delta$. Hence

(9) $$\Sigma_1 \ll \sum_{\rho | P} \sum_j \sum_{\substack{w \leq c_9 N^{2\tau-1} \\ \rho w \equiv j (\mathrm{mod}\, a)}}^{\dagger} |f(w)| \sum_m g^2(am+c) \sum_{\substack{\alpha a^{-1}(\rho w - j) + \beta = xy \\ m^{\tau}/2 < x,y \leq m^{\tau}}} 1.$$

The inner (double-) sum in this bound is at most

(10) $$\sum_{(N/2)^{\tau}/2 < x \leq N^{\tau}} \sum_{\substack{m \leq c_{10} N \\ \alpha(\rho w - j) a^{-1} + \beta \equiv 0 (\mathrm{mod}\, x)}} g^2(\alpha).$$

Here the congruence condition may be written in the form

$$mA\rho w + Aa^{-1}(c\rho w - b) + B \equiv 0 \pmod{x}.$$

Note that $c\rho w \equiv c(ak+j) \equiv b \pmod{a}$. Suppose that there is a solution to this congruence in m. If $r | (A\rho w, x)$ then r does not divide $Aa^{-1}(cw-b) + B$ unless it also divides $Ac\rho w + (aB - Ab)$ and so $aB - Ab$. Thus the highest common factor $(A\rho w, x)$ is bounded independently of x, and the congruence condition is essentially non-trivial, say

$$m \equiv m_0 \pmod{x'},$$

A First Upper Bound

where x' differs from x by a bounded multiple. We now estimate

$$\sum_{\substack{m \le c_{10}N \\ \alpha \equiv A(am_0+c)(\bmod x')}} g^2(\alpha) \ll \sum_{\substack{\alpha \le N \\ \alpha \equiv \alpha_0 (\bmod x')}} \sum_{d \mid \alpha} \mu^2(d) 2^{\omega(d)} d^{-1}$$

$$\ll \sum_{d \le N} \mu^2(d) 2^{\omega(d)} d^{-1} \sum_{\substack{\alpha \le N, \alpha \equiv 0 \,(\bmod d) \\ \alpha \equiv \alpha_0 \,(\bmod x')}} 1,$$

with $\alpha_0 = A(am_0 + c)$. Note that in the situation to hand $(d, x) \mid (\alpha, x) \mid (\alpha, \beta)$, so that (d, x) divides λ $(= A(aB - bA))$. Thus (d, x') is bounded independently of d and x'. Whether or not the congruence conditions are compatible the innermost sum here is

$$\ll \frac{N}{dx'}(d, x') + 1 \ll \frac{N}{dx} + 1.$$

This gives for the (final) double sum an upper bound

$$\ll Nx^{-1} \sum_{d=1}^{\infty} \mu^2(d) 2^{\omega(d)} d^{-2} + (\log N)^2 \ll Nx^{-1}.$$

In view of this result the sum at (10) is $O(N)$, and

(11) $$\Sigma_1 \ll N \sum_{\substack{n \le c_9 N^{2\tau-1} \\ (n, aA\Delta) = 1}} |f(n)|,$$

which will suffice for our present purpose.

The sum Σ_2. This sum does not exceed

$$2 \sum_{r \mid R} \sum_{N^\tau/4 < rx_1 \le N^\tau}'' |f(x_1)| \sum_{N/2 < n \le N} \psi(n) \sum_{\alpha k + \beta = rx_1 y}' 1,$$

where $''$ denotes that x_1 is prime to R, and in particular to $aA\Delta$. We treat the innersum by temporarily fixing k and summing over the values of m, that is considering

$$\sum_{\substack{\alpha \le N \\ \alpha k + \beta \equiv 0 \,(\bmod x)}} g^2(\alpha),$$

where $\alpha = am + c$, $x = rx_1$, and the side condition $x \le N^\tau$ is in force. Note also that (α, x) is bounded. This sum was considered in the treatment of

Σ_2, and there shown to be $\ll Nx^{-1} \ll N^{1-\tau}$. Summing over k we obtain the bound

(12) $$\Sigma_2 \ll N^\tau \sum_{\substack{n \leq N^\tau \\ (n,aA\Delta)=1}} |f(n)|.$$

The First Inductive Proof

From (6), (7), (8), (11) and (12) we have

$$\sum_{N/2 < n \leq N}^\dagger |f(n)| \ll N^{2-2\tau} \sum_{n \leq c_9 N^{2\tau-1}}^\dagger |f(n)| + N^{1-\tau} \sum_{n \leq N^\tau}^\dagger |f(n)|$$

$$+ N \cdot h(c_7 N^{2\tau}) \cdot (\log N)^{34} + NM,$$

or, if we define

$$S(z) = \sum_{\substack{n \leq z \\ (n,aA\Delta)=1}} |f(n)|, \quad z \geq 0,$$

then

(13) $$S(N) \leq S(\tfrac{1}{2}N) + \eta N^{2-2\tau} S(c_9 N^{2\tau-1}) + \eta N^{1-\tau} S(N^\tau)$$
$$+ \eta N \cdot h(c_7 N^{2\tau}) \cdot (\log N)^{34} + \eta NL(c_{10})$$

for some $\eta > 0$ and all $N \geq 2$.

We can now argue by induction.

Suppose that for some $l \geq 0$, $C \geq 1$ and $c_{11} > c_{10}$ we have

(14) $$S(z) \leq Cz(\log z)^l \{h(c_7 z^{2\tau}) + L(c_{11})\} = S_0(z)$$

for $2 \leq z \leq N_0$. If $N_0 < N \leq 2N_0$ and N_0 is sufficiently large that over this range $c_9 N^{2\tau-1} \leq N_0$, then we deduce from (13) that

$$S(N) \leq S_0(N) \left\{ \tfrac{1}{2} + \eta c_9 \left(2\tau - 1 + \frac{\log c_9}{\log N}\right)^l + \eta \tau^l + \eta (\log N)^{34-l} \right\}.$$

We fix $l > 34$ at a value so large that

$$\eta c_9 (2\tau - 1)^l + \eta \tau^l < \tfrac{1}{4}.$$

For N_0 sufficiently large, say $N_0 \geq N_2$, the coefficient of $S_0(N)$ will be less than 1 in value, and the induction will proceed. To complete the proof we choose c_{11} to be the maximum of c_{10} and N_2, and then a sufficiently large value of C so that (14) holds for $2 \leq z \leq N_2$.

Hence the bound (14) holds for all $z \geq 2$, and lemma (3.2) is proved.

Remarks. The argument used in the proof of lemma (3.2) would succeed if we were to try and bound the sum

$$\sum_{\substack{n \leq x \\ (n, a A \Delta) = 1}} |f(n)|$$

in terms of sums of the form

$$\sum_{n \leq x} |f(an+b) - f(An+B)|^\delta$$

for some fixed $\delta > 1$. The value of l in (14) would then depend upon δ.

In terms of our present function $h(x)$, for each value of κ in the range $1 \leq \kappa < 2$ we can find a constant $c(\kappa)$ so that

$$\sum_{\substack{n \leq x \\ (n, a A \Delta) = 1}} |f(n)|^\kappa \ll x (\log x)^{c(\kappa)} \{h(c_{12} x^{2\tau}) + L(c_{13})\}^\kappa$$

for $x \geq 2$. During the estimate of the sum corresponding to Σ_3 we would need an upper bound for a sum of the type

$$\sum_{n'} |f(an'+b) - f(An'+B)|^\kappa d_2(an'+b) d_2(An'+B),$$

where the divisor weighting functions cannot be removed. For $\kappa < 2$ we can apply Hölder's inequality to relate this sum to

$$\sum_{t \leq x} |f(at+b) - f(At+B)|^2$$

and

$$\sum_{t \leq x} (d_2(at+b) d_2(At+b))^{c_{14}}$$

considered separately. For $\kappa = 2$ this is no longer possible, and the above argument fails.

To proceed further we carry out a modified form of the above argument with prime-powers p^l in place of the integers n, and with $f(n)$ replaced by $f(p^l)^2$.

In order not to complicate later matters we first prove

Lemma (3.3). *Let $\delta > 0$ be given. For each constant $\mu > 0$ and integer $r \geq 1$,*

$$\sum_{\substack{\mu y < q_1 \cdots q_r \leq y \\ \min q_i > y^\delta}} \cdots \sum \frac{1}{q_1 \cdots q_r} \ll \frac{1}{\log y}, \qquad y \geq 2,$$

where the q_i are primes and the sum is r-fold.

Proof. We argue by induction on r for all $y \geq 2$.
For $r = 1$ we have

$$\sum_{\mu y < q \leq y} q^{-1} \ll y^{-1} \sum_{q \leq y} 1 \ll (\log y)^{-1}$$

by lemma (1.2).

If $r \geq 2$ then one of the primes q_i, say q_1, must satisfy $q_1^r \leq y$. The sum to be estimated therefore does not exceed

$$r \sum_{y^\delta < q_1 \leq y^{1/r}} q_1^{-1} \sum_{\mu y_1 < q_2 \cdots q_r \leq y_1} \cdots \sum (q_2 \cdots q_r)^{-1},$$

where

$$y_1 = y q_1^{-1} \geq y^{1-1/r},$$

and each prime q_j in the centre (multiple) sum satisfies

$$q_j > y^\delta > y_1^\delta.$$

By the induction hypothesis the innermost sum is

$$\ll (\log y_1)^{-1} \ll (\log y)^{-1},$$

whilst

$$\sum_{y^\delta < q \leq y^{1/r}} q^{-1} \ll \log\left(\frac{(1/r) \log y}{\delta \log y}\right) \ll 1$$

after another application of lemma (1.2).
The induction proceeds and lemma (3.3) is proved.

Lemma (3.4). *For each τ, $\frac{3}{4} < \tau < 1$, there are constants c_{15}, c_{16} and c_{17} so that*

$$\sum_{\substack{p' \leq x \\ (p, a A \Delta) = 1}} p^{-\tau} |f(p')|^2 \ll (\log x)^{c_{15}} \{h(c_{16} x^{2\tau}) + L(c_{17})\}^2$$

uniformly for $x \geq 2$.

The First Inductive Proof

Remark. Apart from the condition $(p, aA\Delta) = 1$ this is the desired result of theorem (3.1).

Proof of lemma (3.4). Confining the integers n to be prime-powers, replacing $f(n)$ by $f(n)^2$ and using lemma (2.2) in place of lemma (2.1) we form analogues of the sums Σ_j. Let us denote them by $\tilde{\Sigma}_j, j = 1, 2, 3, 4$. The analogue of (6) is

(15) $$\sum_{\substack{N/2 < p' \leq N \\ (p, aA\Delta) = 1}} |f(p')|^2 \ll N^{1-2\tau}(\log N)^2 (\tilde{\Sigma}_1 + \tilde{\Sigma}_2 + \tilde{\Sigma}_3 + \tilde{\Sigma}_4),$$

where (in the notation of (6)) $\psi(n)$ is replaced by 1.

The sums $\tilde{\Sigma}_3$ and $\tilde{\Sigma}_4$. The estimation of $\tilde{\Sigma}_3$ is simpler than that of Σ_3. As before

$$\tilde{\Sigma}_3 \ll \sum_{n' \leq c_7 N^{2\tau}} \theta(n')^2 \sum_{k,m} {}' d_2(An' + B),$$

where k, m ($p' = am + c$) are constrained according to the conditions of lemma (2.2). Since $An' + B = xy$ is the product of quasiprimes it has boundedly many factors, and $d_2(An' + B)$ is bounded independently of k, m. For a fixed n'; k, m are constrained by

$$am + c = p', \qquad n' = (ak + j)p',$$

and since n' can have at most one prime-power $p' > N/2$ which divides it exactly ($n' \ll N^{2\tau}$), the double sum over k, m is bounded. Hence

(16) $$\tilde{\Sigma}_3 \ll \sum_{r \leq c_7 N^{2\tau}} \theta(r)^2 \ll N^{2\tau} h(c_7 N^{2\tau})^2.$$

It is clear that

(17) $$\tilde{\Sigma}_4 \ll N^{2\tau}(\log N)^{-3} M^2.$$

The sum $\tilde{\Sigma}_1$. We obtain the analogues of (10) and the inequality (9) which precedes it. The analogue of (10) is

(18) $$\sum_{(N/2)^\tau/2 < x \leq N^\tau} \sum_{\substack{m \leq c_{10} N \\ \alpha(\rho w - j)a^{-1} + \beta \equiv 0 (\bmod x)}} 1,$$

where $am + c = p' \leq N$, and x is a quasiprime. We write the congruence condition in the form

(19) $$Aa^{-1}(\rho w p' - b) + B \equiv 0 \pmod{x}$$

and sum over the prime-powers p^t rather than the integers m. We need an upper bound for the number of prime-powers p^t which satisfy the congruence (19) and for which

$$\{Aa^{-1}(\rho w p^t - b) + B\}x^{-1} (=y)$$

is not divisible by any prime q in the range

$$p_0 < q \le (N/2)^\delta.$$

Since $x \ll N^\tau$ and p^t ranges over an interval of length $N/2$, we can apply Selberg's sieve method, as expressed in lemma (1.6).

Let $0 < \delta_1 < \delta$ and let Q be the product of primes q in the interval

$$p_0 < q \le N^{\delta_1}$$

which do not divide wx. In particular, such primes will not divide $aA\Delta$. Let $I(N, Q)$ denote the number of integers n, not exceeding N, for which

(20) $$A\rho w n + \Delta \equiv 0 \pmod{x}$$

and

$$n(A\rho w n + \Delta)x^{-1}$$

is prime to Q. The number of prime-powers p^t which we need to estimate does not exceed $I(N, Q)$ by more than N^{δ_1}.

We can safely assume that $(A\rho w, x)$ divides Δ, otherwise $I(N, Q)$ is zero. If d divides Q then the number of n, not exceeding N, for which the congruence (20) is satisfied and

$$n(A\rho w n + \Delta)x^{-1} \equiv 0 \pmod{d}$$

is readily estimated to be

$$\frac{N}{x'} \cdot \frac{2^{\omega(d)}}{d} + O(2^{\omega(d)}),$$

where

$$x' = x\{(x, A\rho w)\}^{-1}.$$

In the notation of lemma (1.6)

$$\eta(d) = \frac{2^{\omega(d)}}{d}$$

and (since we seek only an upper bound)

$$S = \sum_{p|Q} \frac{2}{p} \log p \le 2\delta_1 \log N + O(1).$$

We apply that lemma with $k(\ )$ identically one and $z = N^{17\delta_1}$, to obtain the bound

$$I(N, Q) \ll \frac{N}{x'} \prod_{q|Q} \left(1 - \frac{2}{q}\right) + N^{35\delta_1}$$

$$\ll \frac{xw^2}{\varphi(x)^2 \varphi(w)^2} \cdot \frac{N}{(\log N)^2}$$

for a small enough (fixed) value of δ_1. An upper bound of this type can therefore be obtained for the innersum at (18).

Hence the expression at (18) is

$$\ll g(w)^2 N (\log N)^{-2} \sum_{N^\tau/4 < x \le N^\tau} g(x)^2 x^{-1},$$

where x has at most a bounded factor made up of the prime factors of R, and all its remaining prime factors exceed $(N/2)^\delta$. For such values of x, $g(x)$ is bounded, and we can estimate the sum (itself) over x by lemma (3.3) to be $\ll (\log N)^{-1}$.

In this way we obtain for the sum at (18) the upper bound

$$O(g(w)^2 N (\log N)^{-3}),$$

so that (cf. (11))

(21) $$\tilde{\Sigma}_1 \ll N (\log N)^{-3} \sum_{w \le c_9 N^{2\tau-1}}^\dagger |f(w)|^2 g(w)^2.$$

Here (as before) $\dagger$ denotes that the variable of summation is prime to $aA\Delta$.

With a view to a later argument by induction we relate the sum in this last estimate to other sums involving primes.

Let

$$E = \sum_{\substack{p' \le c_9 N^{2\tau-1} \\ (p, aA\Delta) = 1}} p^{-1} f(p').$$

From lemma (3.2), applying an integration by parts, we see that

$$E \ll (\log N)^{c_4+1} \{h(c_5 N^{2\tau}) + L(c_6)\},$$

with possibly a change in the value of c_6.

Let
$$K = \max(M, L(c_{17})),$$
where the constant c_{17}, which will be fixed later, is presently restricted by $c_{17} \geq c_6$.

We apply lemma (1.3), the modified Turán–Kubilius inequality, to the function which coincides with $f(n)$ except on the prime-powers p^t where p divides $aA\Delta$, when it is zero. Then

$$\sum_{r \leq c_9 N^{2\tau-1}} g(r)^2 f(r)^2 \leq 2 \sum_r g(r)^2 \{|f(r) - E|^2 + E^2\}$$

$$\ll N^{2\tau-1} \sum_{p^t \leq c_9 N^{2\tau-1}}^{\dagger} p^{-t} |f(p^t)|^2$$

$$+ N^{2\tau-1} \{h(c_5 N^{2\tau})^2 + K^2\} (\log N)^{2c_4+2}.$$

From (21) we now obtain

(22)
$$\tilde{\Sigma}_1 \ll N^{2\tau} (\log N)^{-3} \sum_{p^t \leq c_9 N^{2\tau-1}}^{\dagger} p^{-t} |f(p^t)|^2$$

$$+ N^{2\tau} (\log N)^{2c_4-1} \{h(c_5 N^{2\tau})^2 + K^2\}.$$

This will suffice for our purpose.

The sum $\tilde{\Sigma}_2$. This is considered in a manner similar to that used in the previous section. We have

$$\tilde{\Sigma}_2 \ll \sum_{r|R} \sum_{N^\tau/4 < rx_1 \leq N^\tau}^{\prime\prime} |f(x_1)|^2 \sum \sum_{\substack{N/2 < p^t \leq N, (p, aA\Delta)=1 \\ \alpha k + \beta = xy}}^{\prime} 1,$$

where $x \, (= x_1 r)$ and y are quasiprimes. Temporarily fixing k we are reduced to estimating the sum

$$\sum_{\substack{p^t \leq N, (p, aA\Delta)=1 \\ \alpha k + \beta \equiv 0 \pmod{x}}} 1,$$

which is of the same form as the innermost sum at (18). It likewise has an upper bound

$$O(g(ak+j)^2 g(x)^2 x^{-1} N (\log N)^{-2}).$$

Summing over k we see that

(23)
$$\tilde{\Sigma}_2 \ll N^{2\tau} (\log N)^{-2} \hat{\sum}_{N^\tau/4R \leq x < N^\tau} x^{-1} |f(x)|^2,$$

The Second Inductive Proof

where $\wedge$ denotes that x is squarefree and if a prime q divides x then $q > (N/2)^\delta$. This is an analogue of the estimate (12).

The Second Inductive Proof

From (15), (16), (17), (22) and (23) we obtain

$$\sum_{N/2 < p' \le N}^\dagger |f(p')|^2 \ll N(\log N)^{-1} \sum_{p' \le c_9 N^{2\tau-1}}^\dagger p^{-t}|f(p')|^2$$

$$+ N \hat{\sum}_{N^\tau/4R < x \le N^\tau} x^{-1}|f(x)|^2$$

$$+ N(\log N)^{c_4}\{h(c_5 N^{2\tau})^2 + K^2\},$$

with a new value for c_4. Setting

$$T(y) = \sum_{\substack{p' \le y \\ (p, aA\Delta) = 1}} |f(p')|^2$$

we obtain the analogue of (13):

$$T(N) \le T(\tfrac{1}{2}N) + \eta N(\log N)^{-1} \sum_{p' \le c_9 N^{2\tau-1}}^\dagger p^{-t}|f(p')|^2$$

(24)
$$+ \eta N \hat{\sum}_{N^\tau/4R < x \le N^\tau} x^{-1}|f(x)|^2 + \eta N(\log N)^{c_4}\{h(c_5 N^{2\tau})^2 + K^2\}$$

for some (new) constant η and all $N \ge 2$.

Once again we argue by induction.

Suppose that

(25) $$T(y) \le Cy(\log y)^v \{h(c_5 y^{2\tau})^2 + K^2\} = T_0(y)$$

holds with some (fixed) $v \ge 0$ for $2 \le y \le N_0$.

Let $N_0 < N \le 2N_0$, and suppose that N_0 is so large that over this range $c_9 N^{2\tau-1} \le N_0$.

If $2 \le z \le N_0$ we integrate by parts:

$$\sum_{p' \le z}^\dagger p^{-t}|f(p')|^2 = z^{-1}T(z) + \int_2^z y^{-2}T(y)\, dy$$

(26)
$$\ll \{h(c_5 z^{2\tau})^2 + K^2\}(\log z)^{v+1}.$$

Thus

$$N(\log N)^{-1} \underset{p^t \leq c_9 N^{2\tau-1}}{\sum^\dagger} p^{-t} |f(p^t)|^2 \ll \{h(c_5 N^{2\tau})^2 + K^2\} N(\log N)^v (2\tau-1)^v$$
(27)
$$\ll (2\tau-1)^v T_0(N).$$

Moreover, we may treat the sum

$$\hat{\sum} x^{-1} |f(x)|^2$$

in a manner similar to that used when considering the proof of lemma (3.3). It will suffice to consider a sum

$$M_r(z) = \sum \cdots \sum (q_1 \cdots q_r)^{-1} |f(q_1 \cdots q_r)|^2,$$

where the q_i are *distinct* primes, each of which exceeds z^δ, restricted by

$$\mu z < q_1 \cdots q_r \leq z, \quad (q, aA\Delta) = 1,$$

say, for some positive constant μ.

We shall show (as a subproof) that

(28) $$M_r(z) \ll z^{-1} T_0(z)$$

uniformly for $2 \leq z \leq N_0$, follows from our temporary hypothesis (25).

Indeed

$$M_1(z) = \underset{\mu z < q \leq z}{\sum^\dagger} q^{-1} |f(q)|^2 \ll z^{-1} T_0(z)$$

follows directly from that hypothesis.

If now $r \geq 2$ then typically

$$|f(q_1 \cdots q_r)|^2 \leq 2|f(q_1)|^2 + 2|f(q_2 \cdots q_r)|^2,$$

so that

$$M_r \leq 2r \sum_{z^\delta < q_1 \leq z^{1/r}} q_1^{-1} \sum \cdots \sum_{\mu z_1 < q_2 \cdots q_r \leq z_1} (q_2 \cdots q_r)^{-1} \{|f(q_1)|^2 + |f(q_2 \cdots q_r)|^2\},$$

say, where $z_1 = zq_1^{-1}$, and in each product $q_2 \cdots q_r$ the q_j are distinct and satisfy $q_j > z^\delta \geq z_1^\delta$. The sum over these products is

$$|f(q_1)|^2 \sum \cdots \sum (q_2 \cdots q_r)^{-1} + M_{r-1}(z_1) \ll |f(q_1)|^2 (\log z)^{-1} + z_1^{-1} T_0(z_1)$$

by lemma (3.3) and our (immediate) induction hypothesis, respectively. Moreover

$$(\log z)^{-1} \sum_{z^\delta < q \leq z^{1/r}}^\dagger q^{-1}|f(q)|^2 \ll z^{-1} T_0(z)$$

by (26). Hence (applying lemma (1.2))

$$M_r(z) \ll z^{-1} T_0(z) + z_1^{-1} T_0(z_1) \sum_{z^\delta < q \leq z^{1/r}} \frac{1}{q} \ll z^{-1} T_0(z),$$

and our inductive subproof is complete.

Using the estimate (28) we have

$$N \hat{\sum}_{N^\tau/4R < x \leq N^\tau} x^{-1}|f(x)|^2 \ll N\{N^{-\tau} T_0(N^\tau)\} \ll \tau^v T_0(N).$$

This bound, together with that of (27), may be applied to the right-hand side of the inequality (24) to give

$$T(N) \leq T_0(N)\{\tfrac{1}{2} + c_{18}((2\tau - 1)^v + \tau^v) + (\log N)^{c_4 - v}\}.$$

It is now straightforward to prove by induction on N (cf. (14) and following) that with suitably chosen constants C and c_{17} (see the definition of K earlier in this proof) the inequality (25) holds for all $y \geq 2$.

The inequality of lemma (3.4) follows directly.

Proof of theorem (3.1). Define additive functions $f_1(n)$, $f_2(n)$ by

$$f_1(p^k) = \begin{cases} f(p^k) & \text{if } p \nmid aA\Delta, \\ 0 & \text{otherwise,} \end{cases}$$

and

$$f_2(n) = f(n) - f_1(n).$$

For notational convenience we set

$$W(x) = (\log x)^{c_{15}} \{h(c_{16} x^{2\tau}) + L(c_{17})\}^2.$$

Let

$$\gamma = \max(a + |b|, A + |B|)$$

and

$$E = \sum_{p^k \leq \gamma x} p^{-k} f_1(p^k).$$

Then applying the Turán–Kubilius inequality (see lemma (1.3) and the remarks following it)

$$\sum_{n \leq x} |f_1(an+b) - f_1(An+B)|^2$$

$$\leq 2 \sum_{n \leq x} |f_1(an+b) - E|^2 + 2 \sum_{n \leq x} |f_1(An+B) - E|^2$$

$$\leq 4 \sum_{m \leq \gamma x} |f_1(m) - E|^2 \ll xW(\gamma x).$$

Since

$$|f_2(an+b) - f_2(An+B)|^2 \leq 2|f(an+b) - f(An+B)|^2$$
$$+ 2|f_1(an+b) - f_1(An+B)|^2,$$

and

$$\sum_{n \leq x} |f(an+b) - f(An+B)|^2 \leq xW(\gamma x),$$

we have

(29) $$\sum_{n \leq x} |f_2(an+b) - f_2(An+B)|^2 \ll xW(\gamma x), \qquad x \geq 2.$$

The function $f_2(n)$ is zero except possibly on the powers of the primes p which divide $aA\Delta$. Let p be such a prime, and let D denote the product of the remaining primes which divide $aA\Delta$.

For each prime divisor q of D we can find a value of j ($=j_q$) and an integer n_q so that neither of

$$an_q + b, \qquad An_q + B$$

is divisible by q^{j+1}. For example, if $q^u \| a$, $q^v \| A$ then

$$an + b \equiv 0 \pmod{q^{u+2}}$$

cannot hold unless $q^u | b$ and n belongs to a certain residue class (mod q^2). Likewise

$$An + B \equiv 0 \pmod{q^{v+2}}$$

cannot hold unless $q^v | b$ and n belongs to a (possibly different) certain residue class (mod q^2). If we set $j = \max(u, v) + 2$ and n_q does not lie in

The Second Inductive Proof

either of these two classes (mod q^2) (since $q^2 \geq 4 > 2$ this can certainly be arranged) then we shall have suitable values.

Let J be the largest of the values j obtained in this manner.

By the Chinese Remainder Theorem there is an n_0 with the property that

$$n \equiv n_0 \pmod{D}, \quad 1 \leq n_0 \leq D,$$

if and only if

$$n \equiv n_q \pmod{q^{j+1}}, \quad j = j_q,$$

for each q dividing D. Denote the arithmetic progression

$$n_0 + mD, \quad m = 0, 1, 2, \ldots$$

by H.

From (29) we obtain for $x \geq 2$

(30) $$\sum_{\substack{t \leq x \\ t \in H}} |f_2(at+b) - f_2(At+B)|^2 \ll xW(\gamma x).$$

Consider now those t for which $at+b$ is exactly divisible by p^k. If $p^r \| a$ and $k \geq r$, then p^r must also divide b. Thus we have

$$a_1 t + b_1 \equiv 0 \pmod{p^{k-r}}, \quad a_1 t + b_1 \not\equiv 0 \pmod{p^{k+1-r}},$$

where

$$a_1 = p^{-r} a, \quad b_1 = p^{-r} b.$$

Suppose further that $p^m \| (a_1 B - A b_1)$ and that $k - r \geq m + 1$. Then we also have

$$a_1(At+B) \equiv -Ab_1 + a_1 B \pmod{p^{k-r}},$$

so that $At+B$ is exactly divisible by p^m.

The number of such t-values which lie in H and do not exceed x may be estimated, using the Chinese Remainder Theorem, to be at least

(31) $$\frac{x}{D} \cdot \frac{1}{p^{k-r}} \left(1 - \frac{1}{p}\right) - 3.$$

Moreover, for each value of t

$$|f_2(at+b) - f_2(At+B) - L(p^k, p^m)|^2 \leq L(D^J)^2.$$

Confining our interest to those prime-powers p^k not exceeding $x/(12D)$ we obtain from (30) and our lower bound (31) that

$$\sum_{\substack{k \\ p^k \le x/(12D)}} \frac{x}{4D} \cdot \frac{1}{p^{k-r}} |L(p^k, p^m)|^2 \ll xW(\gamma x) + xL(D^J)^2$$

and therefore

$$\sum_{\substack{k \\ p^k \le x}} p^{-k} |L(p^k, p^m)|^2 \ll W(12D\gamma x) + L(D^J)^2.$$

We may similarly treat those integers t for which high powers of p divide $At + B$.

Since there are only finitely many possibilities for p, the proof of theorem (3.1) is complete.

Concluding Remarks

The argument of this last section shows that there are positive constants s and c_0, depending at most upon a, A and Δ, so that

$$\sum_{p | aA\Delta} \sum_{p^{\max(i,j)} \le x} p^{-\max(i,j)} |L(p^i, p^j)|^2 \ll x^{-1} \sum_{n \le c_0 x} |f(an+b) - f(An+B)|^2$$
$$+ \sum_{\substack{p^k \le c_0 x \\ (p, aA\Delta)=1}} p^{-k} |f(p^k)|^2 + L(s)^2$$

holds for all $x \ge 1$.

EXAMPLES. Consider an additive function $f(\ )$ which satisfies

$$f(n+1) - f(n) \ll 1$$

for all $n \ge 1$. Then

$$\sum_{n \le x} n^{-1} |f(n+1) - f(n)|^2 \ll \log x,$$

and according to theorem (3.1)

$$\sum_{p^k \le x} p^{-k} |f(p^k)|^2 \ll (\log x)^{c_1+1}, \qquad x \ge 2.$$

Concluding Remarks

This may be compared with the result

$$f(n) \ll \log n$$

and so

$$\sum_{p^k \leq x} p^{-k} |f(p^k)|^2 \ll (\log x)^2$$

essentially obtained by a simple algorithm in the description of the first motive. The higher power of a logarithm which theorem (3.1) gives is compensated by the embracing of a general difference $f(an+b) - f(An+B)$ for which we presently have no simple algorithmic treatment.

The method of the present chapter can be applied directly in an L^∞ sense. Thus, assuming that

$$f(an+b) - f(An+B) \ll 1,$$

holds for all sufficiently large integers n, where $a > 0$, $A > 0$ and

$$\begin{vmatrix} a & b \\ A & B \end{vmatrix} \neq 0,$$

a careful treatment leads to a bound

$$f(n) \ll (\log n)^3$$

for all positive n prime to (a, A).

Chapter 4

Intermezzo: The Group Q^*/Γ

As before, let Q^* denote the multiplicative group of the positive rational numbers.

For each k, let $\Gamma(k)$ be the subgroup of Q^* which is generated by the positive fractions of the form

$$\frac{an+b}{An+B}, \quad \det\begin{pmatrix} a & b \\ A & B \end{pmatrix} \neq 0, \quad n \geq k.$$

I shall prove that *for every k the quotient group*

$$Q^*/\Gamma(k)$$

is finitely generated. With $k = 1$ this justifies an assertion made in the first motive.

This result follows directly from the following

Lemma (4). *Let $a > 0$, $A > 0$, b, B be integers, $aB - bA \neq 0$. Let l be a further integer. Let τ be a real number, $\frac{3}{4} < \tau < 1$.*

Then there is an integer t, so that every positive integer n has a representation

(1) $$n = \frac{r}{s} \cdot \prod_j \left(\frac{an_j + b}{An_j + B}\right)^{\varepsilon_j},$$

where each n_j satisfies

$$l < n_j \leq n^{2\tau},$$

each ε_j has a value $+1$ or -1, there are

(2) $$\ll (\log n)^{c_0}$$

terms in the product, and the integers r and s are made up of primes not exceeding t. Moreover, the upper bound (2) also serves for the multiplicity of each prime divisor of r and s.

Intermezzo: The Group Q^*/Γ

Proof. The proof of this proposition runs along the lines used in the beginning of the proof of theorem (3.1), save that we omit considerations of coprimality, and so on.

If n is a (large enough) integer, prime to a, then we can find $\{k, x, y\}$ by lemma (2.1) so that

$$kn = an' + b,$$

$$An' + B = xy,$$

where it can be arranged that

$$x, y \leq n^\tau < n; \quad k \leq n^{2\tau - 1} < n.$$

We write

$$n = \frac{an' + b}{An' + B} \cdot \frac{xy}{k},$$

where $n' \ll n^{2\tau}$. Speaking group theoretically the element n is multiplicatively congruent to the ratio $xyk^{-1} \pmod{\Gamma(k)}$. Since x, y and k are all less than n we can proceed inductively.

In this manner we obtain the desired representation. Note that after one inductive step we obtain at most one n' which is $\leq c_1 n^{2\tau}$ in size. At the next stage we shall obtain at most three more integers to play similar rôles, each of which is $\leq c_1 (n^\tau)^{2\tau}$ in size; and so on. The total number of n_i which we shall need is therefore not more than

$$1 + 3 + 3^2 + \cdots + 3^m,$$

where m is restricted by

$$c_1 n^{2\tau^{m+1}} \geq 1.$$

Thus, in our representation for n,

$$\ll (\log n)^{c_0}$$

product terms involving the n_i will suffice, and it is possible to take for c_0 the value

$$c_0 = -\frac{\log 3}{\log \tau}.$$

The proof of the lemma is complete.

Remarks. It follows from the lemma that $Q^*/\Gamma(k)$ is a direct product of finitely many cyclic groups.

Some factor of the type r/s must occur in a representation (1) of the present generality. Consider the example $a = A = 3$, $b = 1$, $B = 2$. In this case the group Q^*/Γ cannot be finite. Otherwise some power of 3 could be represented by a product of the form

$$\prod_i \left(\frac{3n_i+1}{3n_i+2}\right)^{\varepsilon_i}, \qquad \varepsilon_i = +1 \text{ or } -1,$$

which is clearly impossible. The group must have at least one infinite generator.

I shall consider the group Q^*/Γ more closely in a later chapter.

Chapter 5

Some Duality

Duality in Finite Spaces

We begin with

Lemma (5.1). *Let c_{ij}, $i = 1, \ldots, m$, $j = 1, \ldots, n$, be mn complex numbers. Let λ be a real number.*
  Then the inequality

$$\sum_{i=1}^{m} \left| \sum_{j=1}^{n} c_{ij} a_j \right|^2 \leq \lambda \sum_{j=1}^{n} |a_j|^2$$

holds for all complex numbers $a_1, \ldots, a_n$, if and only if the inequality

$$\sum_{j=1}^{n} \left| \sum_{i=1}^{m} c_{ij} b_i \right|^2 \leq \lambda \sum_{i=1}^{m} |b_i|^2$$

holds for all complex numbers $b_1, \ldots, b_m$.

Proof. The proof is a judicious application of the Cauchy–Schwarz inequality. Consider the matrix

$$\mathbf{C} = (c_{ij}), \quad 1 \leq i \leq m, 1 \leq j \leq n.$$

Let **a** and **b** denote the (vertical) column vectors with entries $a_1, \ldots, a_n$, and $b_1, \ldots, b_m$, respectively. Then the first of the two inequalities in the statement of lemma (5.1) may be expressed in the form

$$|\mathbf{Ca}|^2 \leq \lambda |\mathbf{a}|^2 \quad \text{for all } \mathbf{a}.$$

Let us assume it to be valid. Clearly $\lambda \geq 0$, and interpreting $0^{1/2}$ to be 0, if necessary,

$$|\mathbf{Ca}| \leq \lambda^{1/2} |\mathbf{a}|.$$

Then for any (m-dimensional) vector **b**, whose transpose we shall denote by $\mathbf{b}^T$, we have

$$|\mathbf{b}^T \mathbf{Ca}| \leq |\mathbf{b}^T||\mathbf{Ca}| \quad \text{(by the Cauchy–Schwarz inequality),}$$

$$\leq \lambda^{1/2}|\mathbf{b}^T||\mathbf{a}| \quad \text{(from the temporary hypothesis).}$$

We set $\mathbf{a} = \bar{\mathbf{C}}^T \bar{\mathbf{b}}$, so that

$$|\mathbf{b}^T \mathbf{C}|^2 \leq \lambda^{1/2}|\mathbf{b}^T||\mathbf{b}^T \mathbf{C}|.$$

Therefore

$$|\mathbf{b}^T \mathbf{C}| \leq \lambda^{1/2}|\mathbf{b}^T|$$

unless perhaps $\mathbf{b}^T \mathbf{C} = \mathbf{0}^T$, when the inequality is trivially valid. This gives the second of the two inequalities in lemma (5.1).

The reverse implication is obtained by interchanging the rôles of i and j. This completes the proof of lemma (5.1).

Before discussing this result we use it to derive the following inequalities, the first of which will play an important part in our treatment of the differences of additive functions in Chapter 8.

Lemma (5.2). *If* $0 \leq \sigma < 1$ *then*

$$\sum_{p^k \leq x} p^k \left| \sum_{\substack{n \leq x \\ p^k \| n}} a_n n^{-\sigma} - \frac{1}{p^k}\left(1 - \frac{1}{p}\right) \sum_{n \leq x} a_n n^{-\sigma} \right|^2 \ll x^{1-\sigma} \sum_{n \leq x} |a_n|^2 n^{-\sigma},$$

and

$$\sum_{p \leq x} p \left| \sum_{\substack{n \leq x \\ n \equiv 0 (\bmod p)}} a_n n^{-\sigma} - \frac{1}{p} \sum_{n \leq x} a_n n^{-\sigma} \right|^2 \ll x^{1-\sigma} \sum_{n \leq x} |a_n|^2 n^{-\sigma}$$

hold uniformly for all complex numbers a_n, $n = 1, 2, \ldots, [x]$, and real $x \geq 2$. The implied constants depend only upon σ, and can be made uniform over any interval of the type $0 \leq \sigma \leq 1 - \delta$ (< 1).

Proof. Define

$$c(p^k, n) = \begin{cases} \left(\dfrac{p^k}{n^\sigma}\right)^{1/2} \left(1 - \dfrac{1}{p^k}\left(1 - \dfrac{1}{p}\right)\right) & \text{if } p^k \| n, \\ -\left(\dfrac{p^k}{n^\sigma}\right)^{1/2} \cdot \dfrac{1}{p^k}\left(1 - \dfrac{1}{p}\right) & \text{otherwise.} \end{cases}$$

If in lemma (1.5) we replace each $f(p^k)$ by $p^{k/2}f(p^k)$ then we obtain

$$\sum_{n \leq x} \left| \sum_{p^k \leq x} f(p^k) c(p^k, n) \right|^2 \ll x^{1-\sigma} \sum_{p^k \leq x} |f(p^k)|^2.$$

Since this inequality holds for all complex numbers $f(p^k)$, by lemma (5.1)

$$\sum_{p^k \leq x} \left| \sum_{n \leq x} c(p^k, n) b_n \right|^2 \ll x^{1-\sigma} \sum_{n \leq x} |b_n|^2$$

holds for all complex numbers b_n. Replacing each b_n by $n^{-\sigma/2} a_n$ gives the first inequality of lemma (5.2).

The second inequality of lemma (5.2) may either be deduced from the first, or more simply by duality from the form of the Turán–Kubilius inequality which appears as (3) in the remarks following the proof of lemma (1.5).

Remarks. The first inequality of lemma (5.2) is convenient for the study of additive or multiplicative arithmetic functions. When $\sigma = 0$ it is the dual of the Turán–Kubilius inequality.

The second inequality is in a form more suitable for strongly additive or multiplicative functions.

This ends the remarks.

Self-adjoint Maps

Let E be a vector space over the complex numbers. A map $E \times E \to \mathbb{C}$ given by

$$(x, y) \mapsto g(x, y),$$

where $g(x, y)$ is linear in the x-variable and satisfies

$$g(y, x) = \overline{g(x, y)}$$

for all pairs (x, y) is said to be *hermitian*.

An hermitian map induces an *hermitian form* $g(x, x)$ on E whose values are real. If $g(x, x) > 0$ for all non-zero vectors x we say that the form is *positive definite*, if $g(x, x) \geq 0$ for all x, then it is *non-negative definite*.

An hermitian map which induces a positive definite form is called an *inner product*.

If E is $\mathbb{C}^m$ for some positive integer m, then the standard inner product is

$$(\mathbf{x}, \mathbf{y}) \mapsto \overline{\mathbf{x}}^T \mathbf{y}.$$

An $m \times m$ matrix $\mathbf{A}$ with complex entries is said to be *hermitian* if it satisfies $\mathbf{A} = \bar{\mathbf{A}}^T$. For such a matrix the map

$$(\mathbf{x}, \mathbf{y}) \mapsto \bar{\mathbf{x}}^T \mathbf{A} \mathbf{y}$$

is hermitian, but the induced form need not be positive definite.

For an inner product [,] we define

$$\|x\| = [x, x]^{1/2} \geq 0.$$

For the standard inner product on $\mathbb{C}^m$,

$$|\mathbf{x}| = (\bar{\mathbf{x}}^T \mathbf{x})^{1/2} \geq 0.$$

The *dual space* of a vector space E is the vector space of linear maps $E \to \mathbb{C}$, and will be denoted by E^*.

Let E be a finite dimensional vector space which is endowed with an inner product [,]. Then given any linear map $E \to \mathbb{C}$ there is an element z in E so that the map is $x \mapsto [x, z]$.

Indeed, if $e_i, i = 1, \ldots, m$ is a basis for E, then $\det([e_i, e_j]) \neq 0$. To represent a linear map $x \mapsto w(x)$ by

$$w(x) = \left[x, \sum_{j=1}^{m} \beta_j e_j \right]$$

we choose the coefficients β_i to satisfy the equations

$$\sum_{j=1}^{m} \bar{\beta}_j [e_i, e_j] = w(e_i), \qquad i = 1, \ldots, m.$$

This determines them uniquely. In particular E^* and E are isomorphic.

Let $T: E \to E$ be a linear map. For each y in E the map

$$x \mapsto [Tx, y]$$

belongs to E^*, and so can be written

$$x \mapsto [x, y^*]$$

for some y^* in E. Since the form $[x, x]$ is positive definite, the map $T^*: E \to E$ given by $y \mapsto y^*$ is linear. T^* is the *adjoint* of T.

A map $T: E \to E$ is said to be *self-adjoint* if $T = T^*$. A map that is self-adjoint with respect to one inner product need not be so with respect to another. However, if $g(x, y)$ is an hermitian function $E \times E \to \mathbb{C}$ and [,]

Self-adjoint Maps 85

is any inner product on E, then an argument of the above type shows that there is a representation

$$g(x, y) = [Tx, y]$$

for some linear $T: E \to E$ which is self-adjoint with respect to [,]. The hermitian functions are the fundamental objects.

If E is $\mathbb{C}^m$, then each linear map T can be represented in the form $\mathbf{x} \mapsto \mathbf{Ax}$ for some matrix $\mathbf{A}$, and it is easy to see that with respect to the standard inner product T is self-adjoint if and only if $\mathbf{A}$ is hermitian.

We shall say that a matrix $\mathbf{A}$ is self-adjoint (with respect to a given inner product) if the corresponding map $\mathbf{x} \mapsto \mathbf{Ax}$ is.

A non-zero vector x is an *eigenvector* of a linear map T if for some scalar λ we have $Tx = \lambda x$. The scalar λ is then called its corresponding *eigenvalue*. An eigenvector of a matrix $\mathbf{A}$ is an eigenvector of the map $\mathbf{x} \mapsto \mathbf{Ax}$, similarly for an eigenvalue.

We now need the spectral theorem:

Let E be a finite dimensional vector space, over the complex numbers, which has an inner product. Let $T: E \to E$ be a self-adjoint linear map. Then E has an orthonormal basis consisting of eigenvectors of T.

An inductive proof of this result may be found in Lang [1], XIV, § 12. The eigenvalues (with multiplicity) that correspond to such a basis we shall call *the* eigenvalues of the operator T. They are determined up to order. Note that if λ is an eigenvalue with associated unit eigenvector x, then

$$\lambda = [\lambda x, x] = [Tx, x] = [x, Tx] = [x, \lambda x] = \bar{\lambda},$$

so that λ is real.

Conversely, let E be a finite dimensional space with a basis, orthogonal with respect to some inner product, made up of eigenvectors of a linear map $T: E \to E$. If the corresponding eigenvalues are real, then T is self-adjoint with respect to that inner product.

Let E be a space of dimension m on which the linear map T is self-adjoint with respect to the inner product [,]. Let z_i, $i = 1, \ldots, m$ be a basis for E consisting of unit eigenvectors of T with associated eigenvalues λ_i. Then for a general member

$$x = \sum_{i=1}^{m} c_i z_i$$

of E we have

$$[Tx, x] = \sum_{i=1}^{m} \lambda_i |c_i|^2.$$

Noting that

$$\sum_{i=1}^{m} |c_i|^2 = \|x\|^2,$$

we see that if λ_i is a largest amongst the eigenvalues, then

$$[Tx, x] \leq \lambda_1 \|x\|^2$$

for all x. Moreover, the choice of z_1 for x effects equality.

A similar characterization can be given for each eigenvalue.

Lemma (5.3). *Let T be self-adjoint with eigenvalues $\lambda_1 \geq \lambda_2 \geq \cdots \geq \lambda_m$. Let $y_1, \ldots, y_{i-1}$ be any collection of $i-1$ elements. Then*

$$\max[Tx, x] \geq \lambda_i$$

when the maximum is taken over those unit vectors x which satisfy $[x, y_j] = 0$, $1 \leq j \leq i-1$. Moreover, a choice of the y_j exists which will effect equality.

Remark. If $i = 1$, the elements y_j are deemed not to exist.

Proof. It is clear that (in the above notation) equality is attained if $y_j = z_j$ for every j. Otherwise we choose a unit vector to satisfy the $m-1$ conditions

$$[x, y_j] = 0, \quad 1 \leq j \leq i-1, \qquad [x, z_j] = 0, \quad i+1 \leq j \leq m.$$

For such a vector

$$[Ax, x] = \sum_{j=1}^{i} \lambda_j |c_j|^2 \geq \lambda_i \sum_{j=1}^{i} |c_j|^2 = \lambda_i \|x\|^2 = \lambda_i,$$

since $c_k = 0$ for $k > i$.

In applications it is sometimes convenient to decompose T into the sum of two maps T_1, T_2 given by

$$T_1 : \sum_{j=1}^{m} c_j z_j \mapsto \sum_{j=1}^{i-1} \lambda_j c_j z_j,$$

$$T_2 : \sum_{j=1}^{m} c_j z_j \mapsto \sum_{j=i}^{m} \lambda_j c_j z_j,$$

so that

$$[T_2 w, w] \leq \lambda_i \|w\|^2.$$

for all w.

Consider the case when E is $\mathbb{C}^m$, with the standard inner product and an hermitian matrix $\mathbf{A}$. Let $\mathbf{U}$ be the matrix whose column vectors, reading left to right, are the (above) eigenvectors $\mathbf{z}_i$ of $\mathbf{A}$. Then

$$\bar{\mathbf{U}}^T \mathbf{A} \mathbf{U} = \mathrm{dg}(\lambda_i)$$

is a diagonal matrix with entries $\lambda_1, \ldots, \lambda_m$ along the main diagonal. Moreover, $\bar{\mathbf{U}}^T \mathbf{U}$ is the identity matrix of order $m \times m$. Corresponding to T_1 will then be the matrix $\mathbf{U} \mathbf{B} \bar{\mathbf{U}}^T$ where the diagonal matrix $\mathbf{B}$ has $\lambda_1, \ldots, \lambda_{i-1}$ for its first $i-1$ entries, and zero for those remaining.

In order to estimate the forms induced by self-adjoint maps one needs bounds for eigenvalues. The following methods offer some help.

Lemma (5.4). *Each eigenvalue λ of the matrix $\mathbf{A} = (a_{ij})$, $1 \leq i, j \leq m$, lies in one of the discs*

$$|\lambda - a_{ii}| \leq \sum_{\substack{j=1 \\ j \neq i}}^{m} |a_{ij}|, \qquad 1 \leq i \leq m.$$

Proof. Let $\mathbf{x}$ be an eigenvector of $\mathbf{A}$ corresponding to λ, and let x_k be one of its coordinates of maximal absolute value. Then from the relation $\mathbf{A}\mathbf{x} = \lambda \mathbf{x}$, we obtain

$$(\lambda - a_{kk}) x_k = \sum_{\substack{j=1 \\ j \neq k}}^{m} a_{kj} x_j,$$

so that

$$|\lambda - a_{kk}| \leq \sum_{\substack{j=1 \\ j \neq k}}^{m} |a_{kj}| |x_j x_k^{-1}|$$

and every $|x_j x_k^{-1}| \leq 1$.

The advantage of this lemma is that one may choose the eigenvalue in advance. A disadvantage is that if the diagonal elements of $\mathbf{A}$ are far apart, then λ has not been properly located.

We define tr **B**, the trace of a square matrix **B**, to be the sum of the elements down its principal diagonal.

If a linear map T is represented by matrices **B**, **C** with respect to different bases, then these are connected by a relation $\mathbf{B} = \mathbf{DCD}^{-1}$ for some non-singular **D**. It is readily checked that tr **B** = tr **C**, and tr T, the *trace* of T, is defined to be the common value.

Lemma (5.5). *Let* **A** *be a matrix which is self-adjoint with respect to some inner product on* $\mathbb{C}^m$. *Then*

$$\lim_{k \to \infty} (\operatorname{tr} \mathbf{A}^{2k})^{1/(2k)} = |\lambda_1|,$$

with λ_1 an eigenvalue of maximal absolute value.

Proof. Let **A** be self-adjoint with respect to the inner product [,]. Let z_i, $i = 1, \ldots, m$ be a basis for $\mathbb{C}^m$, orthonormal with respect to this product, with z_i an eigenvector of **A** with associated eigenvalue λ_i. Then for each positive integer k and typical element

$$\mathbf{x} = \sum_{j=1}^{m} c_j \mathbf{z}_j$$

we have

$$\mathbf{A}^k \mathbf{x} = \sum_{j=1}^{m} \lambda_j^k c_j \mathbf{z}_j.$$

This shows that the eigenvalues of $\mathbf{A}^k$ are the kth-powers of the eigenvalues of **A**.

In computing the trace of $\mathbf{A}^k$ we may assume that it represents a linear map in terms of the basis $\mathbf{z}_j$, $j = 1, \ldots, m$. In such a case it will be diagonal, which gives

$$\operatorname{tr} \mathbf{A}^k = \sum_{j=1}^{m} \lambda_j^m.$$

The assertion of the lemma is now clear.

On the face of it this method will completely determine the eigenvalues. Having obtained $|\lambda_1|$, the next appropriate value, say $|\lambda_2|$, is obtained as

$$\lim_{k \to \infty} (\operatorname{tr} \mathbf{A}^k - \lambda_1^k)^{1/k},$$

the limit being taken through even integers k. Having obtained all the

Self-adjoint Maps 89

absolute values of the λ_i, their signs can be determined by considering the trace of $\mathbf{A}^k$ for the odd and even values of k not exceeding $2m$.

In practice the complications in evaluating high powers of a matrix militate against this.

Note that for a hermitian matrix $\mathbf{A} = (a_{ij})$, $1 \le i, j \le m$,

$$\operatorname{tr} \mathbf{A}^2 = \sum_{i=1}^{m} \sum_{j=1}^{m} |a_{ij}|^2.$$

Lemma (5.6). *Let T be a self-adjoint map. For any vector x and real number α there is an eigenvalue λ of T so that*

$$|\lambda - \alpha| \, \|x\| \le \|Tx - \alpha x\|$$

Proof. Let I denote the identity map on the same space as T. In the notation preceding lemma (5.3)

$$(T - \alpha I)x = \sum_{j=1}^{m} (\lambda_j - \alpha) c_j z_j.$$

Hence

$$\sum_{j=1}^{m} |(\lambda_j - \alpha) c_j|^2 = \|Tx - \alpha x\|^2,$$

and the desired result follows with λ an eigenvalue λ_i for which $|\lambda_i - \alpha|$ is minimal.

Since λ is near to an eigenvalue λ_i one would hope for x to be near a corresponding eigenvector. This requires that λ_i be isolated from the remaining eigenvalues.

Lemma (5.7). *Let T be a linear map, on a finite dimensional space, which is self-adjoint with respect to an inner product $[\ ,\ ]$. Let z_i, $i = 1, \ldots, m$, be an orthonormal basis for the space, comprised of eigenvectors z_i of T with corresponding eigenvalues λ_i. For a given vector x and eigenvalue λ_k define $\delta = \|Tx - \lambda_k x\|$. Then for each positive integer s*

$$\|x - c_k z_k\| \, |\lambda_k| \le \delta \Big(1 + \max_{\substack{1 \le i \le s \\ i \ne k}} |\lambda_i (\lambda_i - \lambda_k)^{-1}|\Big) + \|x - c_k z_k\| \max_{\substack{i > s \\ i \ne k}} |\lambda_i|$$

with $c_k = [x, z_k]$.

Proof. Let
$$x = \sum_{i=1}^{m} c_i z_i.$$

Then
$$Tx - \lambda_k x = \sum_{\substack{i=1 \\ i \neq k}}^{m} c_i(\lambda_i - \lambda_k) z_i,$$

so that
$$\sum_{\substack{i=1 \\ i \neq k}}^{m} |c_i(\lambda_i - \lambda)|^2 = \delta^2.$$

In particular
$$\sum_{\substack{i=1 \\ i \neq k}}^{s} |c_i \lambda_i|^2 = \sum_{\substack{i=1 \\ i \neq k}}^{s} |c_i(\lambda_i - \lambda_k) \lambda_i (\lambda_i - \lambda_k)^{-1}|^2$$

$$\leq \delta^2 \max_{\substack{1 \leq i \leq s \\ i \neq k}} |\lambda_i(\lambda_i - \lambda_k)^{-1}|^2.$$

Moreover,
$$\sum_{\substack{i>s \\ i \neq k}} |c_i \lambda_i|^2 \leq \sum_{\substack{j>s \\ j \neq k}} |c_j|^2 \max_{\substack{i>s \\ i \neq k}} |\lambda_i|^2,$$

where the coefficient of the maximum does not exceed
$$\sum_{\substack{j=1 \\ j \neq k}}^{m} |c_j|^2 = \|x - c_k z_k\|^2.$$

Finally
$$\lambda_k(x - c_k z_k) = \lambda_k x - Tx + Tx - \lambda_k c_k z_k$$

$$= \lambda_k x - Tx + \sum_{\substack{j=1 \\ j \neq k}}^{m} c_j \lambda_j z_j,$$

Self-adjoint Maps

so that by Minkowski's inequality

$$|\lambda_k| \, \|x - c_k z_k\| \le \delta + \left(\sum_{\substack{j=1 \\ j \ne k}}^{m} |c_j \lambda_j|^2 \right)^{1/2}.$$

Combining these inequalities gives the desired bound.

Remark. If several eigenvectors z_j have the same eigenvalue λ_k it is necessary to modify the lemma by replacing $c_k z_k$ by the sum of the $c_j z_j$ taken over all the eigenvectors belonging to λ_k.

An advantage of lemma (5.6) is that any approximate eigenvector of T leads to an estimate for an eigenvalue. A disadvantage is that one does not know where this eigenvalue is in relation to the others.

Examples in the use of lemmas (5.6) and (5.7) will be given in the supplement.

Lemma (5.4) will be applied in Chapter 7, in conjunction with lemma (5.1).

In (the notation of) lemma (5.1) we have

$$\sum_{i=1}^{m} \left| \sum_{j=1}^{n} c_{ij} a_j \right|^2 = \bar{\mathbf{a}}^T \bar{\mathbf{C}}^T \mathbf{C} \mathbf{a},$$

where $\mathbf{C} = (c_{ij})$, $1 \le i \le m$, $1 \le j \le n$, and $\mathbf{a}$ belongs to $\mathbb{C}^m$. Similarly

$$\sum_{j=1}^{n} \left| \sum_{i=1}^{m} c_{ij} b_i \right|^2 = \bar{\mathbf{b}}^T \mathbf{C} \bar{\mathbf{C}}^T \mathbf{b},$$

where $\mathbf{b}$ belongs to $\mathbb{C}^m$. Here $\bar{\mathbf{C}}^T \mathbf{C}$ and $\mathbf{C} \bar{\mathbf{C}}^T$ are clearly hermitian, and give rise to non-negative definite forms. Their eigenvalues are therefore non-negative. Let them be $\mu_1 \ge \mu_2 \ge \cdots \ge \mu_n$ and $\rho_1 \ge \rho_2 \ge \cdots \ge \rho_m$, respectively.

Lemma (5.1) asserts that $\mu_1 = \rho_1$. In fact $\mu_i = \rho_i$ for $i = 1, \ldots, \min(m, n)$, and considerations of rank show that the remaining eigenvalues are zero. For choose vectors $\mathbf{y}_j$, $j = 1, \ldots, i-1$ so that

$$|\mathbf{C} \mathbf{a}| \le \mu_i^{1/2} |\mathbf{a}|$$

for all $\mathbf{a}$ which satisfy $\mathbf{y}_j^T \mathbf{a} = 0$, $1 \le j \le i-1$. Then following the proof of lemma (5.1) we obtain

$$|\mathbf{b}^T \mathbf{C}| \le \mu_i^{1/2} |\mathbf{b}|$$

for all vectors $\mathbf{b}$ which satisfy

$$\mathbf{y}_j^T \bar{\mathbf{C}}^T \mathbf{b} = 0, \quad 1 \le j \le i-1.$$

An application of lemma (5.3) shows that $\rho_i \leq \mu_i$. Since this inequality may also be obtained in the other direction, our assertion is justified.

In particular

$$\sum_{j=1}^{n} \mu_j = \sum_{i=1}^{m} \rho_i = \operatorname{tr} \bar{\mathbf{C}}^T \mathbf{C} = \sum_{i=1}^{m} \sum_{j=1}^{n} |c_{ij}|^2,$$

giving the lower bound

$$\mu_1 \min(m, n) \geq \sum_{i=1}^{m} \sum_{j=1}^{n} |c_{ij}|^2.$$

Applying a result of this type, restricting the range of prime-powers considered to $p^k \leq P$ for a variable $P(\leq x)$, it is possible to show that if the first inequality of lemma (5.2) holds with $\leq \lambda_0$ in place of $\ll x^{1-\sigma}$, then

$$\lambda_0 \geq \frac{x^{1-\sigma}}{1-\sigma} + O(x^{(1-\sigma)^2/(2-\sigma)}(\log x)^2).$$

Only the implied constants in lemma (5.2) (and lemma (1.5)) are susceptible to improvement.

Duality in Hilbert Space

In the remarks following the proof of lemma (1.5) it is shown that for $\sigma > 1$

$$\sum_{n=1}^{\infty} n^{-\sigma} \left| f(n) - \sum_{p^\alpha} p^{-\alpha\sigma}(1 - p^{-\sigma}) f(p^\alpha) \right|^2 \leq \zeta(\sigma) \sum_{p^\alpha} p^{-\alpha\sigma}(1 - p^{-\sigma}) |f(p^\alpha)|^2,$$

with

$$\zeta(\sigma) = \sum_{n=1}^{\infty} n^{-\sigma},$$

Riemann's zeta function. Following the proof of lemma (5.2) we define

$$r(p^\alpha, n) = \begin{cases} \dfrac{p^{\alpha\sigma/2}}{n^{\sigma/2}} \left(1 - \dfrac{1}{p^{\alpha\sigma}}\left(1 - \dfrac{1}{p^\sigma}\right)\right) & \text{if } p^\alpha \| n, \\ -\dfrac{p^{\alpha\sigma/2}}{n^{\sigma/2}} \cdot \dfrac{1}{p^{\alpha\sigma}} \left(1 - \dfrac{1}{p^\sigma}\right) & \text{otherwise,} \end{cases}$$

and obtain

$$\sum_{n=1}^{\infty} \left| \sum_{p^\alpha} r(p^\alpha, n) f(p^\alpha) \right|^2 \leq \zeta(\sigma) \sum_{p^\alpha} |f(p^\alpha)|^2.$$

This inequality is valid provided $\{f(p^\alpha)\}$ belongs to the appropriate Hilbert space of square-summable sequences.

Dualizing we obtain eventually the inequality

(1) $$\sum_{p^\alpha} p^{\alpha\sigma} \left| \sum_{\substack{n=1 \\ p^\alpha \| n}}^{\infty} \frac{a_n}{n^\sigma} - \frac{1}{p^{\alpha\sigma}}\left(1 - \frac{1}{p^\sigma}\right) \sum_{n=1}^{\infty} \frac{a_n}{n^\sigma}\right|^2 \leq \zeta(\sigma) \sum_{n=1}^{\infty} \frac{|a_n|^2}{n^\sigma},$$

which is valid for all complex numbers a_n for which the series on the right-hand side converges. An inequality of this type was obtained as lemma 6 in the author's paper [9]. In the Second Motive we shall discuss the rôle which it played there.

The above argument by duality still applied since the proof of lemma (5.1) depended only upon the Cauchy–Schwarz inequality, and thus the existence of an inner product.

If in our above inequality (1) we set $\sigma = 1 + (\log x)^{-1}$ and $a_n = 0$ for $n > x$, then we can obtain an analogue of lemma (5.2) which is valid when $\sigma = 1$.

Duality in General

Let X be a space, linear over the complex numbers, with norm $\| \ \|$. The space of bounded linear functions f on X into the complex numbers may be given a norm by

$$\|f\| = \sup_{x \neq 0} |f(x)|/\|x\|,$$

x being taken over the elements of X. We call this space the (strong) dual of X, and denote it by X'. It may be considered a linear space by defining

(2) $$(\alpha f_1 + \beta f_2)(x) = \alpha f_1(x) + \beta f_2(x).$$

Let Y be a further normed linear space, with dual Y'. Let T be a bounded linear map on X into Y. To each functional g of Y' we make correspond the functional f on X' which is defined by

$$f(x) = g(T(x)).$$

The map $T': Y' \to X'$ so defined may be viewed as linear and is said to be the *dual* of T.

Giving T' the norm

$$\|T'\| = \sup_{g \neq 0} \|T'(g)\|/\|g\|$$

it follows from the Hahn–Banach theorem that

(3) $$\|T\| = \|T'\|.$$

Suppose now that X is a Hilbert space, with inner product $[\ ,\ ]$. Then we can identify X with its dual space X' by the map J which takes the vector x to the linear map

$$u \mapsto [u, x], \qquad u \in X.$$

This map preserves norms but is not quite linear since for a complex number α, $J(\alpha x) = \bar{\alpha} J(x)$.

Suppose further that Y is a Hilbert space, and let K be its identification with its dual space Y'.

The *adjoint* of a bounded linear map $T: X \to Y$ is then defined to be the map $T^*: Y \to X$ given by

$$T^* = J^{-1} T' K.$$

This ensures that

$$[Tx, y] = [x, T^*y]$$

for all x in X and y in Y, where the inner products are defined on Y and X, respectively. In these circumstances $(T^*)^* = T$.

It is readily checked that for each x in X

$$\|Tx\|^2 = [x, T^*Tx]$$

so that

$$\|T\|^2 = \sup_{\|x\|=1} |[x, T^*Tx]|.$$

Similarly, for each y in Y

$$\|T'Ky\|^2 = [y, TT^*y]$$

from which

$$\|T'\|^2 = \sup_{\|y\|=1} |[y, TT^*y]|.$$

The maps T^*T and TT^* are self-adjoint on the spaces X, Y, respectively, with real spectra contained in a bounded interval of non-negative numbers.

The eigenvectors of T^*T which correspond to non-zero eigenvalues, are taken in a one-to-one manner onto the similar eigenvectors of TT^* by $x \mapsto Tx$. Indeed, if $T^*Tz = \lambda z$, then

$$TT^*(Tz) = \lambda Tz,$$

where $Tz = 0$ can only hold if $\lambda z = 0$. This shows that the map between eigenvectors of T^*T and of TT^* is one-to-one and into. That it is onto follows from a similar argument with T^* in place of every T.

Clearly T^*T and TT^* possess the same positive eigenvalues, and T gives an isomorphism between corresponding eigenspaces. This argument says nothing about whether zero is an eigenvalue of one of these self-adjoint operators, nor whether their spectra contain numbers which are not eigenvalues.

The situation in lemma (5.1) is that $X = \mathbb{C}^n$, $Y = \mathbb{C}^m$ and the map T is given by $\mathbf{a} \mapsto \mathbf{Ca}$. The dual spaces X' and Y' are further copies of $\mathbb{C}^n$ and $\mathbb{C}^m$, respectively. In terms of the standard bases T' is represented by the matrix which is the transpose of $\mathbf{C}$, and T^* by the complex conjugate of the transpose of $\mathbf{C}$.

If we use the appropriate L^2 (mean-square)-norm in each space, then the assertion of lemma (5.1) follows from (3). However, the direct proof of lemma (5.1) does not use the axiom of choice.

To derive the bound (1) we identify X and Y with the square-summable sequences $\{f(p^\alpha)\}$ prime-powers p^α, and $\{d_n\}$ positive integers n, respectively. The map T is represented by the doubly infinite matrix

$$(r(p^\alpha, n)),$$

and the inequalities obtained assert that

$$\|T\| = \|T'\| \leq \zeta(\sigma).$$

Other variants are useful. In Chapter 6, in connection with the inequality of the large sieve, we shall consider a map between a finite space and an infinite dimensional space. In this connection we mention also the possibility of applying a map between a space of sequences and the space of L^2 complex-valued functions on an interval of the line, as indicated in the author's paper [2].

A further application of duality will be made in Chapter 7.

The use of Dirichlet series and the application of Fourier analysis to the study of real numbers (mod 1) may also be viewed as exercises in the application of duality. Indeed, it might be said that much of analytic number theory consists of the study of the discrete by means of the continuous dual.

Second Motive

An additive function $f(n)$ is *strongly-additive* if it satisfies $f(p^m) = f(p)$ for every prime-power p^m. It is then completely determined by its values on the prime numbers.

In the author's paper [9] it was proved that for such a function

(1) $$\sum_{n \leq x} |f(n+1) - f(n)|^2 \ll x$$

holds for all $x \geq 1$ if and only if there is a constant A so that the series

(2) $$\sum p^{-1} |f(p) - A \log p|^2$$

converges. The more difficult part of that argument was the deduction of (2) from (1). It began with the bound

(3) $$\sum_{n \leq x} |f(n)|^2 \ll x (\log x)^2.$$

The problem of deducing a bound of this type from an hypothesis such as (1) was discussed in the First Motive, and dealt with in Chapters 2 and 3.

As the next step in the derivation of (2), the inequality

(4) $$\sum p^\sigma \left| \sum_{\substack{n=1 \\ n \equiv 0 (\text{mod } p)}}^\infty a_n n^{-\sigma} - p^{-\sigma} \sum_{n=1}^\infty a_n n^{-\sigma} \right|^2 \leq \zeta(\sigma) \sum_{n=1}^\infty |a_n|^2 n^{-\sigma}, \qquad \sigma > 1,$$

was obtained in a manner similar to that which gave the inequality (1) of Chapter 5.

In accordance with the philosophy expressed in the introduction to this volume, this inequality was applied with a_n specialized to $f(n+1) - f(n)$, to obtain (after some argument)

(5) $$\sum_p p^{-\sigma} \left| f(p) - \frac{p^\sigma}{\zeta(\sigma)} \sum_{q \neq p} f(q) q^{-\sigma} \Delta(q, p) \right|^2 \ll 1, \qquad 1 < \sigma \leq 2,$$

where

$$\Delta(q, p) = \sum_{\substack{w=1 \\ wq \equiv 1 \pmod p}}^{\infty} \frac{1}{w^\sigma} - \frac{1}{p^\sigma} \zeta(\sigma).$$

An important aspect of this result was that $\Delta(q, p)$ depended only upon the residue class (mod p) to which q belonged, and not upon the size of q itself.

The sum involving the terms

$$f(q) q^{-\sigma} \Delta(q, p)$$

was then treated by means of an inequality

$$(6) \quad \sum_{p \leq Q} (p-1) \sum_{l=1}^{p-1} \left| \sum_{\substack{q \leq x \\ q \equiv l \pmod p}} d_q - \frac{1}{p-1} \sum_{q \leq x} d_q \right|^2 \ll \left(\frac{x}{\log x} + D_1 \right) \sum_{q \leq x} |d_q|^2,$$

where q denotes a prime, and D_1 depends upon Q but not x. This appeared as lemma 9 of the author's paper [9].

After some further argument a bound

$$(7) \quad \sum_{p \leq Q} p^{-\sigma} \left| f(p) - \delta(\sigma, p) \sum_q f(q) q^{-\sigma} \right|^2 \ll Q^{\sigma-1} + Q^{\sigma-1} D_1 \zeta(\sigma)^{-1}$$

was reached, with a certain weighting factor $\delta(\sigma, p)$.

The proof was then completed by letting $\sigma \to 1+$, and $Q \to \infty$.

In order to more accurately measure the control which the differences $f(n+1) - f(n)$ have over the values $f(p)$ of an additive function at the primes p, one implicitly wishes to carry through an argument of the above type starting with an hypothesis which is weaker than (1), say with

$$(8) \quad \sum_{n \leq x} |f(n+1) - f(n)|^2 \ll x (\log x)^d$$

for some fixed $d > 0$. A natural approach is to allow the parameters σ and Q to depend upon x, to try for inequalities of some uniformity in σ, Q. As it stands the above argument leads to inequalities which are in this respect weak, and seems incapable of dealing with an hypothesis such as (8).

Consider the Dirichlet series

$$F(s) = \sum_{n=1}^{\infty} f(n) n^{-s}.$$

In view of (3) the sum function is analytic in the (complex) half-plane

Re(s) > 1. We expect $f(n)$ to behave like $A \log n$ for some constant A, so that if $A \neq 0$ then as $\sigma \to 1+$

$$F(\sigma) \approx A \sum_{n=1}^{\infty} n^{-\sigma} \log n = -A\zeta'(\sigma) \sim A(\sigma-1)^{-2}.$$

Thus $F(s)$ acts as if it has a double-pole at $s = 1$. In letting σ approach 1 in the inequality (7) we are essentially obtaining the residue at this pole. Actually the inequality (7) involves the series

$$\sum_q q^{-\sigma} f(q) = \zeta(s)^{-1} F(s)$$

which has only a simple pole, but the sense of the argument is the same.

This situation may be compared to a standard proof of the prime number theorem. For Re(s) > 1 we have

$$\sum_{p,k} p^{-ks} \log p = -\frac{\zeta'(s)}{\zeta(s)},$$

and the simple pole of $\zeta(s)$ at $s = 1$ ultimately gives rise to the asymptotic estimate

(9) $$\sum_{p^k \leq x} \log p \sim x, \qquad x \to \infty.$$

From this the basic prime number theorem follows. If one seeks an estimate which has a smaller remainder than that implied here, the sum on the left-hand side of (9) is expressed in terms of a contour integral and the integral is moved into the *critical strip* $0 < \text{Re}(s) < 1$. This is possible because $\zeta(s)$ has an analytic continuation over the whole complex s-plane, and is analytic at all points save $s = 1$. A knowledge of the zeros of $\zeta(s)$ now controls the analytic behaviour of the function $\zeta'(s)/\zeta(s)$, and so the sharpening of (9).

We might therefore try to "localize" our treatment of additive arithmetic functions by analytically continuing $F(s)$ to the left of the line Re(s) = 1, and moving our argument into the critical strip $0 < \sigma = \text{Re}(s) < 1$. The existence of such an analytic continuation of $F(s)$ is practically the same as a solution to our general problem, so we compromise by choosing a value of σ, $\frac{1}{2} < \sigma < 1$, and replacing the use of $F(s)$ by that of the truncated series

$$\sum_{n \leq x} \frac{f(n)}{n^\sigma}.$$

Analogues of the inequalities (4) and (5) are readily obtained, but the function $\Delta(q, p)$ is then naturally replaced by another which involves sums of the form

$$S(x, q) = \sum_{\substack{w \leq xq^{-1} \\ wq \equiv 1 (\mathrm{mod}\ p)}} \frac{1}{w^\sigma}.$$

Here the condition $w \leq xq^{-1}$ involves the size of q, and we have lost an important aspect of our earlier treatment.

This disadvantage is overcome by expressing such sums in terms of Mellin transforms:

$$S(x, q) = \frac{1}{2\pi i} \int \sum_{\substack{w \leq x \\ wq \equiv 1 (\mathrm{mod}\ p)}} \frac{1}{w^{\sigma+s}} \cdot \left(\frac{x}{q}\right)^s \frac{ds}{s},$$

the contour being taken over an appropriate line $\mathrm{Re}(s) = \sigma_0 > 0$. As q varies the sum

$$\sum_{\substack{w \leq x \\ wq \equiv 1 (\mathrm{mod}\ p)}} \frac{1}{w^{\sigma+s}}$$

once again depends only upon the value of q (mod p), and the factor $(xq^{-1})^s$ is manageable. There is a little loss of precision in this procedure, but it proves to be more than compensated by the restoration of our old advantage. A detailed account of this argument is given in Chapter 7, at the start of the proof of theorem (7.1).

An appropriate analogue of the inequality (6) is established in Chapter 6. Its rôle in our present treatment of additive functions is somewhat reduced in comparison to that of the corresponding inequality in the author's paper [9].

An analogue to the inequality (7) is obtained in Chapter 8. As might be expected, it is not possible to close our study by choosing σ to be an appropriate function of x. We are, instead, led to the study of certain non-linear approximate functional equations. These we shall consider in the Third Motive.

Chapter 6

Lemmas Involving Prime Numbers

The Large Sieve and Prime Number Sums

For real numbers y, $\|y\|$ denotes the distance of y from the nearest integer.

Lemma (6.1) (The Large Sieve). *Let $x_j, j = 1, \ldots, J$ be real numbers which satisfy*

$$\|x_j - x_k\| \geq \delta > 0$$

for all $j \neq k$. Then the inequality

$$\sum_{j=1}^{J} |S(x_j)|^2 \ll (N + \delta^{-1}) \sum_{n=M}^{M+N} |a_n|^2,$$

with

$$S(x_j) = \sum_{n=M}^{M+N} a_n e^{2\pi i n x_j}$$

holds for all complex numbers a_n, for all positive integers N and integers M. The implied constant is absolute.

Proof. There are several proofs of this result. See, for example, Bombieri [1], Davenport and Halberstam [1], Gallagher [1]. We shall give a proof using an approach suggested by the concluding remarks of the author's paper, Elliott [2].
   Since

$$S(x_j) = e^{2\pi i M x_j} \sum_{n=1}^{N} a_{M+n} e^{2\pi i n x_j}$$

we may without loss of generality assume that $M = 0$.

6. Lemmas Involving Prime Numbers

We look for the smallest value of λ so that

$$\sum_{j=1}^{J} \left| \sum_{n=-\infty}^{\infty} b_n e^{-\pi n^2/N^2 + 2\pi i n x_j} \right|^2 \leq \lambda \sum_{n=-\infty}^{\infty} |b_n|^2 \tag{1}$$

holds for all complex sequences $\{b_n\}$ which belong to the Hilbert space of square-summable doubly infinite sequences.

This has the same value as the least λ for which the dual inequality

$$\sum_{n=-\infty}^{\infty} \left| \sum_{j=1}^{J} c_j e^{-\pi n^2/N^2 + 2\pi i n x_j} \right|^2 \leq \lambda \sum_{j=1}^{J} |c_j|^2$$

is satisfied for all complex numbers c_j, $j = 1, \ldots, J$. Expanding the square and inverting the order of summation gives an hermitian form

$$\sum_{j,k=1}^{J} c_j \bar{c}_k \sum_{n=-\infty}^{\infty} e^{-\pi n^2/N^2 + 2\pi i n(x_j - x_k)}.$$

From the theory of elliptic functions (or Davenport [1], pp. 63–64) we apply the functional equation

$$\sum_{n=-\infty}^{\infty} e^{-\pi n^2 y + 2\pi i n \alpha} = y^{-1/2} \sum_{n=-\infty}^{\infty} e^{-\pi(n+\alpha)^2/y},$$

which is certainly valid for real α and $y > 0$. A typical coefficient in our hermitian form may then be expressed as

$$N \sum_{n=-\infty}^{\infty} e^{-\pi(n+\|x_j - x_k\|)^2 N^2}.$$

Since

$$|c_j \bar{c}_k| \leq (|c_j|^2 + |c_k|^2)/2,$$

we see that

$$\lambda \leq \max_{k} \sum_{j=1}^{J} N \sum_{n=-\infty}^{\infty} e^{-\pi(n+\|x_j - x_k\|)^2 N^2}.$$

For a fixed k the term in this upper bound which has $j = k$ contributes

$$N \sum_{n=-\infty}^{\infty} e^{-\pi n^2 N^2} \leq N + N \int_{-\infty}^{\infty} e^{-\pi w^2 N^2} dw = N + 1.$$

Using the inequality

$$(n+\theta)^2 \geq \theta^2/2 + n^2/8$$

which is valid for all real θ, $|\theta| \leq \frac{1}{2}$, and integers n, it is clear that the remaining terms which bound λ contribute

$$\ll N \sum_{j(\neq k)} e^{-\pi \|x_j - x_k\|^2 N^2/2}$$

$$\ll N \sum_{m=1}^{\infty} e^{-\pi m^2 \delta^2 N^2/2} \ll \delta^{-1}.$$

We now apply the inequality (1), setting

$$b_n = \begin{cases} e^{\pi n^2/N^2} a_n & \text{if } 1 \leq n \leq N, \\ 0 & \text{if } n > N. \end{cases}$$

Note that for the first N positive integers $e^{\pi n^2/N^2} \leq e^{\pi}$. Since

$$\lambda \ll N + \delta^{-1}$$

lemma (6.1) is proved.

Remark. For each k define

$$\delta_k = \min_{j \neq k} \|x_j - x_k\|.$$

Then with minor changes the above method of proof gives

(2) $$\sum_{j=1}^{J} \left(N + \frac{1}{\delta_j} \right)^{-1} |S(x_j)|^2 \ll \sum_{n=M}^{M+N} |a_n|^2,$$

or even

$$\sum_{j=1}^{J} (N + e^{-c_0 N^2} + c_1 \delta_j^{-1} e^{-c_0 \delta_j^2 N^2})^{-1} |S(x_j)|^2 \leq \sum_{n=M}^{M+N} e^{2\pi n^2/N^2} |a_n|^2$$

for certain positive constants c_0 and c_1. We note here only that over the interval

$$x_j - \tfrac{1}{2}\delta_j \leq w \leq x_j + \tfrac{1}{2}\delta_j$$

we have

$$\|w - x_k\| \leq \tfrac{3}{2}\|x_j - x_k\|, \qquad k \neq j,$$

so that for any fixed $c > 0$

$$\sum_{j(\neq k)} \delta_j N e^{-c\|x_j-x_k\|^2 N^2} \leq N \sum_j \int_{x_j-\delta_j/2}^{x_j+\delta_j/2} e^{-4c\|w-x_k\|^2 N^2/9} \, dw$$

$$\leq 2N \int_{\delta_k/2}^{\infty} e^{-4c^2 t^2 N^2/9} \, dt \ll 1.$$

A result of the form (2) was first proved by Montgomery and Vaughan [1], [2]. Their proof of this and the basic large sieve inequality is more complicated, but gives better values for the implied (absolute) constants. This ends our remarks.

Let D be a positive integer and Q a positive real number. Let the x_j in lemma (6.1) run through the fractions

$$\frac{r}{Dd}, \quad 1 \leq r \leq Dd, \quad (r, Dd) = 1, \quad 1 \leq d \leq Q.$$

Any two distinct fractions of this type satisfy

$$\left| \frac{r_1}{Dd_1} - \frac{r_2}{Dd_2} \right| \geq \frac{1}{Dd_1 d_2} \geq \frac{1}{DQ^2},$$

so that

(3) $$\sum_{d \leq Q} \sum_{r=1}^{Dd}{}' \left| S\left(\frac{r}{Dd}\right) \right|^2 \ll (N + DQ^2) \sum_{n=M}^{M+N} |a_n|^2$$

where $'$ indicates that summation is restricted to those integers r which are prime to Dd.

For each prime p

$$\sum_{r=1}^{p-1} \left| S\left(\frac{r}{p}\right) \right|^2 = \sum_{r=0}^{p-1} \left| S\left(\frac{r}{p}\right) \right|^2 - \left| \sum_{n=M}^{M+N} a_n \right|^2$$

$$= p \sum_{\substack{m,n=M \\ m \equiv n(\mathrm{mod}\, p)}}^{M+N} a_m \bar{a}_n - \left| \sum_{n=M}^{M+N} a_n \right|^2$$

$$= p \sum_{s=0}^{p-1} \left| \sum_{\substack{n=M \\ n \equiv s(\mathrm{mod}\, p)}}^{M+N} a_n \right|^2 - \left| \sum_{n=M}^{M+N} a_n \right|^2$$

$$= p \sum_{s=0}^{p-1} \left| \sum_{\substack{n=M \\ n \equiv s(\mathrm{mod}\, p)}}^{M+N} a_n - \frac{1}{p} \sum_{n=M}^{M+N} a_n \right|^2.$$

We deduce from the inequality (3) with $D = 1$ that

$$(4) \quad \sum_{p \leq Q} p \sum_{s=0}^{p-1} \left| \sum_{\substack{n=M \\ n \equiv s \pmod{p}}}^{M+N} a_n - \frac{1}{p} \sum_{n=M}^{M+N} a_n \right|^2 \ll (N + Q^2) \sum_{n=M}^{M+N} |a_n|^2.$$

We shall apply a variant of this result where the (dummy) variable n is confined to prime-powers. To that end we first derive the following theorem of Bombieri and Davenport (see Bombieri [1], Théorème 8, pp. 24–26).

Lemma (6.2). *In the above notation let $a_n = 0$ whenver $(n, d) > 1$ for some d not exceeding Q. Then*

$$\sum_{d \leq Q} \log \frac{Q}{d} \sum^*_{\chi \pmod{d}} \left| \sum_{n=M}^{M+N} a_n \chi(n) \right|^2 \ll (N + Q^2) \sum_{n=M}^{M+N} |a_n|^2,$$

where $*$ denotes that summation is over the Dirichlet characters χ which are primitive with respect to d.

Proof. For each character $\chi \pmod{w}$ we define the Gauss sum

$$\tau(\chi) = \sum_{\substack{r=1 \\ (r,w)=1}}^{w} \chi(r) e^{2\pi i r / w}.$$

If $(n, w) = 1$ then the substitution $r \mapsto rn \pmod{w}$ shows that

$$\tau(\bar{\chi}) \chi(n) = \sum_{r=1}^{w} {}' \bar{\chi}(r) e^{2\pi i r n / w}.$$

It follows at once that in our present circumstances

$$(5) \quad \sum_{\chi \pmod{w}} |\tau(\bar{\chi})|^2 \left| \sum_{n=1}^{N} a_n \chi(n) \right|^2 = \sum_{\chi \pmod{w}} \left| \sum_r {}' \bar{\chi}(r) S\left(\frac{r}{w}\right) \right|^2,$$

where χ is summed over all the characters $\pmod{w}$, including the principal character. Using the fact that

$$\sum_{\chi \pmod{w}} \chi(u) = \begin{cases} \varphi(w) & \text{if } u \equiv 1 \pmod{w}, \\ 0 & \text{otherwise}, \end{cases}$$

the right-hand expression at (5) may be expressed in the form

$$\varphi(w) \sum_{r=1}^{w} {}' \left| S\left(\frac{r}{w}\right) \right|^2.$$

From (3) with $D=1$ we deduce that

(6) $\quad \sum_{w \leq Q} \dfrac{1}{\varphi(w)} \sum_{\chi(\bmod w)} |\tau(\bar{\chi})|^2 \left| \sum_{n=M}^{M+N} a_n \chi(n) \right|^2 \ll (N+Q^2) \sum_{n=M}^{M+N} |a_n|^2.$

Suppose now that $\chi(\bmod w)$ is induced by a (primitive) character $\chi_1(\bmod d)$. Then $d|w$ and

$$|\tau(\bar{\chi})|^2 = \begin{cases} |\mu(wd^{-1})|d & \text{if } (wd^{-1}, d) = 1, \\ 0 & \text{otherwise.} \end{cases}$$

(See Davenport [1], p. 67.) Here $|\mu(n)|$, the absolute value of Möbius' function, is 1 if n has no squared factor greater than 1, and is zero otherwise.

We collect terms in the multiple sum at (6) according to the inducing characters $\chi_1(\bmod d)$. The sum is thus

$$\sum_{d \leq Q} \sum_{\chi_1(\bmod d)}^* \left| \sum_{n=M}^{M+N} a_n \chi_1(n) \right|^2 \sum_{\substack{w \leq Q \\ w \equiv 0(\bmod d)}} \dfrac{1}{\varphi(w)} \sum_{\chi(\bmod w)} |\tau(\bar{\chi})|^2$$

where the inner (double-) sum is taken over those characters $\chi(\bmod w)$ which are induced by $\chi_1(\bmod d)$. Note that if $(n, w) = 1$ (as we presently have) then $\chi(n) = \chi_1(n)$. A typical innersum is

$$\sum_{\substack{w \leq Q \\ w \equiv 0(\bmod d) \\ (wd^{-1},d)=1}} \dfrac{1}{\varphi(w)} \left| \mu\left(\dfrac{w}{d}\right) \right| d = \dfrac{d}{\varphi(d)} \sum_{\substack{t \leq Q/d \\ (t,d)=1}} \dfrac{|\mu(t)|}{\varphi(t)}$$

$$= \sum_{\substack{t \leq Q/d \\ (t,d)=1}} \dfrac{|\mu(t)|}{t} \prod_{p|t}\left(1+\dfrac{1}{p}+\dfrac{1}{p^2}+\cdots\right) \prod_{p|d}\left(1+\dfrac{1}{p}+\dfrac{1}{p^2}+\cdots\right)$$

$$\geq \sum_{t \leq Q/d} \dfrac{1}{t} \geq \int_1^{Q/d} \dfrac{dy}{y} = \log \dfrac{Q}{d}.$$

Lemma (6.2) is proved.

Lemma (6.3). *Let $\varepsilon > 0$ be fixed. Then the inequality*

$$\sum_{p \leq Q} (p-1) \sum_{r=1}^{p-1} \left| \sum_{\substack{q \leq N \\ q \equiv r(\bmod p)}} a_q - \dfrac{1}{p-1} \sum_{q \leq N} a_q \right|^2 \ll \left(\dfrac{N}{\log N} + Q^{2+\varepsilon}\right) \sum_{q \leq N} |a_q|^2$$

holds for all complex numbers a_q, where q runs through the prime-powers not exceeding N, for all integers $N \geq 2$. The implied constant depends only upon ε.

Proof. We first note that if d is a positive integer and $c_n = 0$ unless $(n, d) = 1$, then

$$\varphi(d) \sum_{\substack{r=1 \\ (r,d)=1}}^{d} \left| \sum_{n \equiv r (\bmod d)} c_n - \frac{1}{\varphi(d)} \sum_n c_n \right|^2 = \sum_{\chi (\bmod d)} \left| \sum_n c_n \chi(n) \right|^2,$$

where χ runs through the non-principal Dirichlet characters (mod d). If d is a prime then these will be primitive.

Consider the sum

$$T_1 = \sum_{p \leq Q} (p-1) \sum_{r=1}^{p-1} \left| \sum_{l \equiv r (\bmod p)}'' a_l - \frac{1}{p-1} \sum'' a_l \right|^2,$$

where $''$ indicates that l ranges over the primes in the interval $QN^\varepsilon < l \leq N$. Since each such l is prime to each p not exceeding Q, by lemma (6.2)

$$T_1 \ll \frac{1}{\log N} \sum_{p \leq Q} \log \frac{QN^\varepsilon}{p} \sum_{\chi (\bmod p)}^* \left| \sum'' a_l \chi(l) \right|^2$$

$$\ll \frac{1}{\log N} (N + Q^2 N^{2\varepsilon}) \sum_{q \leq N} |a_q|^2.$$

Now let t run through the prime-powers l^k with $k \geq 2$ which lie in the interval $QN^\varepsilon < l^k \leq N$. By the Cauchy–Schwarz inequality the sum

$$T_2 = \sum_{p \leq Q} (p-1) \sum_{r=1}^{p-1} \left| \sum_{t \equiv r (\bmod p)} a_t - \frac{1}{p-1} \sum_{t \leq N} a_t \right|^2$$

does not exceed $2(T_3 + T_4)$ where

$$T_3 = \sum_{p \leq Q} (p-1) \sum_{r=1}^{p-1} \left| \sum_{t \equiv r (\bmod p)} a_t \right|^2,$$

$$T_4 = \sum_{p \leq Q} (p-1) \sum_{r=1}^{p-1} \frac{1}{(p-1)^2} \left| \sum_{t \leq N} a_t \right|^2.$$

In order to bound T_3 we need an estimate for the sum

$$L(p) = \sum_{\substack{t \leq N \\ t \equiv r (\bmod p)}} 1.$$

This may also be expressed in the form

$$\sum_{k \le \log N/\log 2} \sum_{\substack{l \le N^{1/k} \\ l^k \equiv r \pmod p}} 1.$$

If $k < p$ then the congruence $l^k \equiv r \pmod p$ has at most k solutions in $l \pmod p$. Setting

$$\delta = \min\left(p, \frac{\log N}{\log 2}\right),$$

we see that the terms with $k < \delta$ contribute an amount

$$\ll \sum_{2 \le k < \delta} k\left(\frac{N^{1/k}}{p} + 1\right) \ll (p^{-1} N^{1/2} + 1) \log N.$$

The terms with $k \ge \delta$ are each $\ll N^{1/p}$ in size, and

$$L(p) \ll (p^{-1} N^{1/2} + N^{1/p}) \log N.$$

Applying this bound when $Q > N$ we can readily show that whether this condition is satisfied or not

$$\max_{p \le Q} pL(p) \ll (N^{1/2} + Q) \log N.$$

This together with an application of the Cauchy–Schwarz inequality gives

$$T_3 \le \sum_{p \le Q} p \sum_{r=1}^{p-1} L(p) \sum_{t \equiv r \pmod p} |a_t|^2$$

$$\ll \sum_{r < Q} \sum_{r < p \le Q} pL(p) \sum_{t \equiv r \pmod p} |a_t|^2$$

$$\ll \sum_{r < Q} (N^{1/2} + Q) \log N \sum_{t \le N} |a_t|^2 \sum_{p | (t-r)} 1$$

$$\ll (QN^{1/2} + Q^2)(\log N)^2 \sum_{t \le N} |a_t|^2.$$

Note that in our present circumstance

$$t - r > Q(N^\varepsilon - 1) \ge 0.$$

A simpler argument may be used to bound T_4, and we obtain

$$T_2 \ll (QN^{1/2} + Q^2)(\log N)^2 \sum_{q \le N} |a_q|^2.$$

Moreover, a direct application of (4) shows that

$$(7) \quad \sum_{p \leq Q} (p-1) \sum_{r=1}^{p-1} \left| \sum_{\substack{q \leq QN^{\varepsilon} \\ q \equiv r \pmod{p}}} a_q - \frac{1}{p} \sum_{q \leq QN^{\varepsilon}} a_q \right|^2 \ll (QN^{\varepsilon} + Q^2) \sum_{q \leq N} |a_q|^2,$$

whilst trivially

$$(8) \quad \sum_{p \leq Q} (p-1) \sum_{r=1}^{p-1} \left| \frac{1}{p} \sum_{q \leq QN^{\varepsilon}} a_q - \frac{1}{(p-1)} \sum_{q \leq QN^{\varepsilon}} a_q \right|^2 \ll QN^{\varepsilon} \sum_{q \leq N} |a_q|^2.$$

For $Q < N^{1/2-\varepsilon}$ the bounds for T_1 and T_2 along with those of (7) and (8) give the inequality stated in the lemma. If this condition on Q fails then the argument which gives (7) and (8) will already lead to a similar inequality, since the factor $QN^{\varepsilon} + Q^2$ will be replaced by $N + Q^2$ which will then be $\ll Q^{2+8\varepsilon}$, assuming (as we may) that $4\varepsilon \leq 1$.

Since ε may be fixed at an arbitrarily small value, lemma (6.3) is established.

Remark. I obtained an inequality similar to that of lemma (6.3) (see ref. [18]) using considerations of duality, reducing the problem to the application of estimates for the number of zeros of Dirichlet L-functions in the critical strip. These estimates, of standard type, are generally obtained by the application of inequalities of large sieve type, amongst other ideas. In this way I obtained a result similar to that of the present lemma, but with Q^5 in place of $Q^{2+\varepsilon}$. The possibility of applying lemma (6.2) of Bombieri and Davenport to a similar end was suggested to me by H. Iwaniec.

We shall need the following modified version of lemma (6.3).

Lemma (6.4). *Let $0 \leq \sigma < 1$, $0 < \varepsilon < 1$ hold. Then*

$$\sum_{p \leq Q} (p-1) \sum_{r=1}^{p-1} \left| \sum_{\substack{q \leq N \\ q \equiv r \pmod{p}}} \frac{a_q}{q^{\sigma}} - \frac{1}{p-1} \sum_{q \leq N} \frac{a_q}{q^{\sigma}} \right|^2 \ll \left(\frac{N^{1-\sigma}}{\log N} + Q^{2+\varepsilon} \right) \sum_{q \leq N} \frac{|a_q|^2}{q^{\sigma}}$$

holds for all complex numbers a_q, where q denotes a prime-power, and all integers $N \geq 2$. The implied constant depends only upon σ and ε.

Proof. Assume first that

$$N > Q^{2+2\varepsilon} = N_0.$$

For $0 \leq \sigma < 1$, $N_0 \leq u \leq N$ let

$$F(u, r, p, \sigma) = \sum_{\substack{N_0 < q \leq u \\ q \equiv r \pmod{p}}} \frac{a_q}{q^{\sigma}} - \frac{1}{p-1} \sum_{N_0 < q \leq u} \frac{a_q}{q^{\sigma}}.$$

Integrating by parts we obtain the bound

$$|F(N, r, p, \sigma)| \leq N^{-\sigma}|F(N, r, p, 0)| + \int_{N_0}^{N} \sigma y^{-\sigma-1} |F(y, r, p, 0)| \, dy. \tag{9}$$

We write the integrand in the form

$$\frac{1}{y^{\sigma/2}\sqrt{\log y}} \cdot \frac{\sqrt{\log y}}{y^{\sigma/2}} |F(y, r, p, \sigma)|$$

preparatory to applying the Cauchy–Schwarz inequality for integrals. Squaring both sides of the inequality (9) gives

$$|F(N, r, p, \sigma)|^2 \ll N^{-2\sigma}|F(N, r, p, 0)|^2 + \frac{N^{1-\sigma}}{\log N} \int_{N_0}^{N} \frac{\log y}{y^{\sigma+2}} |F(y, r, p, 0)|^2 \, dy.$$

From the inequality of lemma (6.3)

$$\sum_{p \leq Q} (p-1) \sum_{r=1}^{p-1} |F(y, r, p, 0)|^2 \ll \frac{y}{\log y} \sum_{N_0 < q \leq y} |a_q|^2,$$

since in our present situation

$$Q^{2+\varepsilon} \leq N_0^{1-\varepsilon/4} \ll y(\log y)^{-1}.$$

Moreover, with another integration by parts

$$\int_{N_0}^{N} 1/y^{-\sigma-1} \sum_{N_0 < q \leq y} |a_q|^2 \, dy \ll \sum_{N_0 < q \leq N} \frac{|a_q|^2}{q^\sigma}.$$

Hence

$$\sum_{p \leq Q} (p-1) \sum_{r=1}^{p-1} \left| \sum_{\substack{N_0 < q \leq N \\ q \equiv r(\bmod p)}} \frac{a_q}{q^\sigma} - \frac{1}{p-1} \sum_{N_0 < q \leq N} \frac{a_q}{q^\sigma} \right|^2 \ll \frac{N^{1-\sigma}}{\log N} \sum_{N_0 < q \leq N} \frac{|a_q|^2}{q^\sigma}.$$

In order to estimate the sum similar to that on the left-hand side of this inequality, but with summation over $q \leq N_0$ or $q \leq N \leq N_0$, we apply lemma (6.3) directly and note that

$$\sum_{q \leq N_0} \frac{|a_q|^2}{q^{2\sigma}} \leq \sum_{q \leq N_0} \frac{|a_q|^2}{q^\sigma}.$$

Lemma (6.4) is proved.

Remark. By applying lemma (6.3) more carefully over ranges $U < q \leq 2U$ with $U > Q$, and then arguing directly for $q \leq Q$, we can replace the expression $Q^{2+\varepsilon}$ in the inequality of lemma (6.4) by $Q^2/\log Q$.

The following more standard variant of the Large Sieve in multiplicative form will be applied in our present consideration of Dirichlet series and prime-number sums.

Lemma (6.5). *The inequality*

$$\sum_{d \leq Q} \frac{Dd}{\varphi(Dd)} \sum_{\chi (\mathrm{mod}\, Dd)}^{*} \left| \sum_{n=M}^{M+N} a_n \chi(n) \right|^2 \ll (N + DQ^2) \sum_{n=M}^{M+N} |a_n|^2$$

holds for all complex numbers a_n, $n = M, \ldots, M+N$, integers M and N, $N \geq 1$, and real $Q \geq 1$.

Remark. As earlier, * denotes summation over primitive Dirichlet characters (typically defined with respect to the modulus Dd).

Proof. The relation

$$\tau(\bar{\chi})\chi(n) = \sum_{r=1}^{w}{}' \bar{\chi}(r) e^{2\pi i r n / w}$$

holds not only for n prime to w, as noted earlier, but also *if $\chi(\ )$ is primitive* (mod w), for every n. For such characters we also have

$$|\tau(\bar{\chi})| = w^{1/2}.$$

The analogue of (5) in the proof of lemma (6.2) is

$$\frac{w}{\varphi(w)} \sum_{\chi(\mathrm{mod}\, w)}^{*} \left| \sum_{n=M}^{M+N} a_n \chi(n) \right|^2 = \frac{1}{\varphi(w)} \sum_{\chi(\mathrm{mod}\, w)}^{*} \left| \sum_{n=M}^{M+N} a_n \tau(\bar{\chi})\chi(n) \right|^2$$

$$= \frac{1}{\varphi(w)} \sum_{\chi(\mathrm{mod}\, w)}^{*} \left| \sum_{r=1}^{w}{}' \bar{\chi}(r) S\left(\frac{r}{w}\right) \right|^2$$

$$\leq \frac{1}{\varphi(w)} \sum_{\mathrm{all}\, \chi(\mathrm{mod}\, w)} \left| \sum_{r=1}^{w}{}' \bar{\chi}(r) S\left(\frac{r}{w}\right) \right|^2$$

$$= \sum_{r=1}^{w}{}' \left| S\left(\frac{r}{w}\right) \right|^2.$$

The inequality of lemma (6.5) now follows directly from that of (3).

As an application of this result let

$$R = \sum_{d \leq Q} \frac{Dd}{\varphi(Dd)} \underset{\chi \pmod{Dd}}{\sum\nolimits^*} \max_t \left| \sum_{m=1}^{M} \sum_{\substack{n=1 \\ mn \leq t}}^{N} a_m b_n \chi(mn) \right|.$$

Lemma (6.6). *The inequality*

$$R^2 \ll (M + DQ^2)(N + DQ^2)(\log 2MN)^2 \sum_{m=1}^{M} |a_m|^2 \sum_{n=1}^{N} |b_n|^2$$

holds for all complex numbers a_m, $m = 1, \ldots, M$; b_n, $n = 1, \ldots, N$, positive integers M, N, D, and real $Q \geq 1$.

Proof. For complex $s = \sigma + i\tau$, $\mathrm{Re}(s) = \sigma > 0$, define

$$A(\chi, s) = \sum_{m=1}^{M} a_m \chi(m) m^{-s},$$

$$B(\chi, s) = \sum_{n=1}^{N} b_n \chi(n) n^{-s}.$$

The Cauchy–Schwarz inequality shows that the square of the sum

$$\sum_{d \leq Q} \frac{Dd}{\varphi(Dd)} \underset{\chi \pmod{Dd}}{\sum\nolimits^*} |A(\chi, s) B(\chi, s)|$$

does not exceed the product

$$\sum_{d \leq Q} \frac{Dd}{\varphi(Dd)} \underset{\chi \pmod{Dd}}{\sum\nolimits^*} |A(\chi, s)|^2 \cdot \sum_{d \leq Q} \frac{Dd}{\varphi(Dd)} \underset{\chi \pmod{Dd}}{\sum\nolimits^*} |B(\chi, s)|^2,$$

which by lemma (6.5) is

(10) $$\ll (M + DQ^2) \sum_{m=1}^{M} |a_m|^2 (N + DQ^2) \sum_{n=1}^{N} |b_n|^2.$$

We shall make use of the fact that for real x

$$\frac{1}{2\pi i} \int \frac{x^s}{s} \, ds = \begin{cases} 1 & \text{if } 1 < x, \\ 0 & \text{if } 0 < x < 1, \end{cases}$$

provided that the integration is taken along a line $L: \text{Re}(s) = \sigma_0 > 0$ in the complex s-plane.

In our argument we shall set $\sigma_0 = (\log 2MN)^{-1}$.

If t is not an integer and satisfies $0 < t \leq MN$, then

$$V(t) = \sum_{\substack{m=1 \\ mn \leq t}}^{M} \sum_{n=1}^{N} a_m b_n \chi(mn) = \frac{1}{2\pi i} \int_L A(\chi, s) B(\chi, s) \frac{t^s}{s} ds.$$

Specializing t further to half an odd integer we see that

$$V(y) = V(t)$$

for $t - \frac{1}{4} \leq y \leq t + \frac{1}{4}$. Hence

$$V(t) = 2 \int_{t-1/4}^{t+1/4} V(y) \, dy = \int_L A(\chi, s) B(\chi, s) K(t, s) \, ds$$

with

$$K(t, s) = \frac{1}{\pi i} \int_{t-1/4}^{t+1/4} \frac{y^s}{s} \, ds.$$

By estimating crudely before/after the integration over y in this last integral is carried out, we obtain the bounds

$$|K(t, s)| \ll \begin{cases} \dfrac{1}{|s|}, \\ \dfrac{MN}{|s(s+1)|}, \end{cases}$$

uniformly in the admissible t, and s on L. Note that whether t is an integer or not, $|t^s| = t^{\sigma_0} \leq e$ holds on L. In particular,

$$\int_L \max_t |K(t, s)| \, dt \ll \int_0^{(2MN)^{-1}} \frac{d\tau}{\sigma_0} + \int_{(2MN)^{-1}}^{MN} \frac{d\tau}{\tau} + \int_{MN}^{\infty} \frac{MN}{\tau^2} \, d\tau$$

(11)
$$\ll \log 2MN.$$

The sum R which is to be bounded does not exceed

$$\int_L \sum_{d \leq Q} \frac{Dd}{\varphi(Dd)} \sum_{\chi \pmod{Dd}}^{*} |A(\chi, s) B(\chi, s)| \max_t |K(t, s)| \, d\tau.$$

Writing

$$|K(t,s)| = |K(t,s)|^{1/2} |K(t,s)|^{1/2}$$

and applying the Cauchy–Schwarz inequality for integrals, the inequality of the present lemma is readily obtained from the bounds (10) and (11).

The Method of Vinogradov in Vaughan's Form

During the proof of the next lemma we shall apply Vaughan's form of an argument of Vinogradov. (See Vinogradov [1], Vaughan [1], Vaughan [2], Theorem 3.1, pp. 26–28, Davenport [1].) We give an account of it here.

Let $U \geq 1$, $V \geq 1$ be real numbers, and define

$$F(s) = \sum_{m \leq U} \Lambda(m) m^{-s}, \qquad G(s) = \sum_{d \leq V} \mu(d) d^{-s}.$$

Here $\Lambda(m)$ is von Mangoldt's function, defined to be $\log p$ if m is a power of a prime p, and to be zero otherwise.

Comparing coefficients in the identity

$$-\frac{\zeta'(s)}{\zeta(s)} = F(s) - \zeta(s) F(s) G(s) - \zeta'(s) G(s)$$

$$+ \left(-\frac{\zeta'(s)}{\zeta(s)} - F(s) \right) (1 - \zeta(s) G(s)),$$

we obtain a representation

$$\Lambda(n) = a_1(n) + a_2(n) + a_3(n) + a_4(n),$$

where

$$a_1(n) = \begin{cases} \Lambda(n) & \text{if } n \leq U, \\ 0 & \text{otherwise,} \end{cases}$$

$$a_2(n) = - \sum_{\substack{mdr = n \\ m \leq U, d \leq V}} \Lambda(m) \mu(d),$$

$$a_3(n) = \sum_{\substack{hd = n \\ d \leq V}} \mu(d) \log h,$$

$$a_4(n) = - \sum_{\substack{mk = n \\ m > U \\ k > 1}} \Lambda(m) \sum_{\substack{d \mid k \\ d \leq V}} \mu(d).$$

It follows that for any arithmetic function $f(n)$ whatsoever, and real $y \geq 1$,

$$\sum_{n \leq y} f(n)\Lambda(n) = S_1 + S_2 + S_3 + S_4,$$

where

$$S_1 = \sum_{n \leq U} f(n)\Lambda(n),$$

$$S_2 = -\sum_{t \leq UV} \left(\sum_{\substack{md=t \\ m \leq U, d \leq V}} \mu(d)\Lambda(m) \right) \sum_{r \leq y/t} f(rt),$$

$$S_3 = \sum_{d \leq V} \mu(d) \sum_{h \leq y/d} f(dh) \log h = \sum_{d \leq V} \mu(d) \sum_{h \leq y/d} f(dh) \int_1^h \frac{dw}{w}$$

$$= \int_1^y \left\{ \sum_{d \leq V} \mu(d) \sum_{w \leq h \leq y/d} f(dh) \right\} \frac{dw}{w},$$

and

$$S_4 = \sum_{U < m \leq y/V} \Lambda(m) \sum_{V < k \leq y/m} \left(\sum_{\substack{d \mid k \\ d \leq V}} \mu(d) \right) f(mk),$$

this last since

$$\sum_{\substack{d \mid k \\ d \leq V}} \mu(d) = 0$$

holds for $1 < k \leq V$.

In this way the study of sums

$$\sum_{p \leq X} f(p)$$

is reduced to the study of bilinear forms

$$\sum_{\substack{m=1 \\ mn \leq x}}^{M} \sum_{n=1}^{N} a_m b_n f(mn).$$

Remark. Since

$$\sum_{n \leq y} f(n)\Lambda(n)$$

and

$$S_1 = \sum_{n \le U} f(n)\Lambda(n)$$

have the same form, if $U < y$ there is the possibility of an argument by induction.

For Dirichlet characters χ and real $y \ge \frac{3}{2}$ define

$$\psi(y, \chi) = \sum_{n \le y} \chi(n)\Lambda(n).$$

We can now establish

Lemma (6.7). *The inequality*

$$\sum_{d \le Q} \frac{Dd}{\varphi(Dd)} \sum_{\chi \pmod{Dd}}^* \max_{y \le x} |\psi(y, \chi)| \ll (x + x^{1/2}Q^2 D + x^{5/6}QD^{1/2})(\log xQD)^4$$

holds for all integers $D \ge 1$ and real $x \ge 2$, $Q \ge 1$.

Remark. Our proof of this result will be based upon the method of Vaughan, as presented by Montgomery in Davenport [1]. In their case $D = 1$. In our applications of lemma (6.7) D will sometimes be as large as Q.

Proof. For each character χ we define sums $S_j(\chi)$ by replacing $f(n)$ in the above representations with the character $\chi(\ )$. Then

$$\psi(y, \chi) = \sum_{j=1}^{4} S_j(\chi),$$

and the sum we wish to estimate does not exceed

(12) $$\sum_{j=1}^{4} \sum_{d \le Q} \frac{Dd}{\varphi(Dd)} \sum_{\chi \pmod{Dd}}^* \max_{y \le x} |S_j(\chi)|.$$

To simplify the exposition we shall write, in an obvious notation,

$$\sum_{j=1}^{4} \tilde{\sum} \max_{y \le x} |S_j(\chi)|.$$

If we set

$$b_k = \sum_{\substack{d \mid k \\ d \le V}} \mu(d)$$

and if $1 \leq M \leq x$, then by lemma (6.6)

$$\tilde{\sum}_{y \leq x} \max \Big| \sum_{\substack{U < m \leq y/V \\ M < m \leq 2M}} \sum_{V < k \leq y/m} \Lambda(m) b_k \chi(mk) \Big|$$

$$\ll (M + DQ^2)^{1/2} \Big(\frac{x}{M} + DQ^2\Big)^{1/2} \log x \Big(\sum_{M < m \leq 2M} \Lambda(m)^2\Big)^{1/2} \Big(\sum_{k \leq x/M} |b_k|^2\Big)^{1/2}$$

$$\ll (x + xQ(DM^{-1})^{1/2} + Q(xMD)^{1/2} + DQ^2 x^{1/2})(\log 2M (\log x)^5)^{1/2}.$$

In this last step we use the inequalities

$$\sum_{m \leq 2M} \Lambda(m)^2 \ll \log 2M \sum_{m \leq 2M} \Lambda(m) \ll M \log 2M$$

and (see the remarks following the proof of lemma (1.1))

$$\sum_{k \leq x/M} |b_k|^2 \leq \sum_{k \leq x/M} d(k)^2 \ll x(\log x)^3.$$

Choosing M to be each power 2^r in the range $U/2 < 2^r \leq x/V$ in turn, and adding, we see that the terms involving the $S_4(\chi)$ contribute to (12) an amount which is

(13) $\quad \ll (x + xQ(DU^{-1})^{1/2} + xQ(DV^{-1})^{1/2} + DQ^2 x^{1/2})(\log x)^4.$

We estimate the contribution of the $S_2(\chi)$ by considering separately the ranges $t < T$ and $T \leq t \leq UV$ in the definition of $S_2(\chi)$. The number T will be chosen presently. Call the corresponding sums $S'_2(\chi)$ and $S''_2(\chi)$.

An argument similar to that for $S_4(\chi)$, but with the rôles of m, k, $\Lambda(m)$ and b_k played by t, r,

$$\sum_{\substack{md = t \\ m \leq U, d \leq V}} \mu(d) \Lambda(m)$$

and 1, respectively, leads to a bound

(14) $\quad \tilde{\sum}_{y \leq x} \max |S''_2(\chi)|$

$$\ll (x + xQ(DT^{-1})^{1/2} + Q(xDUV)^{1/2} + DQ^2 x^{1/2})(\log xUV)^3.$$

For the similar sum involving $S'_2(\chi)$ we argue directly

$$|S'_2(\chi)| \leq \sum_{t \leq T} \log t \Big| \sum_{r \leq y/t} \chi(r) \Big|,$$

which by the Pólya–Vinogradov inequality (lemma (1.10)) is

$$\ll T \log 2T (Dd)^{1/2} \log 2Dd.$$

Hence

(15) $$\tilde{\sum} \max_{y \leq x} |S'(\chi)| \ll T(\log 2T) Q^{5/2} D^{3/2} \log 2DQ,$$

so that

(16) $$\tilde{\sum} \max_{y \leq x} |S_2(\chi)| \ll (x + xQ(DT^{-1})^{1/2} + Q(xDUV)^{1/2}$$
$$+ DQ^2 x^{1/2} + TQ^{5/2} D^{3/2})(\log xDQTUV)^3.$$

At this point it is convenient to note that we may ignore the condition $T \leq UV$, since if we ultimately choose a value of T which exceeds UV then the bound (15) will suffice, and no sum $S_2''(\chi)$ need be considered. Moreover, if we choose $T = 1$ then the sums $S_2'(\chi)$ are empty, and we may omit the corresponding term $TQ^{5/2} D^{3/2}$.

If $DQ^{3/2} \leq x$ we set

$$T = \left(\frac{x}{DQ^{3/2}}\right)^{2/3},$$

and the terms in (16) which involve T contribute an amount

$$\ll x^{2/3} Q^{1/2} D^{1/6} \ll x.$$

If $DQ^{2/3} > x$ we set $T = 1$, and in view of our above remark there will be a corresponding contribution of

$$\ll xQD^{1/2} \ll DQ^2 x^{1/2}.$$

We see that in every case the terms in the bound (16) which involve T (disregarding the logarithmic factor) may be omitted in favour of those which remain. The logarithmic factor may be replaced by $(\log xDQUV)^3$.

Moreover, for $\frac{3}{2} \leq y \leq x$,

$$|S_3(\chi)| \leq \int_1^x \left| \sum_{d \leq V} \mu(d) \sum_{w \leq h \leq y/d} \chi(dh) \right| \frac{dw}{w},$$

so that

$$\tilde{\sum} \max_{y \leq x} |S_3(\chi)| \leq \int_1^x \tilde{\sum} \max_{y \leq x} \left| \sum_{d \leq V} \mu(d) \sum_{w \leq h \leq y/d} \chi(dh) \right| \frac{dw}{w}.$$

Here the integrand may be treated in exactly the same way as the corresponding sum involving $S_2(\chi)$, and except for fewer logarithms the bound of (16), with the appropriate T-terms removed and U replaced by 1, still serves.

Let us denote by $L(x)$ the sum, in the statement of lemma (6.7), for which we seek an upper bound. Then

$$\tilde{\sum} \max_{y \leq x} |S_1(\chi)| \leq L(U),$$

and we have proved that

$$L(x) - L(U) \ll (x + xQ(DU^{-1})^{1/2} + xQ(DV^{-1})^{1/2}$$
$$+ DQ^2 x^{1/2} + Q(xDUV)^{1/2})(\log xQDUV)^4.$$

If (ignoring the logarithmic factor) the upper bound here is to have a local minimum, regarded as a function of the variables U, V, then the first partial derivative with respect to U must vanish. This leads to the condition $U = (xV^{-1})^{1/2}$. Choosing $U = V = x^{1/3}$ (which is independent of Q and D) we see that

$$L(x) - L(x^{1/3}) \ll (x + x^{5/6} QD^{1/2} + x^{1/2} Q^2 D)(\log xQD)^4.$$

We do not need to argue by induction, since trivially

$$L(x^{1/3}) \ll Q^2 D x^{1/3},$$

and the proof of lemma (6.7) is complete.

Dirichlet L-Series

For each Dirichlet character χ we define the corresponding L-series

$$L(s, \chi) = \sum_{n=1}^{\infty} \chi(n) n^{-s}.$$

This function may be continued over the whole complex s-plane where it is analytic unless χ is a principal character, when there will be a simple pole at $s = 1$. See, for example, Prachar [1], Kap. VII, Satz (1.1), p. 207. We shall need it only in the half-plane $\text{Re}(s) = \sigma > 0$.

Lemma (6.8). *Let $\delta > 0$ hold. Then*

$$\sum_{1 < k \leq Q} \sum^*_{\chi \pmod k} |L(s, \chi)|^2 \ll Q^2 |s|^{1-\mu}$$

with

$$\mu = \frac{2\sigma - 1}{2\sigma + 1}$$

is satisfied uniformly for $\text{Re}(s) = \sigma \geq \frac{1}{2} + \delta$, $Q \geq 1$.

Remark. The value of μ is poor, but the good dependence of the upper bound upon Q will be the more important in our present applications.

Proof. An integration by parts shows that for any $w > 0$

$$L(s, \chi) = \sum_{n \leq w} \chi(n) n^{-s} + s \int_w^\infty y^{-s-1} \sum_{n \leq y} \chi(n) \, dy.$$

Applying the Pólya–Vinogradov inequality (lemma (1.10)) the integral is seen to be

$$\ll |s| w^{-\sigma} k^{1/2} \log 2k, \qquad \sigma \geq \tfrac{1}{2}.$$

Hence

$$\sum_{k \leq Q} \sum_{\chi \,(\text{mod } k)}^* |L(s, \chi)|^2 \ll \sum_{k \leq Q} \sum_{\chi \,(\text{mod } k)}^* \left| \sum_{n \leq w} \chi(n) n^{-s} \right|^2 + |s|^2 w^{-2\sigma} Q^3 (\log 2Q)^2.$$

An application of lemma (6.5) (with $D = 1$) shows that this last (multiple) sum is

$$\ll (w + Q^2) \sum_{n \leq w} n^{-2\sigma} \ll w + Q^2.$$

With the choice

$$w = (|s|^2 Q^3)^{1/(1 + 2\sigma)}$$

the proof of lemma (6.8) is completed.

Chapter 7

Additive Functions on Arithmetic Progressions with Large Moduli

Additive Functions on Arithmetic Progressions

In this chapter I study the behaviour of additive functions $f(n)$ on short arithmetic progressions which have large differences. I shall make essential use of these results in the next chapter.

Let d, r be integers; c, k positive integers; and σ, y positive real numbers. Define

$$E(y, k, r) = \sum_{\substack{n \leq y \\ n \equiv d \pmod{c} \\ n \equiv r \pmod{k}}} \frac{f(n)}{n^\sigma}\left(1 - \frac{n}{y}\right) - \frac{1}{\varphi(k)} \sum_{\substack{n \leq y \\ n \equiv d \pmod{c} \\ (n,k)=1}} \frac{f(n)}{n^\sigma}\left(1 - \frac{n}{y}\right).$$

In what follows σ, c and d will be regarded as fixed, y, r and k as moving.

Theorem (7.1). *Let $\frac{1}{2} < \sigma < 1$. Then*

$$\sum_{\substack{\log x < p \leq Q \\ (p,c)=1}} (p-1) \max_{(r,p)=1} \max_{y \leq x} |E(y, p, r)|^2$$

$$\ll \left(\frac{x^{1-\sigma}}{\log x} + (Q^{3/2} + x^{5/6-\sigma})Q^{5/2-2\sigma}(\log Q)^5\right) \sum_{q \leq x} \frac{|f(q)|^2}{q^\sigma}$$

holds uniformly for all additive functions $f(n)$, for all $x \geq 2$ and $Q \geq 1$. Here p runs through prime numbers, and q through prime powers.

Remark. If Q is a small enough power of x the leading factor in this upper bound is $\ll x^{1-\sigma}(\log x)^{-1}$. I shall show that apart from the implied constant this is then best possible.

The following variant of theorem (7.1) allows the consideration of primes p as small as $p = 2$.

Theorem (7.2). Let $\frac{1}{2} < \sigma < 1$, $\theta = (3-2\sigma)/(2\sigma+1)$. Then

$$\sum_{\substack{p \leq Q \\ (p,c)=1}} (p-1) \max_{(r,p)=1} \max_{y \leq x} |E(y,p,r)|^2$$

$$\ll \left(\frac{x^{1-\sigma}}{(\log x)^{\theta}} + (Q^{3/2} + x^{5/6-\sigma}) Q^{5/2-\sigma} (\log Q)^5 \right) \sum_{q \leq x} \frac{|f(q)|^2}{q^{\sigma}}$$

holds uniformly for all additive functions $f(n)$, for all $x \geq 2$ and $Q \geq 1$. As before, p denotes primes and q prime-powers.

For application in Chapter 8 two further variants will be derived.

Let d, r be integers; c a positive integer, p a (positive) prime; and σ, y positive real numbers. Define

$$F(y,p,r) = \sum_{\substack{n \leq y \\ n \equiv d \pmod{c} \\ n \equiv r \pmod{p} \\ n \not\equiv r \pmod{p^2}}} \frac{f(n)}{n^{\sigma}} \left(1 - \frac{n}{y}\right) - \frac{1}{p} \sum_{\substack{n \leq y \\ n \equiv d \pmod{c} \\ s(n-r,p)=1}} \frac{f(n)}{n^{\sigma}} \left(1 - \frac{n}{y}\right).$$

In the following two theorems δ will denote any (fixed) positive number which satisfies

$$\delta < \min\left(\frac{1-\sigma}{4-2\sigma}, \frac{1}{15-12\sigma} \right).$$

Theorem (7.3). Let $\frac{1}{2} < \sigma < 1$. Then

$$\sum_{\substack{\log x < p \leq x^{\delta} \\ (p,c)=1}} p \max_{(r,p)=1} \max_{y \leq x} |F(y,p,r)|^2 \ll \frac{x^{2(1-\sigma)} \log \log 2x}{\log x} \cdot \sum_{q \leq x} \frac{|f(q)|^2}{q}$$

holds for all additive functions $f(n)$, for all $x \geq 2$.

This is an analogue of theorem (7.1), and there is an analogue of theorem (7.2).

Theorem (7.4). Let $\frac{1}{2} < \sigma < 1$. Then

$$\sum_{\substack{p \leq x^{\delta} \\ (p,c)=1}} p \max_{(r,p)=1} \max_{y \leq x} |F(y,p,r)|^2 \ll \frac{x^{2(1-\sigma)}}{(\log x)^{1/4}} \sum_{q \leq x} \frac{|f(q)|^2}{q}$$

holds for all additive functions $f(n)$, for all $x \geq 2$.

In theorems (7.3) and (7.4), as in the earlier results, p denotes primes, and q prime-powers.

Remark. If in theorem (7.4) we replace $F(y, p, r)$ by

$$F(y, p, r) + \frac{1}{(p-1)} \cdot \frac{y^{1-\sigma}}{(1-\sigma)(2-\sigma)} \cdot \sum_{p^k \leq y} \frac{f(p^k)}{p^k},$$

then the power of $\log x$ in the upper bound may be increased.
After proving these four theorems I shall discuss their nature.

It will be convenient, for the duration of this chapter to adopt the convention that whenever q is a prime-power, q_0 is the prime which generates it.

Let k and t be positive integers, $1 \leq t \leq k$.
Define

$$A(x, t, z) = A(x, k, t, z) = \sum_{\substack{m \leq x \\ m \equiv t \pmod{k}}} m^{-z} - k^{-z} \sum_{m \leq xk^{-1}} m^{-z} - t^{-z}$$

for complex numbers $z = \beta + i\tau$, $\beta = \mathrm{Re}(z)$, $0 < \beta < 1$.

Lemma (7.5). *Let $\nu = 2\beta - 1$. Then*

$$\sum_{t=1}^{k} |A(x, t, z)|^2 \ll k^{-\nu} |z|^{1-\nu} + k x^{-2\beta}$$

uniformly for all $k \geq 1$, $x \geq 1$ and $\beta_0 \leq \beta \leq 1 - \beta_0$.

Remark. Here β_0 may be any positive constant not exceeding $\frac{1}{2}$ in value.

Proof. Consider first the sum over the integers $m \equiv t \pmod{k}$. Setting $m = t + uk$ we see that

$$\sum_{\substack{m \leq x \\ m \equiv t \pmod{k}}} m^{-z} = \sum_{1 \leq u \leq xk^{-1}} (t + uk)^{-z} + t^{-z} + O(x^{-\beta}).$$

For $u \geq 2$

$$(t + uk)^{-z} = (uk)^{-z} \{1 + tu^{-1}k^{-1}\}^{-z} = (uk)^{-z} + O(|z| k^{-\beta} u^{-1-\beta}).$$

Hence

$$\sum_{|z| < u \leq xk^{-1}} \{(t + uk)^{-z} - (uk)^{-z}\} \ll |z| k^{-\beta} \sum_{u > |z|} u^{-1-\beta} \ll k^{-\beta} |z|^{1-\beta}.$$

Moreover, estimating crudely

$$\sum_{u \le |z|} \{(t+uk)^{-z} - (uk)^{-z}\} \ll k^{-\beta} \sum_{u \le |z|} u^{-\beta} \ll k^{-\beta}|z|^{1-\beta}.$$

Thus

$$A(x, t, z) \ll k^{-\beta}|z|^{1-\beta} + x^{-\beta},$$

from which the lemma follows trivially.

Remark. At this stage we note that for $0 < \sigma < 1$

$$\sum_{m \le xk^{-1}} m^{-z} \ll (xk^{-1})^{1-\beta} + 1$$

so that

$$\sum_{\substack{m \le x \\ m \equiv t \pmod{k}}} m^{-z} \ll x^{1-\beta}k^{-1} + 1.$$

A straightforward argument gives

$$\sum_{\substack{n \le y \\ n \equiv n_0 \pmod{pc}}} \frac{f(n)}{n^\sigma}\left(1 - \frac{n}{y}\right) = \sum_{q \le y} \frac{f(q)}{q^\sigma} \sum_{\substack{m \le yq^{-1} \\ qm \equiv n_0 \pmod{pc} \\ q \| m}} \frac{1}{m^\sigma}\left(1 - \frac{mq}{y}\right)$$

$$\ll \sum_{q \le y} \frac{|f(q)|}{q^\sigma}\left(\left(\frac{y}{q}\right)^{1-\sigma}\frac{1}{p} + 1\right)$$

$$\ll \frac{y^{1-\sigma}}{p} \sum_{q \le y} \frac{|f(q)|}{q^{\sigma/2} \cdot q^{1-\sigma/2}} + \sum_{q \le y} \frac{|f(q)|}{q^{\sigma/2} \cdot q^{\sigma/2}},$$

and, applying the Cauchy–Schwarz inequality,

$$|E(y, p, r)|^2 \ll \left(\frac{y^{2(1-\sigma)}}{p^2} + y^{1-\sigma}\right) \sum_{q \le y} \frac{|f(q)|^2}{q^\sigma}.$$

Here we have made use of the condition $\sigma < 1$. Hence

$$\sum_{\substack{p \le Q \\ (p,c)=1}} (p-1) \max_{(r,p)=1} \max_{y \le x} |E(y, p, r)|^2$$

$$\ll (x^{2(1-\sigma)} \log \log 4Q + x^{1-\sigma}Q^2) \sum_{q \le x} \frac{|f(q)|^2}{q^\sigma},$$

and we obtain theorems (7.1) and (7.2) trivially if $x \le Q^2$.
This ends the remark.

Additive Functions on Arithmetic Progressions

To continue with our argument we apply the integral

(1) $$\frac{1}{2\pi i}\int_L \frac{y^s}{s(s+1)}\,ds = \begin{cases} 1-1/y & \text{if } y \geq 1, \\ 0 & \text{if } 0 < y < 1, \end{cases}$$

where the integration is taken along the line $\text{Re}(s) = \sigma_0 = (\log x)^{-1}$ in the complex s-plane. Since $x \geq 2$ we have $\sigma_0 > 0$.

For primes p and integers l, u, $1 \leq l \leq p-1$, $1 \leq u \leq c$, define

$$B_1^u(l, p, s) = \sum_{\substack{q \leq x \\ q \equiv u(\bmod c) \\ q \equiv l(\bmod p)}} \frac{f(q)}{q^{\sigma+s}} - \frac{1}{p-1} \sum_{\substack{q \leq x \\ q \equiv u(\bmod c) \\ (q,p)=1}} \frac{f(q)}{q^{\sigma+s}},$$

$$B_2^u(l, p, s) = \sum_{\substack{qq_0 \leq x \\ qq_0 \equiv u(\bmod c) \\ qq_0 \equiv l(\bmod p)}} \frac{f(q)}{(qq_0)^{\sigma+s}} - \frac{1}{p-1} \sum_{\substack{qq_0 \leq x \\ qq_0 \equiv u(\bmod c) \\ (q,p)=1}} \frac{f(q)}{(qq_0)^{\sigma+s}},$$

and

$$B^u(l, p, s) = B_1^u(l, p, s) - B_2^u(l, p, s).$$

For $0 < y \leq x$ define the integral

$$J(y, p, r) = \frac{1}{2\pi i}\int_L \sum_{u=1}^{c} \sum_{l=1}^{p-1} B^u(l, p, s) A(x, pc_1, t_l(u), \sigma+s) \frac{y^s}{s(s+1)}\,ds.$$

Here the notation of lemma (7.5) is considered still in force, and $t = t_l(u)$ is determined by the conditions

$$ut \equiv d \pmod{c}, \qquad lt \equiv r \pmod{p}, \qquad 1 \leq t \leq pc_1$$

with $c_1 = c(c, u)^{-1}$. These conditions will, of course, not have any solutions t unless (c, u), the highest common factor of c and u, divides d.

Since for each u

$$\sum_{l=1}^{p-1} B^u(l, p, s) = 0,$$

the integral has the alternative representation

$$\sum_{u=1}^{c} \frac{1}{2\pi i}\int_L \sum_{l=1}^{p-1} B^u(l, p, s) \left\{ \sum_{\substack{m \leq x \\ m \equiv t_l(u)(\bmod pc_1)}} \frac{1}{m^{\sigma+s}} - \frac{1}{t_l(u)^{\sigma+s}} \right\} \frac{y^s}{s(s+1)}\,ds.$$

7. Additive Functions on Arithmetic Progressions with Large Moduli

Interchanging the order of summation and integration, and applying (1), the integral

$$Y(u) = \frac{1}{2\pi i} \int_L \sum_{l=1}^{p-1} B^u(l,p,s) \cdot \sum_{\substack{m \leq x \\ m \equiv t_l(u) \,(\text{mod } pc_1)}} \frac{1}{m^{\sigma+s}} \cdot \frac{y^s}{s(s+1)} ds$$

is readily evaluated as

$$\sum_{l=1}^{p-1} \sum_{\substack{q \leq y \\ q \equiv u \,(\text{mod } c) \\ q \equiv l \,(\text{mod } p)}} \frac{f(q)}{q^\sigma} \sum_{\substack{m \leq yq^{-1} \\ um \equiv d \,(\text{mod } c) \\ lm \equiv r \,(\text{mod } p)}} \frac{1}{m^\sigma} \left(1 - \frac{m}{yq^{-1}}\right)$$

$$- \sum_{l=1}^{p-1} \sum_{\substack{qq_0 \leq y \\ qq_0 \equiv u \,(\text{mod } c) \\ qq_0 \equiv l \,(\text{mod } p)}} \frac{f(q)}{(qq_0)^\sigma} \sum_{\substack{m_1 \leq yq^{-1}q_0^{-1} \\ um_1 \equiv d \,(\text{mod } c) \\ lm_1 \equiv r \,(\text{mod } p)}} \frac{1}{m_1^\sigma} \left(1 - \frac{m_1}{yq^{-1}q_0^{-1}}\right)$$

$$- \frac{1}{p-1} \sum_{\substack{q \leq y \\ q \equiv u \,(\text{mod } c) \\ (q,p)=1}} \frac{f(q)}{q^\sigma} \sum_{\substack{m \leq yq^{-1} \\ um \equiv d \,(\text{mod } c) \\ (m,p)=1}} \frac{1}{m^\sigma} \left(1 - \frac{m}{yq^{-1}}\right)$$

$$+ \frac{1}{p-1} \sum_{\substack{qq_0 \leq y \\ qq_0 \equiv u \,(\text{mod } c) \\ (q,p)=1}} \frac{f(q)}{(qq_0)^\sigma} \sum_{\substack{m_1 \leq yq^{-1}q_0^{-1} \\ um_1 \equiv d \,(\text{mod } c) \\ (m,p)=1}} \frac{1}{m_1^\sigma} \left(1 - \frac{m_1}{yq^{-1}q_0^{-1}}\right).$$

By writing $m = m_1 q_0$ when q_0 divides m, these four sums may be expressed in the form

$$\sum_{l=1}^{p-1} \sum_{\substack{q \leq y \\ q \equiv u \,(\text{mod } c) \\ q \equiv l \,(\text{mod } p)}} \frac{f(q)}{q^\sigma} \sum_{\substack{mq \leq y \\ mq \equiv d \,(\text{mod } c) \\ mq \equiv r \,(\text{mod } p)}} \frac{1}{m^\sigma} \left(1 - \frac{mq}{y}\right)$$

$$- \sum_{l=1}^{p-1} \sum_{\substack{qq_0 \leq y \\ qq_0 \equiv u \,(\text{mod } c) \\ qq_0 \equiv l \,(\text{mod } p)}} \frac{f(q)}{q^\sigma} \sum_{\substack{mq \leq y \\ mq \equiv d \,(\text{mod } c) \\ mq \equiv r \,(\text{mod } p) \\ m \equiv 0 \,(\text{mod } q_0)}} \frac{1}{m^\sigma} \left(1 - \frac{mq}{y}\right)$$

$$- \frac{1}{p-1} \sum_{\substack{q \leq y \\ q \equiv u \,(\text{mod } c) \\ (q,p)=1}} \frac{f(q)}{q^\sigma} \sum_{\substack{mq \leq y \\ mq \equiv d \,(\text{mod } c) \\ (m,p)=1}} \frac{1}{m^\sigma} \left(1 - \frac{mq}{y}\right)$$

$$+ \frac{1}{p-1} \sum_{\substack{qq_0 \leq y \\ qq_0 \equiv u \,(\text{mod } c) \\ (q,p)=1}} \frac{f(q)}{q^\sigma} \sum_{\substack{mq \leq y \\ mq \equiv d \,(\text{mod } c) \\ (m,p)=1 \\ m \equiv 0 \,(\text{mod } q_0)}} \frac{1}{m^\sigma} \left(1 - \frac{mq}{y}\right).$$

Additive Functions on Arithmetic Progressions

The condition $mq \equiv r \pmod{p}$ ensures that only prime-powers q which satisfy $(q, p) = 1$ give a contribution to the first two of these sums, and

$$\sum_{u=1}^{c} Y(u) = \sum_{\substack{q \leq y}} \frac{f(q)}{q^\sigma} \sum_{\substack{mq \leq y \\ mq \equiv d \pmod{c} \\ mq \equiv r \pmod{p} \\ (m,q)=1}} \frac{1}{m^\sigma}\left(1 - \frac{mq}{y}\right)$$

$$- \frac{1}{p-1} \sum_{\substack{q \leq y \\ (q,p)=1}} \frac{f(q)}{q^\sigma} \sum_{\substack{mq \leq y \\ mq \equiv d \pmod{c} \\ (m,pq)=1}} \frac{1}{m^\sigma}\left(1 - \frac{mq}{y}\right)$$

$$= \sum_{\substack{n \leq y \\ n \equiv d \pmod{c} \\ n \equiv r \pmod{p}}} \frac{f(n)}{n^\sigma}\left(1 - \frac{n}{y}\right) - \frac{1}{p-1} \sum_{\substack{n \leq y \\ n \equiv d \pmod{c} \\ (n,p)=1}} \frac{f(n)}{n^\sigma}\left(1 - \frac{n}{y}\right)$$

$$= E(y, p, r).$$

Similarly

$$\sum_{u=1}^{c} \frac{1}{2\pi i} \int_L \sum_{l=1}^{p-1} B^u(l, p, s) \cdot t_l(u)^{-\sigma-s} \cdot \frac{y^s}{s(s+1)} ds$$

$$= \sum_{u=1}^{c} \sum_{l=1}^{p-1} \frac{1}{t_l(u)^\sigma} \left\{ \sum_{\substack{qt_l(u) \leq y \\ q \equiv l \pmod{p}}} \frac{f(q)}{q^\sigma}\left(1 - \frac{qt_l(u)}{t}\right) \right.$$

$$\left. - \frac{1}{(p-1)} \sum_{\substack{qt_l(u) \leq y \\ (q,p)=1}} \frac{f(q)}{q^\sigma}\left(1 - \frac{qt_l(u)}{y}\right) \right\}$$

$$- \sum_{u=1}^{c} \sum_{l=1}^{p-1} \frac{1}{t_l(u)^\sigma} \left\{ \sum_{\substack{qq_0 t_l(u) \leq y \\ qq_0 \equiv l \pmod{p}}} \frac{f(q)}{(qq_0)^\sigma}\left(1 - \frac{qq_0 t_l(u)}{y}\right) \right.$$

$$\left. - \frac{1}{(p-1)} \sum_{\substack{qq_0 t_l(u) \leq y \\ (q,p)=1}} \frac{f(q)}{(qq_0)^\sigma}\left(1 - \frac{qq_0 t_l(u)}{y}\right) \right\}$$

$$= C(y, p, r),$$

say.

Hence, by means of the Cauchy–Schwarz inequality,

(2) $$\sum_{\substack{w < p \leq Q \\ (p,c)=1}} (p-1) \max_{(r,p)=1} \max_{y \leq x} |E(y, p, r)|^2 = \Lambda_1 + O((\Lambda_1 \Lambda_2)^{1/2} + \Lambda_2),$$

with

$$\Lambda_1 = \sum_{\substack{w<p\leq Q \\ (p,c)=1}} (p-1) \max_{(r,p)=1} \max_{y\leq x} |C(y,p,r)|^2$$

$$\Lambda_2 = \sum_{\substack{w<p\leq Q \\ (p,c)=1}} (p-1) \max_{(r,p)=1} \max_{y\leq x} |J(y,p,r)|^2.$$

Tail first; the estimation of Λ_2.

Lemma (7.6). *For each fixed* $\varepsilon > 0$ *the inequality*

(3) $$\Lambda_2 \ll (w^{-1}\log x)^\eta \left(\frac{x^{1-\sigma}}{\log x} + Q^{2+\varepsilon}\right) \sum_{q\leq x} \frac{|f(q)|^2}{q^\sigma},$$

with $\eta = 2\sigma - 1$, *holds uniformly for* $\tfrac{1}{2} \leq \sigma \leq 1$ *and* $2 \leq Q \leq x/\log x$, $x \geq 2$.

Remark. If w is appreciably larger than $\log x$, then Λ_2 is smaller than the upper bound given in the statement of theorem (7.1). To this extent the sum Λ_2 is a genuine error term.

Proof. Let

$$Z(p,r) = \int_L \left|\sum_{l=1}^{p-1} s^{-1} B^u(l,p,s)\right| \cdot |(s+1)^{-1} A(x, pc_1, t_l(u), \sigma+s)| \, d\tau,$$

where $\tau = \mathrm{Im}(s)$. Since $|y^s| \leq e$ holds uniformly for $1 \leq y \leq x$ and all s on L, it will be enough if we show that for each value of u,

(4) $$\sum_{\substack{w<p\leq Q \\ (p,c)=1}} p \max_{(r,p)=1} |Z(p,r)|^2$$

has an upper bound of the type given at (3).

Applying the Cauchy–Schwarz inequality for integrals and then sums gives

$$|Z(p,r)|^2 \leq \int_L \sum_{l=1}^{p-1} |B^u(l,p,s)|^2 \frac{d\tau}{|s|^2} \cdot \int_L \sum_{l=1}^{p-1} |A(x, pc_1, t_l(u), \sigma+s)|^2 \frac{d\tau}{|s+1|^2}.$$

The second of these integrals is, after lemma (7.5),

$$\ll \int_L (p^{-\eta}|s|^{1-\eta} + px^{-2\beta}) \frac{d\tau}{|s|^2} \ll (p^{-1}\log x)^\eta + px^{-2\beta}\log x,$$

Additive Functions on Arithmetic Progressions   129

the last step by means of a substitution $\tau = \theta(\log x)^{-1}$. This and an application of lemma (6.4) give for the sum at (4) the upper bound

$$\ll \int_L \sum_{w<p\leq Q} \{p^{1-\eta}(\log x)^\eta + p^2 x^{-2\beta} \log x\} \sum_{l=1}^{p-1} |B^u(l,p,s)|^2 \frac{d\tau}{|s+1|^2}$$

$$\ll \{(w^{-1} \log x)^\eta + Qx^{-2\beta} \log x\} \int_L \left(\frac{x^{1-\sigma}}{\log x} + Q^{2+\varepsilon}\right) \sum_{q\leq x} \frac{|f(q)|^2}{q^\sigma} \cdot \frac{d\tau}{|s+1|^2}$$

$$\ll (w^{-1} \log x)^\eta \left(\frac{x^{1-\sigma}}{\log x} + Q^{2+\varepsilon}\right) \sum_{q\leq x} \frac{|f(q)|^2}{q^\sigma}$$

uniformly for $2\sigma \geq 1$ and $2 \leq w \leq Q \leq x/\log x$. The second of these steps has been somewhat wasteful. Note that

$$\sum_{p\leq Q} (p-1) \sum_{l=1}^{p-1} \left|\frac{1}{p-1} \sum_{\substack{q\leq x \\ (q,p)=1}} \frac{f(q)}{q^\sigma} - \frac{1}{p-1} \sum_{q\leq x} \frac{f(q)}{q^\sigma}\right|^2$$

$$= \sum_{p\leq Q} \left|\sum_r \frac{f(p^r)}{p^{r\sigma}}\right|^2 \ll \sum_{p\leq Q} \frac{1}{p^\sigma} \sum_r \frac{|f(p^r)|^2}{p^{r\sigma}}$$

$$\ll \sum_{q\leq x} \frac{|f(q)|^2}{q^\sigma}.$$

Lemma (7.6) is proved.

Remark. The success of this lemma, and so in part of the proof of theorems (7.1) and (7.2), depends upon interpreting sums of the type

$$\sum_{n\leq y} f(n) n^{-\sigma}$$

as (Dirichlet) convolutions. A suitable Fourier analysis (we have used a Mellin transform) then reduces us to the study of *products* of functions of a complex variable (here s), so introducing a degree of independence between the various parameters and arithmetic functions involved.

Now the head; the estimation of Λ_1.

In this section we, so-to-speak, begin again.

Lemma (7.7). *Let $\frac{1}{2} < \sigma < 1$. The inequality*

(5) $$\Lambda_1 \ll \left(\frac{x^{1-\sigma}}{\log x} + (Q^{3/2} + x^{5/6-\sigma}) Q^{5/2-2\sigma} (\log Q)^5\right) \sum_{q\leq x} \frac{|f(q)|}{q^\sigma}$$

holds uniformly for $x \geq 2$ and $Q \geq w \geq 1$.

Proof. Let $\sum'$ denote summation over the set K of primes p, $w < p \le Q$, $(p, c) = 1$. For each prime p in K let an integer r_p and a real number y_p be chosen which maximize $|C(y, p, r)|$, subject to the restraints $1 \le r_p \le p-1$, $1 \le y_p \le x$.

We seek an upper bound for the sum

$$\Gamma = \sum_p{}' (p-1) |C(y_p, p, r_p)|^2.$$

Define

$$C^u(y, p, r) = \sum_{l=1}^{p-1} \frac{1}{t_l(u)^\sigma} \left\{ \sum_{\substack{qt_l(u) \le y \\ q \equiv l \pmod p}} \frac{f(q)}{q^\sigma} \left(1 - \frac{qt_l(u)}{y}\right) \right.$$

$$\left. - \frac{1}{(p-1)} \sum_{\substack{qt_l(u) \le y \\ (q,p)=1}} \frac{f(q)}{q^\sigma} \left(1 - \frac{qt_l(u)}{y}\right) \right\}$$

and

$$C_u(y, p, r) = \sum_{l=1}^{p-1} \frac{1}{t_l(u)^\sigma} \left\{ \sum_{\substack{qq_0 t_l(u) \le y \\ q \equiv l \pmod p}} \frac{f(q)}{(qq_0)^\sigma} \left(1 - \frac{qq_0 t_l(u)}{y}\right) \right.$$

$$\left. - \frac{1}{(p-1)} \sum_{\substack{qq_0 t_l(u) \le y \\ (q,p)=1}} \frac{f(q)}{(qq_0)^\sigma} \left(1 - \frac{qq_0 t_l(u)}{y}\right) \right\}$$

so that

$$C(y, p, r) = \sum_{u=1}^{c} \{C^u(y, p, r) - C_u(y, p, r)\}.$$

We first show that the sums C_u may be neglected. For

$$\sum_{q \le x} \frac{1}{q^\sigma q_0^{2\sigma}} \ll \sum \frac{1}{q_0^{3\sigma}} \ll 1, \qquad \sigma > \tfrac{1}{3},$$

so that applying the Cauchy–Schwarz inequality several times

$$|C_u(y, p, r)|^2 \ll \sum_{l=1}^{p-1} \frac{1}{t_l(u)^{2\sigma}} \sum_{l=1}^{p-1} \left\{ \sum_{\substack{q \le x \\ q \equiv l \pmod p}} \frac{|f(q)|^2}{q^\sigma} + \frac{1}{p^2} \sum_{q \le x} \frac{|f(q)|^2}{q^\sigma} \right\}$$

$$\ll \sum_{q \le x} \frac{|f(q)|^2}{q^\sigma},$$

uniformly for $1 \le y \le x$ and r prime to p. Here we have made use of the fact that for distinct values of $l \pmod p$, the $t_l(u)$ are distinct $\pmod{pc}$. Hence

$$\sum_{l=1}^{p-1} \frac{1}{t_l(u)^{2\sigma}} \le \sum_{m=1}^{pc} \frac{1}{m^{2\sigma}} \ll 1, \qquad \sigma > \tfrac{1}{2}.$$

In particular

(6) $$\sum_{p}' (p-1)|C_u(y_p, p, r_p)|^2 \ll \frac{Q^2}{\log Q} \sum_{q \le x} \frac{|f(q)|^2}{q^\sigma},$$

a result which is certainly as strong as we shall need in order to obtain (5). In what follows we shall therefore study the sum

$$\Gamma^u = \sum_{p}' (p-1)|C^u(y_p, p, r_p)|^2.$$

We shall regard u as fixed, and write t_l in place of $t_l(u)$.

For p in K, and $1 \le l \le p-1$, define the arithmetic function

$$\theta(l, p, n) = \begin{cases} \dfrac{(p-1)^{1/2}}{t_l^\sigma n^{\sigma/2}} \left(1 - \dfrac{nt_l}{y_p}\right)\left(1 - \dfrac{1}{p-1}\right) & \text{if } n \equiv l \pmod p \text{ and } nt_l \le y_p; \\[2ex] \dfrac{-(p-1)^{1/2}}{t_l^\sigma n^{\sigma/2}} \left(1 - \dfrac{nt_l}{y_p}\right) \dfrac{1}{p-1} & \text{if } (n, p) = 1, \, nt_l \le y_p, \\ & \text{but } n \not\equiv l \pmod p; \end{cases}$$

and to be zero for all other primes p and integers l, n. The integers t_l are determined by the conditions

$$ut_l \equiv d \pmod c, \qquad lt_l \equiv r_p \pmod p, \qquad 1 \le t_l \le pc(c, u)^{-1}.$$

We look for the least value of λ so that the inequality

$$\sum_{p}' \left| \sum_{l=1}^{\infty} \sum_{q \le x} \theta(l, p, q) f(q) \right|^2 \le \lambda \sum_{q \le x} |f(q)|^2$$

holds for all complex numbers $f(q)$.

We dualize this inequality and investigate the (same) best bound λ for the inequality

$$\sum_{q \le x} \left| \sum_{l=1}^{\infty} \sum_{p}' \theta(l, p, q) b(p) \right|^2 \le \lambda \sum_{p}' |b(p)|^2$$

for all complex numbers $b(p)$.

In fact it is convenient to consider a slightly modified form. Let $\Lambda(n)$ denote von Mangoldt's function. We consider the expression

$$\hat{\Gamma} = \sum_{n \leq x} \Lambda(n) \left| \sum_{l=1}^{\infty} {\sum_{p}}' \theta(l, p, n) b(p) \right|^2.$$

Expanding the square and inverting the order of summation gives an hermitian form

$$G = \sum_i \sum_j b(p_i) \overline{b(p_j)} a_{ij},$$

where

(7) $$a_{ij} = \sum_{l_1=1}^{\infty} \sum_{l_2=1}^{\infty} \sum_{n \leq x} \theta(l_1, p_i, n) \theta(l_2, p_j, n) \Lambda(n).$$

In accordance with our remarks on duality, in Chapter 5, the least value of λ_1 so that

$$|G| \leq \lambda_1 {\sum_p}' |b(p)|^2$$

holds for all $b(p)$, is a maximal eigenvalue of the matrix (a_{ij}). Moreover, every eigenvalue ν of this matrix lies in a Gershgorin disc

$$|\nu - a_{ii}| \leq \sum_{j \neq i} |a_{ij}|$$

for some i.

Consider a diagonal term a_{ii}. Writing p for p_i this term may be expressed in the form

$$\sum_{n \leq x} \Lambda(n) \sum_{l_1=1}^{\infty} \sum_{l_2=1}^{\infty} \theta(l_1, p, n) \theta(l_2, p, n).$$

For each n considered in the outside sum the summand will be zero unless $(n, p) = 1$. If then $n \equiv l \pmod{p}$ with $(l, p) = 1$, we shall obtain a contribution of $\Lambda(n)$ weighted with a factor

$$\frac{p-1}{t_l^{2\sigma} n^{\sigma}} \left(1 - \frac{nt_l}{y_p}\right)^2 \left(1 - \frac{1}{p-1}\right)^2$$

$$- \frac{2(p-1)}{t_l^{\sigma} n^{\sigma}} \left(1 - \frac{nt_l}{y_p}\right) \left(1 - \frac{1}{p-1}\right) \frac{1}{(p-1)} \sum_{\substack{r=1 \\ r \neq l}}^{p-1} \frac{1}{t_r^{\sigma}} \left(1 - \frac{nt_r}{y_p}\right)$$

$$+ \frac{(p-1)}{n^{\sigma}} \frac{1}{(p-1)^2} \left(\sum_{\substack{r=1 \\ r \neq l}}^{p-1} \frac{1}{t_r^{\sigma}} \left(1 - \frac{nt_r}{y_p}\right) \right)^2,$$

where it is to be understood that any term in this expression which has a factor

$$\left(1 - \frac{nt_r}{y_p}\right)$$

in it is to be replaced by zero if $nt_r > y_p$. As r varies over the interval $1 \leq r \leq p-1$, t_r assumes distinct values (mod pc). In particular

$$\sum_{r=1}^{p-1} \frac{1}{t_r^\sigma} \leq \sum_{m=1}^{p-1} \frac{1}{m^\sigma} \ll (p-1)^{1-\sigma}.$$

The above weighting factor is therefore

$$\ll \frac{p-1}{t_l^{2\sigma} n^\sigma},$$

and

$$a_{ii} \ll (p-1) \sum_{l=1}^{p-1} \frac{1}{t_l^{2\sigma}} \sum_{\substack{n \leq y_p \\ n \equiv l \pmod{p}}} \frac{\Lambda(n)}{n^\sigma}.$$

We can estimate the innersum here by means of a form of the Brun–Titchmarsh theorem:

$$\sum_{\substack{n \leq z \\ n \equiv n_0 \pmod{k}}} \Lambda(n) \ll \frac{z}{\varphi(k)}.$$

For each fixed $\varepsilon > 0$, this inequality is valid uniformly for n_0 prime to k, $1 \leq k \leq z^{1-\varepsilon}$ and $z \geq 2$. See, for example, Halberstam and Richert [1], Theorem (3.8), p. 110; Prachar [1] Satz (4.1), p. 44; or Elliott [11], Volume I, Chapter 2.

If $x \leq p^{1+\varepsilon}$, $0 < \varepsilon < \sigma$, then

$$\sum_{\substack{n \leq x \\ n \equiv l \pmod{p}}} \frac{\Lambda(n)}{n^\sigma} \ll \log p \left\{ \frac{1}{l^\sigma} + \frac{1}{(l+p)^\sigma} + \cdots \right\} \ll \log p.$$

Otherwise we integrate by parts:

$$\sum_{\substack{p^{1+\varepsilon} < n \leq x \\ n \equiv l \pmod{p}}} \frac{\Lambda(n)}{n^\sigma} \ll \frac{x^{1-\sigma}}{p-1} + \sigma \int_{p^{1+\varepsilon}}^{x} y^{-\sigma-1} \frac{y}{\varphi(p)} dy \ll \frac{x^{1-\sigma}}{p-1}.$$

A typical diagonal term thus satisfies

(8)
$$a_{ii} \ll \sum_{l=1}^{p-1} \frac{1}{t_l^{2\sigma}} \{x^{1-\sigma} + p \log p\} \ll \sum_{r=1}^{p-1} \frac{1}{r^{2\sigma}} \{x^{1-\sigma} + p \log p\}$$
$$\ll x^{1-\sigma} + p^2 \log p,$$

since $2\sigma > 1$.

Remark. This estimate is best possible in the following sense. A closer examination of the above argument shows that the weighting factor is

$$\frac{p-1}{t_l^{2\sigma} n^\sigma} \left(1 - \frac{nt_l}{y_p}\right)^2 + O\left(\frac{p-1}{t_l^\sigma n^\sigma p^\sigma}\right)$$

so that

$$a_{ii} \geq \sum_{l=1}^{p-1} \sum_{\substack{nt_l \leq y_p \\ n \equiv l \pmod{p}}} \frac{\Lambda(n)}{n^\sigma} \left\{ \frac{p-1}{t_l^{2\sigma}} \left(1 - \frac{nt_l}{y_p}\right)^2 + O\left(\frac{p-1}{t_l^\sigma p^\sigma}\right) \right\}.$$

There is a value of l, say l_0, for which $1 \leq t_l \leq c$. Since

$$\sum_{l=1}^{p-1} \sum_{\substack{n \leq y_p \\ n \equiv l \pmod{p}}} \frac{\Lambda(n)}{n^\sigma} \cdot \frac{p-1}{t_l^\sigma p^\sigma} \ll p^{1-\sigma} \sum_{l=1}^{p-1} \frac{1}{t_l^\sigma} \left(\frac{x^{1-\sigma}}{p-1} + \log p\right)$$
$$\ll p^{1-2\sigma} x^{1-\sigma} + p^{2(1-\sigma)} \log p,$$

we have

$$a_{ii} \geq \frac{p-1}{c^{2\sigma}} \sum_{\substack{n \leq y_p c^{-1} \\ n \equiv l_0 \pmod{p}}} \frac{\Lambda(n)}{n^\sigma} \left(1 - \frac{nc}{y_p}\right)^2 + O(p^{1-2\sigma} x^{1-\sigma} + p^{2(1-\sigma)} \log p).$$

By using results on the distribution of primes in short arithmetic progressions (see, for example, Elliott [7]) one may show that if Q (and so p) does not exceed a certain fixed power of x, and if $y_p = x$, then for each p in the range $Q/2 < p \leq Q$, $(p, c) = 1$, with at most one exception, there is a uniform lower bound

$$a_{ii} \gg x^{1-\sigma}.$$

This ends our remark.

Additive Functions on Arithmetic Progressions 135

Consider now a typical term a_{ij} with $i \neq j$. A typical innermost sum in (7), apart from the factor

$$(t_{l_1} t_{l_2})^{-\sigma} \{(p_i - 1)(p_j - 1)\}^{1/2},$$

becomes the sums

$$\sum_{\substack{n \leq N \\ n \equiv l_1 \pmod{p_i} \\ n \equiv l_2 \pmod{p_j}}} \gamma(n) \Lambda(n) n^{-\sigma} - \frac{1}{\varphi(p_i)} \sum_{\substack{n \leq N \\ n \equiv l_2 \pmod{p_j} \\ (n, p_i) = 1}} \gamma(n) \Lambda(n) n^{-\sigma}$$

$$- \frac{1}{\varphi(p_j)} \sum_{\substack{n \leq N \\ n \equiv l_1 \pmod{p_i} \\ (n, p_j) = 1}} \gamma(n) \Lambda(n) n^{-\sigma} + \frac{1}{\varphi(p_i p_j)} \sum_{\substack{n \leq N \\ (n, p_i p_j) = 1}} \gamma(n) \Lambda(n) n^{-\sigma},$$

where it is understood that in all four sums

$$N = \min\left(\frac{y_{p_i}}{t_{l_1}}, \frac{y_{p_j}}{t_{l_2}}\right), \quad \gamma(n) = \left(1 - \frac{nt_{l_1}}{y_{p_i}}\right)\left(1 - \frac{nt_{l_2}}{y_{p_j}}\right).$$

If we define h by

$$h \equiv l_1 \pmod{p_i} \quad \text{and} \quad h \equiv l_2 \pmod{p_j}$$

then these four sums combine to be

$$\frac{1}{\varphi(p_i p_j)} \sum_{\chi \pmod{p_i p_j}}^{*} \bar{\chi}(h) \sum_{n \leq N} \chi(n) \gamma(n) \Lambda(n) n^{-\sigma},$$

where * denotes summation over primitive characters (mod $p_i p_j$). Note that if χ is a non-principal character (mod $p_i p_j$) then it is either primitive (mod $p_i p_j$), or of the form $\chi_1 \chi_2$ where χ_1 is the principal character (mod p_i) and χ_2 is a non-principal character (mod p_j), or of the form $\chi_3 \chi_4$ where χ_4 is the principal character (mod p_j) and χ_3 is a non-principal character (mod p_i).

Using the estimate

$$\sum_{r=1}^{p-1} t_r^{-\sigma} \ll (p-1)^{1-\sigma}$$

several times we obtain the bounds

$$a_{ij} \ll (p_i p_j)^{1/2 - \sigma} \sum_{\chi \pmod{p_i p_j}}^{*} \left| \sum_{n \leq N} \chi(n) \gamma(n) \Lambda(n) n^{-\sigma} \right|$$

and

$$\text{(9)} \quad \sum_{j \neq i} |a_{ij}| \ll p_i^{1/2-\sigma} \sum_{w < p_j \leq Q} p_j^{1/2-\sigma} \sum_{\chi \pmod{p_i p_j}}^* \left| \sum_{n \leq N} \chi(n) \gamma(n) \Lambda(n) n^{-\sigma} \right|.$$

We shall estimate this multiple sum using a result of Chapter 6. If δ is fixed and non-negative, then an integration by parts shows that

$$\sum_{n \leq z} \left(\frac{n}{z}\right)^\delta \chi(n) \Lambda(n) n^{-\sigma} \ll \max_{u \leq z} \left| \sum_{n \leq u} \chi(n) \Lambda(n) n^{-\sigma} \right|.$$

With a further integration by parts (and in the notation of lemma (6.7))

$$\sum_{n \leq u} \chi(n) \Lambda(n) n^{-\sigma} = u^{-\sigma} \psi(u, \chi) + \sigma \int_{3/2}^{u} y^{-\sigma-1} \psi(y, \chi) \, dy.$$

The upper bound in (9) therefore does not exceed a constant multiple of

$$\text{(10)} \quad \begin{aligned} & p_i^{1/2-\sigma} \sum_{w < p_j \leq Q} p_j^{1/2-\sigma} \sum_{\chi \pmod{p_i p_j}}^* \max_{3/2 \leq u \leq x} u^{-\sigma} |\psi(u, \chi)| \\ & + p_i^{1/2-\sigma} \int_{3/2}^{x} y^{-\sigma-1} \sum_{w < p_j \leq Q} p_j^{1/2-\sigma} \sum_{\chi \pmod{p_i p_j}}^* |\psi(y, \chi)| \, dy. \end{aligned}$$

If $P \leq Q$ and $H \leq x$ then from lemma (6.7)

$$\sum_{P < p \leq 2P} p^{1/2-\sigma} \sum_{\chi \pmod{p_i p}}^* \max_{H < u \leq 2H} u^{-\sigma} |\psi(u, \chi)|$$

$$\ll (P^{1/2-\sigma} H^{1-\sigma} + H^{1/2-\sigma} P^{5/2-\sigma} p_i + H^{5/6-\sigma} P^{3/2-\sigma} p_i^{1/2}) (\log xQ)^4.$$

We set $P = 2^{-k} Q$ with $k = 1, 2, \ldots$, restricted by $2^{-k} Q \geq w/2$; and $H = 2^{-m} x$, $m = 1, 2, \ldots$, and add, to obtain that the first multiple sum at (10) is

$$\ll p_i^{1/2-\sigma} (w^{1/2-\sigma} x^{1-\sigma} + Q^{5/2-\sigma} p_i + Q^{3/2-\sigma} p_i^{1/2} x^{5/6-\sigma}) (\log xQ)^5.$$

Here we have made use of the condition $\tfrac{1}{2} < \sigma \leq 1$. A similar argument gives, apart from the implied constant, the same bound for the integral at (10). We note only that

$$\int_{3/2}^{x} y^{-\sigma-1/6} \, dy \ll (x^{5/6-\sigma} + 1) \log x$$

holds uniformly for $\tfrac{1}{2} \leq \sigma \leq 1$.

Hence
$$\max_{w<p_i\leq Q}\sum_{j\neq i}|a_{ij}| \ll (w^{1-2\sigma}x^{1-\sigma}+Q^{4-2\sigma}+x^{5/6-\sigma}Q^{5/2-2\sigma})(\log xQ)^5.$$

Define
$$w_0 = (\log xQ)^{6/(2\sigma-1)}.$$

Under the condition $w \geq w_0$ we see that our eigenvalues ν satisfy
$$|\nu - a_{ii}| \ll \frac{x^{1-\sigma}}{\log xQ} + (Q^{4-2\sigma}+x^{5/6-\sigma}Q^{5/2-2\sigma})(\log xQ)^5,$$

so that (see (8))

(11) $\qquad \lambda_1 \ll x^{1-\sigma} + (Q^{4-2\sigma}+x^{5/6-\sigma}Q^{5/2-2\sigma})(\log xQ)^5.$

For every fixed $\mu > 0$
$$\sum_{x^\mu < q \leq x}\left|\sum_{l=1}^{\infty}\sideset{}{'}\sum_{p} \theta(l,p,q)b(p)\right|^2 \leq \frac{\hat{\Gamma}}{\mu \log x} \leq \frac{\lambda_1}{\mu \log x}\sideset{}{'}\sum |b(p)|^2$$

for all complex $b(p)$, where *for the moment q* denotes a prime number. Dualizing:
$$\sideset{}{'}\sum_p \left|\sum_{l=1}^{p-1}\sum_{x^\mu<q\leq x}\theta(l,p,q)f(q)\right|^2 \leq \frac{\lambda_1}{\mu\log x}\sum_{x^\mu<q\leq x}|f(q)|^2.$$

Let $''$ indicate that q is (now) either a prime not exceeding x^μ, or a power t^m, $m \geq 2$, of a prime t. We estimate
$$\tilde{\Gamma} = \sideset{}{'}\sum_p \left|\sum_{l=1}^{p-1}\sideset{}{''}\sum_{q\leq x}\theta(l,p,q)f(q)\right|^2$$

crudely. Applying the Cauchy–Schwarz inequality several times:
$$\left|\sum_{l=1}^{p-1}t_l^{-\sigma/2}\cdot t_l^{\sigma/2}\sideset{}{''}\sum_{q\leq x}\theta(l,p,q)f(q)\right|^2$$
$$\ll p^{1-\sigma}\sum_{l=1}^{p-1}t_l^\sigma\left|\sideset{}{''}\sum_{q\leq x}\theta(l,p,q)f(q)\right|^2$$
$$\ll p^{1-\sigma}\sum_{l=1}^{p-1}t_l^\sigma\sideset{}{''}\sum_{q\leq x}\theta(l,p,q)^2\sideset{}{''}\sum_{q\leq x}|f(q)|^2.$$

The sum which involves the function θ is

$$\ll \sum_{\substack{q \leq x \\ q \equiv l (\bmod p)}}^{\prime\prime} pt_l^{-2\sigma} q^{-\sigma} + \sum_{\substack{q \leq x \\ q \not\equiv l (\bmod p)}}^{\prime\prime} p^{-1} t_l^{-2\sigma} q^{-\sigma}$$

$$\ll pt_l^{-2\sigma}(1 + p^{-1} x^{\mu(1-\sigma)}) + p^{-1} t_l^{-2\sigma} x^{\mu(1-\sigma)}$$

$$\ll t_l^{-2\sigma}(p + x^{\mu(1-\sigma)}).$$

Hence

$$\tilde{\Gamma} \ll \sum_{p \leq Q} (p + x^{\mu(1-\sigma)}) p^{2(1-\sigma)} \sum_{q \leq x}^{\prime\prime} |f(q)|^2$$

$$\ll (Q^{4-2\sigma} + x^{\mu(1-\sigma)} Q^{3-2\sigma}) \sum_{q \leq x}^{\prime\prime} |f(q)|^2.$$

If $x \leq Q$ then with $\mu = 1$ and each $f(q)$ replaced by $f(q) q^{-\sigma/2}$, this bound for $\tilde{\Gamma}$ already leads to the inequality in the statement of lemma (7.7). For each value of u, $1 \leq u \leq c$, we apply it to estimate Γ^u from above, restricting the primes q by $q \equiv u \pmod{c}$. We then appeal to (6).

Suppose now that $Q < x \leq Q^{1/\mu}$. Then our bound (11) and the argument which follows it show that

(12)
$$\sum_{\substack{w_1 < p \leq Q \\ (p,c)=1}} (p-1) |C(y_p, p, r_p)|^2$$
$$\ll \left(\frac{x^{1-\sigma}}{\log x} + (Q^{4-2\sigma} + x^{5/6-\sigma} Q^{5/2-2\sigma}) (\log Q)^5 \right) \sum_{q \leq x}^{\prime\prime} |f(q)|^2,$$

where q runs through prime-powers and $w_1^{2\sigma-1} = (2 \log x)^6$. This will hold (with a different implied constant) for each value of μ not exceeding 1. However, if $x > Q^{1/\mu}$ then, by fixing μ at a value small enough in terms of σ,

$$(Q^{4-2\sigma} + x^{5/6-\sigma} Q^{5/2-2\sigma})(\log Q)^5 + x^{\mu(1-\sigma)} Q^{3-2\sigma} \ll x^{(1-\sigma)/2},$$

so that our inequality (12) remains valid.

To complete our proof of lemma (7.7) it will suffice to show that for each (fixed) positive value of h

$$\sum_{\substack{p \leq (\log x)^h \\ (p,c)=1}} (p-1) |C(y_p, p, r_p)|^2 \ll \frac{x^{1-\sigma}}{\log x} \sum_{q \leq x}^{\prime\prime} |f(q)|^2.$$

We can follow the treatment of $\hat{\Gamma}$ but omit the factor $\Lambda(n)$ and deal directly with the prime-power values of n. The argument proceeds similarly save that a factor $(\log x)^{-1}$ appears in the upper bound for the diagonal terms a_{ii}. At the analogue of (9) we estimate sums

$$\sum_{q \leq y} \chi(q)\gamma(q)q^{-\sigma}, \qquad \tfrac{3}{2} < y \leq x,$$

by means of the Siegel–Walfisz theorem (see, for example, Prachar [1], Satz (8.3), p. 44), since the modulus to which χ is defined does not exceed a fixed power of $\log x$. These sums are then (uniformly in y)

$$\ll x^{1-\sigma} \exp(-c_1 \sqrt{\log x})$$

for some $c_1 > 0$, which is more than enough for our purpose.

Lemma (7.7) is proved.

Remark. From the remark following the upper bound (8) and the inequality preceding that at (11), we see that if every $y_p = x$ then

$$\lambda_1 \gg x^{1-\sigma}$$

holds for all Q not exceeding a certain fixed positive power of x. There are then values of $b(p)$, not all zero, so that

$$\sum_{q \leq x} \left| \sum_{l=1}^{\infty} \sum_p{}' \theta(l, p, q) b(p) \right|^2 \geq \frac{\hat{\Gamma}}{\log x} \gg \frac{x^{1-\sigma}}{\log x} \sum_p{}' |b(p)|^2.$$

The eigenvalue λ is therefore bounded below by a constant multiple of $x^{1-\sigma}/\log x$, and we can find values of $f(q)$, not all zero, giving

$$\sum_{\substack{Q/2 < p \leq Q \\ (p,c)=1}} (p-1) \max_{(r,p)=1} |C(x, p, r)|^2 \gg \frac{x^{1-\sigma}}{\log x} \sum_{q \leq x} \frac{|f(q)|^2}{q^\sigma}.$$

From this we see that, over a wide range of values of Q, the term $x^{1-\sigma}/\log x$ which appears in the upper bound (factor) of lemma (7.7) is essentially best possible.

Proof of theorem (7.1). In view of the remark which follows the proof of lemma (7.5) we may assume that $Q < \sqrt{x}$. The desired inequality now follows from those of lemmas (7.6) and (7.7).

By taking $w = Q/2$ in the inequality of lemma (7.6) and restricting Q to be a small enough power of x, we see from (2) and the remark which follows the proof of lemma (7.7) that an additive function $f(n)$ may be found, not identically zero over the interval $1 \leq n \leq x$, so that

$$\sum_{\substack{Q/2<p\leq Q \\ (p,c)=1}} (p-1) \max_{(r,p)=1} \max_{y\leq x} |E(y,p,r)|^2 > \frac{c_0 x^{1-\sigma}}{\log x} \sum_{q\leq x} \frac{|f(q)|^2}{q^\sigma}$$

holds for some positive constant c_0. This justifies the remark which follows the statement of theorem (7.1).

In order to prove theorem (7.2) we treat the small primes differently.

Lemma (7.8). *Let $\frac{1}{2}<\sigma<1$. With $x\geq 2$ let P be a fixed power of $\log x$. Then*

$$\sum_{\substack{p\leq P \\ (p,c)=1}} (p-1) \max_{(r,p)=1} \max_{y\leq x} |E(y,p,r)|^2 \ll \frac{x^{1-\sigma}P^2}{\log x} \sum_{q\leq x} \frac{|f(q)|^2}{q^\sigma}$$

holds for all additive functions $f(n)$.

Proof. Define

$$E_1(y,p,r) = \sum_{q\leq y} f(q) \sum_{\substack{n\leq y \\ n\equiv d(\bmod c) \\ n\equiv r(\bmod p) \\ n\equiv 0(\bmod q)}} \frac{1}{n^\sigma}\left(1-\frac{n}{y}\right)$$

$$- \frac{1}{(p-1)} \sum_{q\leq y} f(q) \sum_{\substack{n\leq y \\ n\equiv d(\bmod c) \\ (n,p)=1 \\ n\equiv 0(\bmod q)}} \frac{1}{n^\sigma}\left(1-\frac{n}{y}\right),$$

and

$$E_2(y,p,r) = \sum_{qq_0\leq y} f(q) \sum_{\substack{n\leq y \\ n\equiv d(\bmod c) \\ n\equiv r(\bmod p) \\ n\equiv 0(\bmod qq_0)}} \frac{1}{n^\sigma}\left(1-\frac{n}{y}\right)$$

$$- \frac{1}{(p-1)} \sum_{qq_0\leq y} f(q) \sum_{\substack{n\leq y \\ n\equiv d(\bmod c) \\ (n,p)=1 \\ n\equiv 0(\bmod qq_0)}} \frac{1}{n^\sigma}\left(1-\frac{n}{y}\right),$$

so that

$$E(y,p,r) = E_1(y,p,r) - E_2(y,p,r).$$

We shall prove that

(13) $$\sum_{\substack{p \leq P \\ (p,c)=1}} p \max_{(r,p)=1} \max_{y \leq x} |E_j(y, p, r)|^2 \ll \frac{x^{1-\sigma} P^2}{\log x} \sum_{q \leq x} \frac{|f(q)|^2}{q^\sigma}$$

for $j = 1, 2$. From this result lemma (7.8) will follow at once.

Let $(c, d) = l$, $c = c_1 l$, $d = d_1 l$. For each prime-power q let $q_1 = q(q, l)^{-1}$, so that $(q_1, l) = 1$. Consider the sum

$$\sum_{\substack{n \leq y \\ n \equiv d \pmod{c} \\ n \equiv r \pmod{p} \\ n \equiv 0 \pmod{q}}} \frac{1}{n^\sigma}\left(1 - \frac{n}{y}\right) = \frac{1}{(lq_1)^\sigma} \sum_{\substack{m \leq y(lq_1)^{-1} \\ q_1 m \equiv d_1 \pmod{c_1} \\ lq_1 m \equiv r \pmod{p}}} \frac{1}{m^\sigma}\left(1 - \frac{m}{y(lq_1)^{-1}}\right), \quad (p, rc) = 1.$$

The sum on the right-hand side we express by the contour integral

$$\frac{1}{(lq_1)^\sigma} \frac{1}{2\pi i} \int \frac{1}{\varphi(pc_1)} \sum_{\chi \pmod{pc_1}} \bar{\chi}(t) L(\sigma + s, \chi) \frac{(yl^{-1}q_1^{-1})^s}{s(s+1)} \, ds,$$

where the integration is carried out over the line $\operatorname{Re}(s) = 2$ (for the moment) and t is the residue class $\pmod{pc_1}$ determined by

$$q_1 t \equiv d_1 \pmod{c_1}, \qquad lq_1 t \equiv r \pmod{p}.$$

We consider, separately, the contribution which arises from the principal character χ_0. Moving the corresponding integral to the line $\operatorname{Re}(s) = \frac{1}{2} - \sigma$ we pass over the pole at $s = 1 - \sigma$ to obtain

(14) $$\frac{1}{(lq_1)^\sigma \varphi(pc_1)} \operatorname*{Res}_{s=1-\sigma} L(\sigma + s, \chi_0) \frac{(yl^{-1}q_1^{-1})^s}{s(s+1)}$$
$$+ \frac{1}{(lq_1)^\sigma \varphi(pc_1)} \cdot \frac{1}{2\pi i} \int_{1/2-i\infty}^{1/2+i\infty} L(z, \chi_0) \left(\frac{y}{lq_1}\right)^{z-\sigma} \frac{dz}{(z-\sigma)(z+1-\sigma)}.$$

Since

$$L(z, \chi_0) = \zeta(z) \prod_{g | pc_1} (1 - g^{-z}),$$

where g runs through the prime divisors of pc_1, the bound

$$L(z, \chi_0) \ll |z|^{1/2}$$

holds on the line $\operatorname{Re}(z) = \frac{1}{2}$. (See Titchmarsh [1] Chapter 5, or Prachar [1],

Satz (5.4), p. 115.) The integral at (14) is then seen to be

(15) $$O\left(\frac{y^{1/2-\sigma}}{pq^{1/2}}\right).$$

Similarly

$$\frac{1}{\varphi(p)} \sum_{\substack{n \leq y \\ n \equiv d \pmod{c} \\ n \equiv 0 \pmod{q} \\ (n,p)=1}} \frac{1}{n^\sigma}\left(1-\frac{n}{y}\right) = \frac{1}{(lq_1)^\sigma} \cdot \frac{1}{\varphi(p)} \frac{1}{2\pi i} \int \frac{1}{\varphi(c_1)} \sum_{\chi \pmod{c_1}} \bar{\chi}(t')$$

$$\times \left(1-\frac{\chi(p)}{p^{\sigma+s}}\right) L(\sigma+s, \chi) \frac{(yl^{-1}q_1^{-1})^s}{s(s+1)} ds,$$

where t' is determined by $q_1 t' \equiv d_1 \pmod{c_1}$. We treat the contribution arising from the principal character (mod c_1) in the same manner that we treated that of the character $\chi_0 \pmod{pc_1}$. In the expression corresponding to (14) the residue term has the same value, and the corresponding integral is estimated by (15).

From these results

$$E_1(y, p, r) = \sum_{q \leq x} \frac{f(q)}{q^\sigma}\left(I_1(q, p) + I_2(q, p) + O\left(\frac{x^{1/2-\sigma}}{pq^{1/2}}\right)\right)$$

uniformly for $1 \leq y \leq x$, where

$$I_1(q, p) = \frac{(q, l)^\sigma}{l^\sigma \varphi(pc_1)} \cdot \frac{1}{2\pi i} \int_{2-i\infty}^{2+i\infty} \tilde{\sum}_{\chi \pmod{pc_1}} \bar{\chi}(t) L(\sigma+s, \chi) \frac{(yl^{-1}q_1^{-1})^s}{s(s+1)} ds,$$

$$I_2(q, p) = \frac{(q, l)^\sigma}{l^\sigma \varphi(pc_1)} \cdot \frac{1}{2\pi i} \int_{2-i\infty}^{2+i\infty} \tilde{\sum}_{\chi \pmod{c_1}} \bar{\chi}(t') L(\sigma+s, \chi) \frac{(yl^{-1}q_1^{-1})^s}{s(s+1)} ds,$$

and $\tilde{\sum}$ denotes summation over non-principal characters to the modulus under consideration.

We apply the Cauchy–Schwarz inequality, with an obvious notation:

$$|E_1(y, p, r)|^2 \leq \sum_{q \leq x} \frac{1}{q^\sigma}\left|I_1 + I_2 + O\left(\frac{x^{1/2-\sigma}}{pq^{1/2}}\right)\right|^2 \sum_{q \leq x} \frac{|f(q)|^2}{q^\sigma}.$$

Here

$$\sum_{q \leq x} \frac{1}{q^\sigma} \left(\frac{x^{1/2-\sigma}}{pq^{1/2}} \right)^2 \ll \frac{x^{1-2\sigma}}{p^2}.$$

The integral I_1 is treated by moving the contour to a line $\text{Re}(s) = -(\sigma - \frac{1}{2} - \varepsilon)$, where ε is a positive number not exceeding $\frac{1}{4}$, chosen so that $-\sigma + \frac{1}{2} + \varepsilon$ is negative. In this manner we pass over a pole at $s = 0$, where there is a residue of

$$\frac{1}{l^\sigma \varphi(pc_1)} \sum_{\chi (\text{mod } pc_1)}^{\sim} \bar{\chi}(t) L(\sigma, \chi).$$

Along the line $\text{Re}(s) = -\sigma + \frac{1}{2} + \varepsilon$, with $\tau = \text{Im}(s)$, the integral is

$$\ll \frac{1}{\varphi(pc_1)} \left(\frac{x}{q} \right)^{-\sigma+1/2+\varepsilon} \int_{-\infty}^{\infty} \sum_{\chi (\text{mod } pc_1)}^{\sim} |L(\tfrac{1}{2}+\varepsilon+i\tau, \chi)| \frac{d\tau}{|s(s+1)|}.$$

Hence, applying the Cauchy–Schwarz inequality for sums and for integrals,

$$|I_1|^2 \ll \frac{1}{\varphi(pc_1)} \sum_{\chi (\text{mod } pc_1)}^{\sim} |L(\sigma, \chi)|^2$$

$$+ \left(\frac{x}{q} \right)^{-2\sigma+1+2\varepsilon} \int_{-\infty}^{\infty} \frac{1}{\varphi(pc_1)} \sum_{\chi (\text{mod } pc_1)}^{\sim} |L(\tfrac{1}{2}+\varepsilon+i\tau, \chi)|^2 \frac{d\tau}{|s+1|^{5/2}}.$$

Similarly

$$|I_2|^2 \ll \frac{1}{\varphi(pc_1)^2} \sum_{\chi (\text{mod } c_1)}^{\sim} |L(\sigma, \chi)|^2$$

$$+ \frac{1}{\varphi(pc_1)^2} \left(\frac{x}{q} \right)^{-2\sigma+1+2\varepsilon} \int_{-\infty}^{\infty} \sum_{\chi (\text{mod } c_1)}^{\sim} |L(\tfrac{1}{2}+\varepsilon+i\tau, \chi)|^2 \frac{d\tau}{|s+1|^{5/2}}.$$

Note that these upper bounds only involve q to the extent of a power of q as a factor in the integrals.

Altogether, therefore,

$$\sum_{\substack{p \leq P \\ (p,c)=1}} (p-1) \max_{(r,p)=1} \max_{y \leq x} |E_1(y, p, r)|^2 \ll (L_1 + L_2 + L_3) \sum_{q \leq x} \frac{|f(q)|^2}{q^\sigma},$$

where

$$L_1 = \frac{x^{1-\sigma}}{\log x} \sum_{\substack{p \leq P \\ (p,c)=1}} \widetilde{\sum_{\chi(\bmod pc_1)}} |L(\sigma, \chi)|^2$$

$$+ \frac{x^{1-\sigma}}{\log x} \int_{-\infty}^{\infty} \sum_{\substack{p \leq P \\ (p,c)=1}} \widetilde{\sum_{\chi(\bmod pc_1)}} |L(\tfrac{1}{2}+\varepsilon+i\tau, \chi)|^2 \frac{d\tau}{|s+1|^{5/2}},$$

$$L_2 = \frac{x^{1-\sigma}}{\log x} \sum_{\substack{p \leq P \\ (p,c)=1}} \frac{1}{\varphi(p)} \widetilde{\sum_{\chi(\bmod c_1)}} |L(\sigma, \chi)|^2$$

$$+ \frac{x^{1-\sigma}}{\log x} \int_{-\infty}^{\infty} \sum_{\substack{p \leq P \\ (p,c)=1}} \frac{1}{\varphi(p)} \widetilde{\sum_{\chi(\bmod c_1)}} |L(\tfrac{1}{2}+\varepsilon+i\tau, \chi)|^2 \frac{d\tau}{|s+1|^{5/2}},$$

and

$$L_3 = x^{1-2\sigma} \sum_{\substack{p \leq P \\ (p,c)=1}} \frac{1}{p}.$$

Note that our restrictions upon ε ensure that $2\varepsilon < \sigma$ and

$$\sum_{q \leq x} \frac{1}{q^\sigma} \left(\frac{x}{q}\right)^{-2\sigma+1+2\varepsilon} = x^{-2\sigma+1+2\varepsilon} \sum q^{-1-2\varepsilon+\sigma}$$

$$\ll x^{-2\sigma+1+2\varepsilon} \frac{x^{\sigma-2\varepsilon}}{\log x} \ll \frac{x^{1-\sigma}}{\log x}.$$

We now show that

(16) $$\sum_{\substack{p \leq P \\ (p,c)=1}} \widetilde{\sum_{\chi(\bmod pc_1)}} |L(s, \chi)|^2 \ll P^2 |s|$$

holds in every half-plane $\operatorname{Re}(s) \geq \tfrac{1}{2} + \delta > \tfrac{1}{2}$.

Each non-principal character $\chi \pmod{pc_1}$ is induced by a primitive non-principal character $\chi_1 \pmod{k}$, where k divides pc_1. Moreover

$$L(s, \chi) = L(s, \chi_1) \prod_{g|k} (1 - \chi_1(g)g^{-s}),$$

where g runs through the prime divisors of k. In this situation if p divides

k then χ_1 uniquely determines χ, and the corresponding terms in the sum at (16) may be estimated by lemma (6.8) to contribute

$$\ll \sum_{k \leq Pc_1} \sum_{\chi_1 \pmod{k}}^* |L(s, \chi_1)|^2 \ll P^2 |s|.$$

However, if k divides c_1 and $\chi \pmod{pc_1}$ is induced by $\chi_1 \pmod{k}$, then once again applying lemma (6.8), now with $Q = c_1$,

$$|L(s, \chi)|^2 \ll |L(s, \chi_1)|^2 \ll |s|,$$

the implied constants depending upon c_1 but not upon p.
The inequality (16) is now clear.

With the aid of this inequality it is easy to show that $L_1 \ll x^{1-\sigma}/\log x$. A smaller bound may be so obtained for L_2, and with the trivial bound for L_3, inequality (13) is established for $j = 1$.

The argument for $E_2(y, p, r)$ proceeds in a very similar manner. The only change of any consequence is that q_1 is now to be defined as $qq_0(qq_0, l)^{-1}$.
This concludes the proof of lemma (7.8).

Proof of theorem (7.2). From lemmas (7.6), (7.7) and (7.8)

$$\sum_{\substack{p \leq Q \\ (p,c)=1}} (p-1) \max_{(r,p)=1} \max_{y \leq x} |E(y, p, r)|^2$$

$$\ll \left(\left\{ \left(\frac{\log x}{P} \right)^{2\sigma-1} + P^2 \right\} \frac{x^{1-\sigma}}{\log x} + (Q^{3/2} + x^{5/6-\sigma}) Q^{5/2-2\sigma} (\log Q)^5 \right) \sum_{q \leq x} \frac{|f(q)|^2}{q^\sigma},$$

provided that P is a power of $\log x$. With the choice $P = (\log x)^{(2\sigma-1)/(2\sigma+1)}$ theorem (7.2) is proved.

Remark. There is a result of Wolke [1] which considers a large sieve inequality when the moduli are primes. If one argues as in Elliott [3] his result can be applied to improve by a power of $\log \log x$ the term $x^{1-\sigma}(\log x)^{-\theta}$ which appears in the statement of theorem (7.2).

Proof of theorem (7.3). It is readily checked that $F(y, p, r)$ differs from $E(y, p, r)$ by

$$\left(\frac{1}{p} \sum_{n \equiv r \pmod{p}} - \sum_{n \equiv r \pmod{p^2}} + \frac{1}{p(p-1)} \sum_n - \frac{1}{p-1} \sum_{n \equiv 0 \pmod{p}} \right) \frac{f(n)}{n^\sigma} \left(1 - \frac{n}{y}\right)$$

$$= \Sigma_1 - \Sigma_2 + \Sigma_3 - \Sigma_4,$$

say, where an obvious notation is in force, and in every summation condition it is to be understood that $n \equiv d \pmod{c}$ and $n \le y$ hold.

Let D be a positive integer. Let $(r, D) = 1$, and for each w prime to D define $t(w)$ by $wt(w) \equiv r \pmod{D}$, $1 \le t(w) \le D$.

We shall first show that

(17) $$\sum_{q \le x} \frac{1}{q^{2\sigma-1} t(q)^{2\sigma}} \ll \frac{x^{2(1-\sigma)}}{\varphi(D) \log x} + 1.$$

Indeed, the sum to be estimated may be written in the form

(18) $$\sum_{\substack{w=1 \\ (w,D)=1}}^{D} \frac{1}{t(w)^{2\sigma}} \sum_{\substack{q \le x \\ q \equiv w \pmod{D}}} \frac{1}{q^{2\sigma-1}}.$$

We already considered a sum

$$\sum_{\substack{n \le x \\ n \equiv l \pmod{p}}} \frac{\Lambda(n)}{n^{\sigma}}$$

during the proof of lemma (7.7), and the argument given there may be readily adapted to show that a typical innersum at (18) is

$$\ll \frac{x^{2(1-\sigma)}}{\varphi(D) \log x} + 1.$$

Since the map $w \mapsto t(w)$ induces a permutation of the reduced residue classes $\pmod{D}$

$$\sum_{\substack{w=1 \\ (w,D)=1}}^{D} \frac{1}{t(w)^{2\sigma}} \le \sum_{m=1}^{D} \frac{1}{m^{2\sigma}} \ll 1, \quad 2\sigma > 1,$$

and (17) is clearly valid.

Arguing much as in the remark which follows the proof of lemma (7.5)

(19) $$\sum_{\substack{n \le y \\ n \equiv r \pmod{D}}} \frac{|f(n)|}{n^{\sigma}} \left(1 - \frac{n}{y}\right) \le \sum_{q \le y} \frac{|f(q)|}{q^{\sigma}} \sum_{\substack{m \le yq^{-1} \\ qm \equiv r \pmod{D}}} \frac{1}{m^{\sigma}}.$$

From the estimate for $A(yq^{-1}, t(q), \sigma)$ which is established during the proof of lemma (7.5), a typical innersum here is

$$\ll \frac{(yq^{-1})^{1-\sigma}}{D} + t(q)^{-\sigma}.$$

The majorizing expression at (19) is therefore

$$\ll \frac{y^{1-\sigma}}{D} \sum_{q \leq y} \frac{|f(q)|}{q} + \sum_{q \leq y} \frac{|f(q)|}{(qt(t))^{\sigma}}.$$

Applying the Cauchy–Schwarz inequality to each of these sums in turn, and then applying to the bound (17), we see that

$$\left\{ \sum_{\substack{n \leq y \\ n \equiv r \pmod{D}}} \frac{|f(n)|}{n^{\sigma}} \left(1 - \frac{n}{y}\right) \right\}^2$$

$$\ll \left(\frac{x^{2(1-\sigma)} \log \log 2x}{D^2} + \frac{x^{2(1-\sigma)}}{\varphi(D) \log x} + 1 \right) \sum_{q \leq x} \frac{|f(q)|^2}{q}$$

uniformly for $1 \leq y \leq x$.

It is now easy to obtain the bounds

(20) $\quad \sum_{P < p \leq x^{\delta}} p \max_{(r,p)=1} \max_{y \leq x} |\Sigma_j|^2 \ll x^{2(1-\sigma)} \left\{ \frac{\log \log 2x}{P^2 \log P} + \frac{\log \log 2x}{\log x} \right\} \sum_{q \leq x} \frac{|f(q)|^2}{q}$

uniformly for $2 \leq P \leq x^{\delta}$ and $1 \leq j \leq 3$.

Moreover,

$$\sum_{\substack{n \leq y \\ n \equiv 0 \pmod{p}}} \frac{f(n)}{n^{\sigma}} \left(1 - \frac{n}{y}\right) = \sum_{\substack{q \leq yp^{-1} \\ (q,p)=1}} \frac{f(q)}{(pq)^{\sigma}} \sum_{m \leq y(pq)^{-1}} \frac{1}{m^{\sigma}} \left(1 - \frac{m}{y(pq)^{-1}}\right)$$

(21) $\qquad\qquad + \sum_{q = p^k \leq y} \frac{f(q)}{q^{\sigma}} \sum_{m \leq yq^{-1}} \frac{1}{m^{\sigma}} \left(1 - \frac{m}{yq^{-1}}\right)$

$$\ll \frac{x^{1-\sigma}}{p} \sum_{q \leq x} \frac{|f(q)|}{q} + x^{1-\sigma} \sum_{q = p^k \leq x} \frac{|f(q)|}{q}$$

uniformly for $2 \leq y \leq x$. Since, by the Cauchy–Schwarz inequality,

$$\left(\sum_{q = p^k \leq x} \frac{|f(q)|}{q} \right)^2 \ll \sum_{k=1}^{\infty} \frac{1}{p^k} \sum_{q \leq x} \frac{|f(q)|^2}{q} \ll \frac{1}{p} \sum_{q \leq x} \frac{|f(q)|^2}{q},$$

we obtain readily the fourth bound

(22) $\quad \sum_{P < p \leq x^{\delta}} p \max_{(r,p)=1} \max_{y \leq x} |\Sigma_4|^2 \ll x^{2(1-\sigma)} \left\{ \frac{\log \log 2x}{P^2 \log P} + \frac{1}{P \log P} \right\} \sum_{q \leq x} \frac{|f(q)|^2}{q}.$

In connection with the remark which follows the statement of theorem (7.4), we note that here the "worst" error factor $(P \log P)^{-1}$ in (22) arises from the consideration of the q in (21) which are powers of the prime p.

With $P = \log x$, theorem (7.3) follows directly from theorem (7.1) and the inequalities at (20) and (22).

These same inequalities, in combination with lemmas (7.6) and (7.7), give

$$\sum_{\substack{P < p \leq x^\delta \\ (p,c)=1}} p \max_{(r,p)=1} \max_{y \leq x} |F(y,p,r)|^2$$

(23)
$$\ll \left\{ \frac{\log \log 2x}{P^2 \log P} + \frac{1}{P \log P} + \left(\frac{\log x}{P}\right)^{2\sigma - 1} \frac{1}{\log x} \right.$$

$$\left. + \frac{\log \log 2x}{\log x} \right\} x^{2(1-\sigma)} \sum_{q \leq x} \frac{|f(q)|^2}{q}$$

uniformly for $2 \leq P \leq x^\delta$.

Proof of theorem (7.4). The proof of lemma (7.8) may be readily adapted to give

(24) $$\sum_{\substack{p \leq P \\ (p,c)=1}} p \max_{(r,p)=1} \max_{y \leq x} |F(y,p,r)|^2 \ll \frac{x^{2(1-\sigma)}}{\log x} (P^3 + \log \log 2x) \sum_{q \leq x} \frac{|f(q)|^2}{q}$$

as long as P does not exceed a fixed power of $\log x$.

For this purpose $F(y, p, r)$ is better represented in the form

(25)
$$\left(1 + \frac{1}{p}\right) \sum_{\substack{n \leq y \\ n \equiv d \pmod{c} \\ n \equiv r \pmod{p}}} \frac{f(n)}{n^\sigma}\left(1 - \frac{n}{y}\right) - \sum_{\substack{n \leq y \\ n \equiv d \pmod{c} \\ n \equiv r \pmod{p^2}}} \frac{f(n)}{n^\sigma}\left(1 - \frac{n}{y}\right)$$

$$- \frac{1}{p} \sum_{\substack{n \leq y \\ n \equiv d \pmod{c}}} \frac{f(n)}{n^\sigma}\left(1 - \frac{n}{y}\right).$$

Of these three sums it will suffice to estimate the third directly, omitting the factor p^{-1}, by

$$\sum_{q \leq y} \frac{f(q)}{(q_1 l)^\sigma} \sum_{\substack{m \leq y(q_1 l)^{-1} \\ q_1 m \equiv d_1 \pmod{c_1}}} \frac{1}{m^\sigma}\left(1 - \frac{m}{y(q_1 l)^{-1}}\right)$$

$$- \sum_{q q_0 \leq y} \frac{f(q)}{(q_2 l)^\sigma} \sum_{\substack{m \leq y(q_1 l)^{-1} \\ q_2 m \equiv d_1 \pmod{c_1}}} \frac{1}{m^\sigma}\left(1 - \frac{m}{y(q_2 l)^{-1}}\right),$$

Algebraicanalytic Inequalities 149

where $q_2 = qq_0(qq_0, l)^{-1}$. Here typical innersums (see lemma (8.4) of the next chapter) may be estimated by

$$\frac{(y(q_j l)^{-1})^{1-\sigma}}{(1-\sigma)(2-\sigma)c_1} + O(1), \qquad j = 1, 2.$$

This gives

(26)
$$\sum_{\substack{n \leq y \\ n \equiv d \pmod{c}}} \frac{f(n)}{n^\sigma}\left(1 - \frac{n}{y}\right) = \frac{y^{1-\sigma}}{(1-\sigma)(2-\sigma)c_1} \sum_{q \leq y} \frac{f(q)}{l}\left(\frac{1}{q_1} - \frac{1}{q_2}\right)$$

$$+ O\left(\sum_{q \leq y} \frac{|f(q)|}{q^\sigma}\right).$$

The two remaining sums at (25) can be treated exactly as the sums $E_j(y, p, r)$ were during the proof of lemma (7.8). Taking into consideration the estimate (26), the "main" terms of the three sums at (25) cancel, and we are left with "error" terms involving the same sums L_i as before, together with corresponding extra terms when

$$\sum_{\substack{p \leq P \\ (p,c)=1}} \widetilde{\sum_{\chi \pmod{pc_1}}} |L(s, \chi)|^2$$

for various values of s, is replaced (in an obvious notation) by

$$\sum_{\substack{p \leq P \\ (p,c)=1}} p^{-1} \widetilde{\sum_{\chi \pmod{p^2 c_1}}} |L(s, \chi)|^2.$$

The analogue to the bound (16) is easily seen to be here $\ll P^3 |s|$, and we readily obtain the inequality (24). The factor involving $\log \log 2x$ arises from the error term at (26).

Theorem (7.4) now follows from the inequalities (23) and (24) with $P = (\log x)^{1/4}$.

Algebraicanalytic Inequalities

In order to obtain a result

$$\sum_{\log x < p \leq x^\delta} (p-1) \max_{(r,p)=1} \max_{y \leq x} |E(y, p, r)|^2 \ll \frac{x^{1-\sigma}}{\log x} \sum_{q \leq x} \frac{|f(q)|^2}{q^\sigma}$$

with δ as large as possible we apply theorem (7.1) with

$$\delta < \min\left(\frac{1-\sigma}{4-2\sigma}, \frac{\frac{1}{6}}{\frac{5}{2}-2\sigma}\right), \quad \tfrac{1}{2} < \sigma < 1.$$

When $\sigma = (25 - \sqrt{97})/24$ this condition becomes $\delta < (\sqrt{97} - 5)/36$. In particular $\delta = \tfrac{1}{8}$ is permissible.

This may be compared with results of Bombieri–Vinogradov, and of Wolke. Let

$$R(y, D, l) = \pi(y, D, l) - \frac{1}{\varphi(D)} \int_2^y \frac{dw}{\log w},$$

where $\pi(y, D, l)$ denotes the number of primes p not exceeding y which satisfy $p \equiv l \pmod{D}$. Then the theorem of Bombieri–Vinogradov (see Bombieri [1], Théorème 17, p. 57, or Davenport [1], Chapter 28, p. 161) asserts that for any constant $A > 0$ there is a further constant B so that

$$\sum_{D \leq x^{1/2}(\log x)^{-B}} \max_{(l,D)=1} \max_{y \leq x} |R(y, D, l)| \ll x(\log x)^{-A}.$$

This result may be put in a form more like that of theorem (7.1) by employing the Brun–Titchmarsh estimate

$$\pi(x, D, l) \ll \frac{x}{\varphi(D) \log x},$$

valid over the present range of D, to obtain

(27) $$\sum_{D \leq x^{1/2}(\log x)^{-B}} \varphi(D) \max_{(l,D)=1} \max_{y \leq x} |R(y, D, l)|^2 \ll x^2 (\log x)^{-A-1}.$$

Moreover, from a result of this last type an application of the Cauchy–Schwarz inequality enables one to retrieve a form of the Bombieri–Vinogradov theorem.

A crude upper bound for the sum which appears in (27) would be

$$O\left(\sum_{D \leq x^{1/2}} \varphi(D) \left(\frac{x}{\varphi(D) \log x}\right)^2\right) \ll \frac{x^2}{\log x}.$$

In comparison with this estimate the inequality (27) saves "essentially" an arbitrary power of $\log x$.

More generally, let the multiplicative function $g(n)$ satisfy

$$|g(p^a)| \leq D_1 a^{D_2}$$

for all prime powers p^a, and

$$\sum_{p \leq y} |g(p) - \tau| \ll y(\log y)^{-A_1}, \qquad y \geq 2,$$

where D_1, D_2, τ are constants, and A_1 may be fixed at any positive value. Then Wolke [2] showed that for each $A > 0$ we can find a B such that

$$\sum_{k \leq x^{1/2}(\log x)^{-B}} \max_{(l,k)=1} \max_{y \leq x} \left| \sum_{\substack{n \leq y \\ n \equiv l (\bmod k)}} g(n) - \frac{1}{\varphi(k)} \sum_{\substack{n \leq y \\ (n,k)=1}} g(n) \right| \ll x(\log x)^{-A}.$$

From this result applied to a power of the divisor function $d(n)$ we see that

$$\sum_k \max_l \max_y \sum_{\substack{n \leq y \\ n \equiv l (\bmod k)}} d(n)^{c_0} \ll x(\log x)^{-A} + x(\log x) \sum_{n \leq x} d(n)^{c_0}$$

$$\ll x(\log x)^{c_1}$$

for some constant c_1, this last step by the example following the proof of lemma (1.2). Since for some c_0

$$|g(n)| \leq d(n)^{c_0},$$

a crude estimate for the general sum considered by Wolke would also be $\ll x(\log x)^{c_1}$. Once again an arbitrary power of a logarithm has been saved.

What in our situation corresponds to a "crude" estimate?

Let k_1, k_2, k_3 denote respectively the number of prime-powers not exceeding x, the number of integers n in the interval $x/2 < n \leq x$, and the number of pairs (r, p) with $1 \leq r \leq p$ and p a prime not exceeding Q. Let

$$G(p, r) = \sum_{\substack{x/2 < n \leq x \\ n \equiv r (\bmod p)}} \frac{f(n)}{n^\sigma} - \frac{1}{p} \sum_{x/2 < n \leq x} \frac{f(n)}{n^\sigma}.$$

To obtain an upper bound for the sum

$$S = \sum_{p \leq Q} p \sum_{r=1}^{p} |G(p, r)|^2$$

define

$$H(p, r) = \sum_{\substack{x/2 < n \leq x \\ n \equiv r (\bmod p)}} \frac{1}{n^\sigma} (f(n) - A) - \frac{1}{p} \sum_{x/2 < n \leq x} \frac{1}{n^\sigma} (f(n) - A),$$

where

$$A = \sum_{q \leq x} \frac{f(q)}{q}.$$

By the Cauchy–Schwarz inequality and an application of lemma (1.2)

$$|A|^2 \ll \log \log x \sum_{q \leq x} \frac{|f(q)|^2}{q}, \quad x \geq 3,$$

and since

$$\sum_{\substack{x/2 < n \leq x \\ n \equiv r \pmod{p}}} \frac{1}{n^\sigma} - \frac{1}{p} \sum_{x/2 < n \leq x} \frac{1}{n^\sigma} \ll 1,$$

we have

$$S \ll \sum_{p \leq Q} p \sum_{r=1}^{p} |H(p,r)|^2 + \frac{Q^3 \log \log x}{\log Q} \sum_{q \leq x} \frac{|f(q)|^2}{q}$$

for $Q \geq 2$. We estimate the sum involving the $H(p, r)$ by considering two linear maps τ_1, τ_2 and their composition $\tau_2 \circ \tau_1$:

$$\begin{array}{ccc} & \mathbb{R}^{k_2} & \\ {}^{\tau_1}\nearrow & & \searrow^{\tau_2} \\ \mathbb{R}^{k_1} & \xrightarrow{\tau_2 \circ \tau_1} & \mathbb{R}^{k_3}. \end{array}$$

Here τ_1 takes a vector with typical component $q^{-1/2} f(q)$ to a vector with typical component

$$n^{-\sigma}(f(n) - A).$$

The map τ_2 takes a vector with typical component a_n to a vector with typical component

$$p^{1/2}\left(\sum_{\substack{x/2 < n \leq x \\ n \equiv r \pmod{p}}} a_n - \frac{1}{p} \sum_{x/2 < n \leq x} a_n\right).$$

Then, using $\| \ \|$ to denote the appropriate L^2-norm(s),

$$(28) \qquad S \ll \left(\|\tau_2 \circ \tau_1\|^2 + \frac{Q^3 \log \log x}{\log Q}\right) \sum_{q \leq x} \frac{|f(q)|^2}{q}.$$

Algebraicanalytic Inequalities

From inequality (4) of Chapter 6, a form of the Large Sieve, we see that

$$\|\tau_2\|^2 \ll x + Q^2,$$

whilst from lemma (1.5), a form of the Turán–Kubilius inequality,

$$\|\tau_1\|^2 \ll x^{1-2\sigma}.$$

By means of the elementary inequality

$$\|\tau_2 \circ \tau_1\| \leq \|\tau_2\| \cdot \|\tau_1\|$$

we obtain in the bound for S at (28) a factor which is

$$\ll x^{2-2\sigma} + Q^2 x^{1-2\sigma} + \frac{Q^3 \log \log x}{\log Q}.$$

In particular, if $Q^3 \leq x^{1/2}$, $\frac{1}{2} < \sigma < 1$, then

(29) $$\sum_{p \leq Q} p \max_{(r,p)=1} |G(p,r)|^2 \ll x^{2(1-\sigma)} \sum_{q \leq x} \frac{|f(q)|^2}{q}.$$

However, if Q is a small enough power of x then a similar sum may be estimated by theorem (7.1) to be

(30) $$\ll \frac{x^{1-\sigma}}{\log x} \sum_{q \leq x} \frac{|f(q)|^2}{q^\sigma} \ll \frac{x^{2(1-\sigma)}}{\log x} \sum_{q \leq x} \frac{|f(q)|^2}{q}.$$

The definition of the $G(p, r)$ can readily be modified so as to make this comparison more accurate in detail.

With (29) as a standard we may view theorem (7.1) as saving a logarithm in comparison with a straightforward estimate. The results of Bombieri and Wolke save an arbitrary power of $\log x$, but the improvement is purchased by quite strong (implicit or explicit) assumptions concerning the growth of the arithmetic functions under consideration. *In our theorem (7.1) no assumptions are made concerning $f(n)$ beyond that it be additive,* and in this generality, as we have already indicated, the saving of one $\log x$ is largely best possible.

The application of the bound $\|\tau_2\|^2 \ll x + Q^2$ in the "crude" treatment of S wastes the fact that the $a_n = f(n) - A$ have an algebraic structure. Our theorem (7.1) is obtained by treating the map $\tau_2 \circ \tau_1$ directly. In this way we obtain an inequality which takes into account the algebraic nature of the functions involved. Such inequalities may be regarded as having both an algebraic and an analytic nature. They are analytic in so far as one is estimating the norm of an operator, and algebraic in so far that they operate upon elements which have a coordinate representation *where the coordinates have an algebraic structure.*

From this point of view it is more appropriate that we seek estimates in mean-square (or perhaps higher powers) rather than in an L^1 sense, because one expects the dual of an L^2 space to be another L^2 space.

It may occur that theorem (7.1) implicitly saves $(\log x)^2$ rather than $\log x$. For example, if $f(q) = \log q$ then (see (29))

$$\sum_{q \leq x} \frac{|f(q)|^2}{q} \sim \tfrac{1}{2}(\log x)^2, \qquad x \to \infty.$$

The second upper bound at (30) then gives, apart from the factor $x^{2(1-\sigma)}$,

$$\ll \frac{1}{\log x} \sum_{q \leq x} \frac{|f(q)|^2}{q} \ll \log x.$$

However, a direct application of the bound which comes from theorem (7.1) gives

$$\ll \frac{x^{1-\sigma}}{\log x} \sum_{q \leq x} \frac{(\log q)^2}{q^\sigma} \ll x^{2(1-\sigma)}$$

effecting a saving of $(\log x)^2$ over the bound at (29).

One can write

$$E(y, p, r) = \frac{1}{y} \int_1^y \left\{ \sum_{\substack{n \leq w \\ n \equiv d \pmod{c} \\ n \equiv r \pmod{p}}} \frac{f(n)}{n^\sigma} - \frac{1}{\varphi(p)} \sum_{\substack{n \leq w \\ n \equiv d \pmod{c} \\ (n,p)=1}} \frac{f(n)}{n^\sigma} \right\} dw.$$

Perhaps a form of theorem (7.1) holds without this integration.

One might also hope to improve the dependence upon Q. Arguments of the type considered in Chapter 5 show that it cannot be reasonably hoped to improve the factor in the bound of theorem (7.1) beyond

$$\frac{x^{1-\sigma}}{\log x} + Q^{2(1-\sigma)}.$$

There would be no difficulty in obtaining analogues of the theorems in the present chapter with the parameter σ replaced by a complex number. This concludes our remarks.

Chapter 8

The Loop

Theorem (8.1). Let $\frac{1}{2} < \sigma < 1$. Then there is a positive constant δ so that the inequality

$$\sum_{\substack{\log x < p \leq x^\delta \\ (p, aA\Delta)=1}} \frac{1}{p} \left| f(p) - E(x) + E\left(\frac{x}{p}\right) \right|^2$$

$$\ll \frac{\log \log x}{\log x} \sum_{q \leq Ax + |\Delta|} \frac{|f(q)|^2}{q} + \frac{1}{x^{1-\sigma}} \sum_{n \leq (x-b)/a} \frac{|f(an+b) - f(An+B)|^2}{n^\sigma},$$

with

$$E(x) = \sum_{\substack{p^k \leq x \\ (p, aA\Delta)=1}} \frac{f(p^k)}{p^k}\left(1 - \frac{1}{p}\right),$$

holds for all additive functions $f(n)$, for all $x \geq 4(aA\Delta)^2$.

Remarks. As in previous chapters p denotes a prime, q a prime-power. The integers $a > 0$, $A > 0$, b, B are assumed to satisfy $\Delta = aB - Ab \neq 0$.

It will transpire that for δ we may take any value which satisfies

(1) $$\delta < \min\left(\frac{1-\sigma}{4-2\sigma}, \frac{1}{15-12\sigma}\right).$$

In particular, the choice $\sigma = (25 - \sqrt{97})/24$, $\delta = \frac{1}{8}$ is possible.

If the factor $\log \log x / \log x$, which appears in the upper bound, is reduced to $(\log x)^{-1/4}$, then the summation condition $\log x < p$ may be removed.

We shall need a variant of theorem (8.1) which treats more carefully the contribution towards $f(n)$ arising from the powers of the primes which divide $aA\Delta$. We adopt the notation L_1, L_2 and $L(p^i, p^j)$ from the beginning of Chapter 3. It will also be convenient to use p_{ij} to denote the maximum of p^i and p^j.

Theorem (8.2). *Let the notation of theorem* (8.1) *be in force, and let δ satisfy* (1). *Then*

$$\sum_{\substack{p \leq x^\delta \\ (p, aA\Delta)=1}} \frac{1}{p} \left| f(p) - E(x) + E\left(\frac{x}{p}\right) \right|^2$$

$$\ll \frac{1}{x^{1-\sigma}} \sum_{n \leq (x-b)/a} \frac{|f(an+b) - f(An+B)|^2}{n^\sigma} + \frac{1}{(\log x)^{1/4}} \sum_{\substack{q \leq Ax+|\Delta| \\ (q, aA\Delta)=1}} \frac{|f(q)|^2}{q}$$

$$+ \frac{1}{x^{1-\sigma-\delta}} \left(\sum_{p|aA\Delta} \sum_{p_{ij} \leq Ax+|\Delta|} \frac{|L(p^i, p^j)|^2}{p_{ij}} + \sum_{t=1}^{2} |L_t|^2 \right)$$

holds with the same uniformities as in theorem (8.1).

Remarks. The pairs p^i, p^j are restricted to those for which an integer n exists so that $p^i \| (an+b)$ and $p^j \| (An+B)$. Otherwise $L(p^i, p^j)$ is not defined. Note that if $p^w \| \Delta$ then $\min(i, j) \leq w$ must hold.

If only one fixed-divisor pair exists (see Chapter 3) the term involving $|L_2|^2$ is to be omitted.

The implied constants in theorem (8.1) and (8.2) depend only upon a, b, A, B, δ and σ.

The quality of these theorems is much enhanced by the introduction of a parameter. The following extension of theorem (8.2) will be employed in Chapter 10. A similar extension of theorem (8.1) may be constructed.

Theorem (8.3). *Let $\frac{1}{2} < \sigma < 1$, and let δ satisfy the condition* (1). *Then*

$$\sum_{\substack{p \leq x^\delta \\ (p, aA\Delta)=1}} \frac{1}{p} \left| f(p) - U \log p - E^*(x) + E^*\left(\frac{x}{p}\right) \right|^2$$

$$\ll \frac{1}{x^{1-\sigma}} \sum_{n \leq (x-b)/a} \frac{|f(an+b) - f(An+B)|^2}{n^\sigma} + \frac{1}{(\log x)^{1/4}} \sum_{\substack{q \leq Ax+|\Delta| \\ (q, aA\Delta)=1}} \frac{|f^*(q)|^2}{q}$$

$$+ \frac{1}{x^{1-\sigma-\delta}} \left(\sum_{p|aA\Delta} \sum_{p_{ij} \leq Ax+|\Delta|} \frac{|L^*(p^i, p^j)|^2}{p_{ij}} + \sum_{t=1}^{2} |L_t^*|^2 \right) + \frac{|U|^2}{x^{2(1-\sigma-\delta)}}$$

*holds uniformly for all additive functions $f(n)$, for all complex U, for all real $x \geq 4(aA\Delta)^2$. Here * denotes that the definitions of theorems* (8.1) *and* (8.2) *(and of Chapter 3) are in force, but with $f(q)$ everywhere replaced by $f(q) - U \log q$.*

Following the proofs of these results we give an example in the application of theorem (8.1).

The next two lemmas will be useful many times.

Lemma (8.4). *Let $0<\sigma<1$ hold. Then*

$$\sum_{t\le y}\frac{1}{t^\sigma}=\frac{y^{1-\sigma}}{1-\sigma}+O(1)$$

and

$$\sum_{\substack{m\le y\\m\equiv r(\bmod D)}}\frac{1}{m^\sigma}\left(1-\frac{m}{y}\right)=\frac{y^{1-\sigma}}{(1-\sigma)(2-\sigma)D}+O(1)$$

holds uniformly for all integers r, D with $D\ge 1$, and all real $y>0$.

Proof. Integrating by parts

$$\sum_{t\le y}\frac{1}{t^\sigma}=\int_{1-}^{y+}z^{-\sigma}d[z]=[y+]y^{-\sigma}+\sigma\int_1^y [z]z^{-\sigma-1}\,dz,$$

where $[z]$ denotes the greatest integer not exceeding z. With the estimate $[z]=z+O(1)$ the first assertion of lemma (8.4) is readily obtained. As a corollary

(2) $$\sum_{t\le w}\frac{1}{t^\sigma}\left(1-\frac{t}{w}\right)=\frac{1}{w}\int_1^w\sum_{t\le y}\frac{1}{t^\sigma}\,dy=\frac{w^{1-\sigma}}{(1-\sigma)(2-\sigma)}+O(1).$$

For the second assertion of lemma (8.4) we note that without loss of generality $1\le r\le D$, and $y>D$. The sum to be estimated can then be represented by

$$\sum_{1\le s\le y/D}\frac{1}{(sD+r)^\sigma}\left(1-\frac{sD+r}{y}\right)+O(1).$$

Here a typical summand is

$$\frac{1}{(sD)^\sigma}\left(1+\frac{r}{sD}\right)^{-\sigma}\left(1-\frac{sD+r}{y}\right)=\frac{1}{(sD)^\sigma}\left(1+O\!\left(\frac{1}{s}\right)\right)\left(1-\frac{sD}{y}+O\!\left(\frac{D}{y}\right)\right)$$

$$=\frac{1}{(sD)^\sigma}\left(1-\frac{sD}{y}\right)+O\!\left(\frac{D^{1-\sigma}}{ys^\sigma}+\frac{1}{D^\sigma s^{\sigma+1}}\right).$$

Since
$$\frac{D^{1-\sigma}}{y} \sum_{s \le y/D} \frac{1}{s^\sigma} \ll \frac{D^{1-\sigma}}{y}\left(\frac{y}{D}\right)^{1-\sigma} \ll 1$$

and
$$\frac{1}{D^\sigma} \sum_{s \le y/D} \frac{1}{s^{\sigma+1}} \ll \frac{1}{D^\sigma} \ll 1,$$

the desired result follows from (2) with $w = y/D$.

Lemma (8.5). *If* $0 < \sigma < 1$ *then*

$$\sum_{n \le x} \frac{|f(n)|}{n^\sigma} \ll x^{1-\sigma} \left(\sum_{q \le x} \frac{|f(q)|^2}{q} \cdot \log\log x \right)^{1/2},$$

$$\sum_{\substack{n \le x \\ n \equiv 0 (\bmod p)}} \frac{|f(n)|}{n^\sigma} \ll x^{1-\sigma} \sum_{k} \frac{|f(p^k)|}{p^k} + \frac{x^{1-\sigma}}{p}\left(\sum_{q \le x} \frac{|f(q)|^2}{q} \cdot \log\log x \right)^{1/2}$$

hold for all additive functions $f(n)$, *primes* p *and real* $x \ge 4$. *In the second of these inequalities k ranges over the positive integers for which* $p^k \le x$.

Proof. In the notation of lemma (1.5), applying the Cauchy–Schwarz inequality,

$$\sum_{n \le x} \frac{1}{n^\sigma} |f(n) - E(x)| \ll \{x^{1-\sigma} \cdot x^{1-\sigma} D(x)^2\}^{1/2}.$$

However, by another application of the Cauchy–Schwarz inequality, and lemma (1.2),

$$E(x)^2 \le D(x)^2 \sum_{q \le x} \frac{1}{q} \ll D(x)^2 \log\log x,$$

so that
$$\sum_{n \le x} \frac{|E(x)|}{n^\sigma} \ll x^{1-\sigma} D(x) \sqrt{\log\log x}.$$

The first inequality of lemma (8.5) is now evident.

The second of the sums to be majorized in lemma (8.5) does not exceed

$$\sum_k \sum_{\substack{n\leq x \\ p^k \| n}} \frac{|f(n)|}{n^\sigma} \leq \sum_k \frac{1}{p^{k\sigma}} \sum_{r\leq x/p^k} \frac{|f(p^k)|+|f(r)|}{r^\sigma}$$

$$\ll \sum_k \frac{1}{p^{k\sigma}} \left\{ |f(p^k)| \left(\frac{x}{p^k}\right)^{1-\sigma} + \sum_{r\leq x/p^k} \frac{|f(r)|}{r^\sigma} \right\}.$$

By means of the first inequality of lemma (8.5)

$$\sum_k \frac{1}{p^{k\sigma}} \cdot \sum_{r\leq x/p^k} \frac{|f(r)|}{r^\sigma} \ll x^{1-\sigma} D(x) \sqrt{\log\log x} \sum_k \frac{1}{p^k}$$

$$\ll p^{-1} x^{1-\sigma} D(x) \sqrt{\log\log x}$$

and lemma (8.5) is proved.

It will be convenient to continue with the convention, adopted in the previous chapter, that whenever q denotes a prime-power, q_0 will be the prime which generates it.

We shall also write $n \stackrel{\scriptscriptstyle\|}{\equiv} r \pmod{q}$ when the congruence $n \equiv r \pmod{q}$ is satisfied, but $n \equiv r \pmod{qq_0}$ is not. An alternative notation would therefore be $q \| (n-r)$.

Proof of theorem (8.1). We begin with the inequality

(3) $$\sum_{p\leq x} p \left| \sum_{\substack{n\leq x \\ p\|n}} a_n n^{-\sigma} - \frac{1}{p} \sum_{\substack{n\leq x \\ (n,p)=1}} a_n n^{-\sigma} \right|^2 \ll x^{1-\sigma} \sum_{n\leq x} |a_n|^2 n^{-\sigma}.$$

This may be derived from the first inequality of lemma (5.2) if we note that

$$\sum_{p\leq x} p \left| \frac{1}{p^2} \sum_{n\leq x} a_n n^{-\sigma} \right|^2$$

and

$$\sum_{p\leq x} p \left| \frac{1}{p} \sum_{\substack{n\leq x \\ n\equiv 0 \pmod{p}}} a_n n^{-\sigma} \right|^2$$

are both majorized by the upper bound at (3).

In the inequality (3) we set

$$a_n = \begin{cases} z_m \left(\dfrac{am+b}{m} \right)^\sigma & \text{if } n \text{ has the form } am+b \text{ for some integer } m, \\ 0 & \text{otherwise.} \end{cases}$$

This gives

$$\sum_{p \leq x} p \left| \sum_{\substack{0 < am+b \leq x \\ am+b \equiv 0 (\bmod p)}} z_m m^{-\sigma} - \frac{1}{p} \sum_{\substack{0 < am+b \leq x \\ (am+b,p)=1}} z_m m^{-\sigma} \right|^2 \ll x^{1-\sigma} \sum_{0 < am+b \leq x} |z_m|^2 m^{-\sigma}.$$

Specializing further, set

$$z_m = \{f(am+b) - f(Am+B)\} \left(\frac{m}{Am+B} \right)^\sigma \left(1 - \frac{a(Am+B)}{Ax+\Delta} \right)$$

if $Am+B$ is positive, and $z_m = 0$ otherwise. We assume x large enough that $Ax+\Delta$ is positive, and restrict the primes to those which do not divide $aA\Delta$. To simplify the exposition let $x_1 = (Ax+\Delta)/a$, and define

$$R = \frac{x^{2(1-\sigma)} \log \log 2x}{\log x} \sum_{q \leq Ax+|\Delta|} \frac{|f(q)|^2}{q} + x^{1-\sigma} \sum_{an+b \leq x} \frac{|f(an+b) - f(An+B)|^2}{n^\sigma}.$$

Then

$$\sum_{\substack{p \leq x \\ (p,aA\Delta)=1}} p \left| \sum_{\substack{am+b \leq x \\ am+b \equiv 0(\bmod p)}} \frac{f(am+b)}{(Am+B)^\sigma} \left(1 - \frac{Am+B}{x_1} \right) \right.$$

$$- \frac{1}{p} \sum_{\substack{am+b \leq x \\ (am+b,p)=1}} \frac{f(am+b)}{(Am+B)^\sigma} \left(1 - \frac{Am+B}{x_1} \right)$$

$$\left. - \left\{ \sum_{\substack{n \leq x_1 \\ n \equiv B(\bmod A) \\ an \equiv \Delta(\bmod p)}} \frac{f(n)}{n^\sigma} \left(1 - \frac{n}{x_1} \right) - \frac{1}{p} \sum_{\substack{n \leq x_1 \\ n \equiv B(\bmod A) \\ (an-\Delta,p)=1}} \frac{f(n)}{n^\sigma} \left(1 - \frac{n}{x_1} \right) \right\} \right|^2 \ll R.$$

In an obvious notation we write

(4) $$\sum_{\substack{p \leq x \\ (p,aA\Delta)=1}} p |\Sigma_1 - \Sigma_2|^2 \ll R.$$

The Loop

Note that in (4) summation conditions $Am + B > 0$, or $am + b > 0$ as the case may be, have been omitted from the sums Σ_j. This is permissible since it introduces an error which is

$$\ll \sum_{p \leq x} p \left(\sum_{\substack{n \leq |\Delta| \\ n \equiv 0 \pmod{p}}} |f(n)| + \frac{1}{p} \sum_{n \leq |\Delta|} |f(n)| \right)^2.$$

Applying the Cauchy–Schwarz inequality we see that this expression is in turn

$$\ll \sum_{p \leq |\Delta|} pD(x)^2 + \sum_{p \leq x} \frac{1}{p} D(x)^2 \ll D(x)^2 \log \log x,$$

an amount which is negligible in comparison with R.

We now further restrict the primes p by $\log x < p \leq x^\delta$, where δ is any positive number which satisfies the bound (1). Then from theorem (7.3)

$$\sum_{\substack{\log x < p \leq x^\delta \\ (p, a A \Delta) = 1}} p \left| \sum_{\substack{n \leq x_1 \\ n \equiv B \pmod{A} \\ an \equiv \Delta \pmod{p}}} \frac{f(n)}{n^\sigma} \left(1 - \frac{n}{x_1}\right) - \frac{1}{p} \sum_{\substack{n \leq x_1 \\ n \equiv B \pmod{A} \\ (an - \Delta, p) = 1}} \frac{f(n)}{n^\sigma} \left(1 - \frac{n}{x_1}\right) \right|^2 \ll R.$$

In the notation of theorem (7.3), the relevant value of r for a typical $F(y, p, r)$ is that given by $ar \equiv \Delta \pmod{p}$.

It follows from this and (4) that

(5) $$\sum_{\substack{\log x < p \leq x^\delta \\ (p, a A \Delta) = 1}} p |\Sigma_1|^2 \ll R.$$

An argument similar to that used during the proof of the second estimate of lemma (8.4) gives

$$\frac{1}{(Am+B)^\sigma} \left(1 - \frac{Am+B}{x_1}\right) = \frac{1}{(Am)^\sigma} \left(1 - \frac{Am}{x_1}\right) + O\left(\frac{1}{m^{\sigma+1}} + \frac{1}{m^\sigma x}\right).$$

This, together with the like estimate obtained when A, B, x_1 are replaced by a, b, $x + \Delta A^{-1}$, respectively, shows that

$$\frac{1}{(Am+B)^\sigma} \left(1 - \frac{Am+B}{x_1}\right) = \left(\frac{a}{A}\right)^\sigma \frac{1}{(am+b)^\sigma} \left(1 - \frac{am+b}{x}\right) + O\left(\frac{1}{m^{\sigma+1}}\right)$$

holds uniformly for $\max(-B/A, -b/a) < m \leq Ax + |\Delta|$.

This allows us to deduce from the inequality (5) another of more convenient form:

(6)
$$\sum_{\substack{\log x < p \leq x^\delta \\ (p,a A\Delta)=1}} p \left| \sum_{\substack{am+b \leq x \\ am+b \equiv 0 (\bmod p)}} \frac{f(am+b)}{(am+b)^\sigma}\left(1 - \frac{am+b}{x}\right) \right.$$

$$\left. - \frac{1}{p} \sum_{\substack{am+b \leq x \\ (am+b,p)=1}} \frac{f(am+b)}{(am+b)^\sigma}\left(1 - \frac{am+b}{x}\right) \right|^2 \ll R.$$

This step is readily justified by the appropriate application of lemma (8.5). For example, using an integration by parts

$$\sum_{n \leq x} \frac{|f(n)|}{n^{\sigma+1}} \ll \left(\sum_{q \leq x} \frac{|f(q)|^2}{q} \log \log x \right)^{1/2},$$

so that

$$\sum_{\log x < p \leq x^\delta} \frac{1}{p} \left| \frac{1}{p} \sum_{am+b \leq x} O\left(\frac{|f(am+b)|}{m^{\sigma+1}} \right) \right|^2 \ll \sum_{q \leq x} \frac{|f(q)|^2}{q} (\log \log x)^2,$$

which is much smaller than R.

We now turn to the innersums of (6).

Inverting the order of summation we have

(7)
$$\sum_{\substack{am+b \leq x \\ am+b \equiv 0 (\bmod p)}} \frac{f(am+b)}{(am+b)^\sigma}\left(1 - \frac{am+b}{x}\right) = \sum_{q \leq x} \frac{f(q)}{q^\sigma} \sum_{\substack{t \leq x/q \\ qt \equiv b (\bmod a) \\ qt \equiv 0 (\bmod p) \\ (t,q)=1}} \frac{1}{t^\sigma}\left(1 - \frac{qt}{x}\right).$$

Suppose first that $(q, ap) = 1$. Then a typical innersum on the right-hand side may be given the representation

$$\frac{1}{p^\sigma} \sum_{\substack{s \leq x(pq)^{-1} \\ qps \equiv b(\bmod a)}} \frac{1}{s^\sigma}\left(1 - \frac{s}{x(pq)^{-1}}\right) - \frac{1}{(pq_0)^\sigma} \sum_{\substack{r \leq x(pqq_0)^{-1} \\ qq_0 pr \equiv b(\bmod a)}} \frac{1}{r^\sigma}\left(1 - \frac{r}{x(p^2 qq_0)^{-1}}\right)$$

$$- \frac{1}{p^{2\sigma}} \sum_{\substack{s \leq x(p^2 q)^{-1} \\ qp^2 s \equiv b(\bmod a)}} \frac{1}{s^\sigma}\left(1 - \frac{s}{x(p^2 q)^{-1}}\right)$$

$$+ \frac{1}{(p^2 q_0)^\sigma} \sum_{\substack{r \leq x(p^2 qq_0)^{-1} \\ qq_0 p^2 r \equiv b(\bmod a)}} \frac{1}{r^\sigma}\left(1 - \frac{r}{x(p^2 qq_0)^{-1}}\right).$$

By lemma (8.4) these together are

$$\frac{(x/q)^{1-\sigma}}{(1-\sigma)(2-\sigma)a}\frac{1}{p}\left(1-\frac{1}{p}\right)\left(1-\frac{1}{q_0}\right)+O\left(\frac{1}{p^\sigma}\right).$$

Corresponding to such prime-powers q, which cannot exceed x/p, we have towards the sum at (7) an amount

$$\frac{x^{1-\sigma}}{(1-\sigma)(2-\sigma)a}\frac{1}{p}\left(1-\frac{1}{p}\right)\sum_{\substack{q\leq x/p \\ (q,pa)=1}}\frac{f(q)}{q}\left(1-\frac{1}{q_0}\right)+O(S_1),$$

where

$$S_1 = p^{-\sigma}\sum_{q\leq x/p}q^{-\sigma}|f(q)|.$$

Corresponding to the q which are powers of the prime p we have a contribution

$$\frac{x^{1-\sigma}}{(1-\sigma)(2-\sigma)a}\frac{f(p)}{p}\left(1-\frac{1}{p}\right)+O(S_2),$$

where

$$S_2 = p^{-\sigma}|f(p)|.$$

Suppose now that q is a power of a prime q_0 which divides a, so that q_0 cannot be p. Let

$$q_0^{h_1}\|a, \quad q_0^{h_2}\|b, \qquad h_1 \geq 1.$$

Then $q = q_0^k\|(am+b)$ is possible for some integer m only under the following circumstances:

(8)
    (i) $h_2 < h_1$ and $k = h_2$

    (ii) $h_2 \geq h_1$ and $k \geq h_1$.

In the first of these two cases $am+b$ is exactly divisible by q for every m. In the second it is so if and only if m belong to a certain set of (q_0-1) residue classes (mod q^{k-h_1+1}).

Corresponding to a prime-power q of the type (8(ii)) we will have toward the sum at (7) a contribution which is

$$\frac{f(q)}{q^\sigma}\sum_{\substack{t\leq xq^{-1} \\ q_1 t \equiv b_1 \pmod{a_1} \\ t \equiv 0 \pmod p \\ (t,q)=1}}\frac{1}{t^\sigma}\left(1-\frac{t}{xq^{-1}}\right),$$

where $q_1 = q(q, a)^{-1}$, $a_1 = a(q, a)^{-1}$, $b_1 = b(q, a)^{-1}$. By lemma (8.4) this is

$$\frac{x^{1-\sigma}}{(1-\sigma)(2-\sigma)}\frac{1}{p}\left(1-\frac{1}{p}\right)\frac{f(q)}{qa_1}\left(1-\frac{1}{q_0}\right)+O\left(\frac{|f(q)|}{(qp)^\sigma}\right).$$

Note that in these circumstances $(q, a) = (qq_0, a) = q_0^{h_1}$.

When q is a prime-power of the type (8(i)) we gain toward the sum at (7) a contribution

$$\frac{x^{1-\sigma}}{(1-\sigma)(2-\sigma)}\frac{1}{p}\left(1-\frac{1}{p}\right)\frac{f(q)}{qa_1}+O\left(\frac{|f(q)|}{(qp)^\sigma}\right),$$

since in this case a congruence such as $qq_0 pr \equiv b \pmod{a}$ has no solution.

Define

$$\eta(q) = \begin{cases} (a, q) & \text{if } q \text{ is of type (8(ii))}, \\ (a, q)(1-1/q_0)^{-1} & \text{if } q \text{ is of type (8(i))}. \end{cases}$$

Then the total contribution of the prime-powers q which satisfy (8) is

$$\frac{x^{1-\sigma}}{(1-\sigma)(2-\sigma)a}\frac{1}{p}\left(1-\frac{1}{p}\right)\sum_{\substack{q \leq x/p \\ (q,a)>1}}\frac{f(q)}{q}\eta(q)\left(1-\frac{1}{q_0}\right)+O(S_1).$$

Putting these results together

$$\sum_{\substack{am+b \leq x \\ am+b \equiv 0 \pmod{p}}}\frac{f(am+b)}{(am+b)^\sigma}\left(1-\frac{am+b}{x}\right)$$

$$=\frac{x^{1-\sigma}}{(1-\sigma)(2-\sigma)a}\frac{1}{p}\left(1-\frac{1}{p}\right)\left\{f(p)+\sum_{\substack{q \leq x/p \\ (q,pa)=1}}\frac{f(q)}{q}\left(1-\frac{1}{q_0}\right)\right.$$

$$\left.+\sum_{\substack{q \leq x/p \\ (q,a)>1}}\frac{f(q)}{q}\eta(q)\left(1-\frac{1}{q_0}\right)\right\}+O\left(\sum_{j=1}^{2}S_j\right).$$

A similar but simpler argument gives

$$\frac{1}{p}\sum_{\substack{am+b \leq x \\ (am+b,p)=1}}\frac{f(am+b)}{(am+b)^\sigma}\left(1-\frac{am+b}{x}\right)$$

$$=\frac{x^{1-\sigma}}{(1-\sigma)(2-\sigma)a}\frac{1}{p}\left(1-\frac{1}{p}\right)\left\{\sum_{\substack{q \leq x \\ (q,pa)=1}}\frac{f(q)}{q}\left(1-\frac{1}{q_0}\right)\right.$$

$$+ \sum_{\substack{q \le x \\ (q,a)>1}} \frac{f(q)}{q} \eta(q) \left(1 - \frac{1}{q_0}\right)\right\} + O(S_3),$$

where

$$S_3 = \frac{1}{p} \sum_{q \le x} \frac{|f(q)|}{q^\sigma}.$$

The expression inside the square at (6) therefore may be estimated by

$$\frac{x^{1-\sigma}}{(1-\sigma)(2-\sigma)a} \frac{1}{p}\left(1 - \frac{1}{p}\right)\left\{f(p) - E(x) + E\left(\frac{x}{p}\right)\right\} + O\left(\sum_{j=1}^{4} S_j\right),$$

with

$$S_4 = \frac{x^{1-\sigma}}{p} \sum_{\substack{x/p < q \le x \\ (q,a\Delta)>1}} \frac{|f(q)|}{q}.$$

In order to deduce the inequality of theorem (8.1) from this estimate and the inequality (6) it will suffice to prove that

$$\sum_{p \le x^\delta} p S_j^2 \ll R$$

holds for each value of j, $1 \le j \le 4$.

For example, applying the Cauchy–Schwarz inequality

$$S_1^2 = \left(p^{-\sigma} \sum_{q \le x/p} q^{-\sigma/2} |f(q)| q^{-\sigma/2}\right)^2$$

$$\le p^{-2\sigma} \sum_{q \le x/p} q^{-\sigma} |f(q)|^2 \sum_{q \le x/p} q^{-\sigma}$$

$$\ll p^{-2\sigma} \sum_{q \le x/p} q^{-\sigma} |f(q)|^2 \frac{(x/p)^{1-\sigma}}{\log x},$$

so that

$$\sum_{p \le x^\delta} p S_1^2 \ll \frac{x^{1-\sigma}}{\log x} \sum_{q \le x} \frac{|f(q)|^2}{q^\sigma} \sum_{p \le x/q} \frac{1}{p^\sigma}$$

$$\ll \frac{x^{1-\sigma}}{\log x} \sum_{q \le x} \frac{|f(q)|^2}{q^\sigma} \left(\frac{x}{q}\right)^{1-\sigma} \ll R.$$

As another example

$$S_4^2 \ll \frac{x^{2(1-\sigma)}}{p^2} \sum_{q \le x} \frac{|f(q)|^2}{q} \sum_{\substack{x^{1-\delta} \le q \le x \\ (q,a\Delta)>1}} \frac{1}{q}.$$

Since

$$\sum_{\substack{x^{1-\delta} \le q \le x \\ (q,a\Delta)>1}} \frac{1}{q} = \sum_{q_0 | a\Delta} \sum_{q_0^k > x^{1-\delta}} \frac{1}{q_0^k} \ll \sum_{q_0 | a\Delta} \frac{1}{x^{1-\delta}},$$

in fact

$$\sum_{p \le x^\delta} pS_4^2 \ll x^{-1/2} R.$$

The remaining S_j^2 are more simply treated, and theorem (8.1) is proved.

To obtain the larger range $p \le x^\delta$, $(p, a A \Delta) = 1$, with the attendant change of the factor $\log \log x / \log x$ to $(\log x)^{-1/4}$, we apply theorem (7.4) in place of theorem (7.3).

Proof of theorem (8.2). If $f(q) = 0$ whenever $(q, aA\Delta) > 1$ then the desired inequality follows from that of theorem (8.1) taking the third remark into account.

More generally we follow the proof of theorem (8.1) except that we separate off the contribution towards the z_m coming from those $f(q)$ with $(q, aA\Delta) > 1$. The $f(q)$ with $(q, aA\Delta) = 1$ are treated as before.

For each prime p let

$$Y_p = \sum_{\substack{0 < am+b \le x \\ am+b \equiv 0 \pmod p}} y_m m^{-\sigma} - \frac{1}{p} \sum_{\substack{0 < am+b \le x \\ (am+b,p)=1}} y_m m^{-\sigma},$$

where

$$y_m = \{f(am+b) - f(Am+B)\} \left(\frac{m}{Am+B}\right)^\sigma \left(1 - \frac{a(Am+B)}{Ax+\Delta}\right),$$

save that *we now assume that* $f(q) = 0$ *whenever* $(q, aA\Delta) = 1$.

It will be convenient to use g to denote a prime number, so that

$$g_{ij} = \max(g^i, g^j).$$

Let

$$T = \sum_{g_{ij} \leq Ax+|\Delta|} \frac{|L(g^i, g^j)|^2}{g_{ij}} + \sum_{t=1}^{2} |L_t|^2,$$

where the first sum is taken over all pairs g^i, g^j for which an integer m can be found so that $g^i \| (am+b)$ and $g^j \| (Am+B)$.

To complete the proof of theorem (8.2) it will be enough to show that

$$\sum_{\substack{p \leq x^\delta \\ (p, aA\Delta)=1}} p|Y_p|^2 \ll x^{1-\sigma+\delta} T.$$

For each t ($= 1, 2$) let

$$Z_t(g^i, g^j) = \sum_{\substack{0 < am+b \leq x \\ am+b \equiv 0 \pmod{p}}} \frac{1}{(Am+B)^\sigma} \left(1 - \frac{a(Am+B)}{Ax+\Delta}\right)$$

$$- \frac{1}{p} \sum_{\substack{0 < am+b \leq x \\ (am+b, p)=1}} \frac{1}{(Am+B)^\sigma} \left(1 - \frac{a(Am+B)}{Ax+\Delta}\right)$$

where, besides the summation conditions indicated, m must run through those integers for which $am+b \equiv 0 \pmod{g^i}$, $Am+B \equiv 0 \pmod{g^j}$; and $am+b \equiv 0 \pmod{\alpha}$, $Am+B \equiv 0 \pmod{\beta}$ where (α, β) is the fixed-divisor pair which features in the definition of L_t. (For notation see Chapter 3.)

Here the summation condition $am+b \leq x$ is the same as $Am+B \leq (Ax+\Delta)/a$.

Consider for the moment the first of the two sums which appear in this definition. Without $am+b \equiv 0 \pmod{p}$ the congruence conditions require that m belong to a certain bounded number of residue classes (mod $Dg_{ij}g$), where D is a bounded integer, depending only upon the prime divisors of $aA\Delta$, which serves to guarantee the conditions $am+b \equiv 0 \pmod{\alpha}$ and $Am+B \equiv 0 \pmod{\beta}$. Since p does not divide $aA\Delta$ we may employ the Chinese Remainder Theorem to split this sum into the difference of two others whose summation conditions require that m belong to certain residue classes (mod $pDg_{ij}g$) and (mod $p^2Dg_{ij}g$) respectively.

In this way we can relate $Z_t(g^i, g^j)$ to a bounded number of sums each of which may be estimated by lemma (8.4). Their leading terms cancel and we obtain the estimate (uniform in i, j, p, t)

$$Z_t(g^i, g^j) \ll 1.$$

Whilst this is satisfactory for small values of i and j we need something better when g_{ij} is large.

Suppose that $i \leq j$. Then arguing crudely

$$Z_t(g^i, g^j) \ll \sum_{\substack{am+b \leq x \\ am+b \equiv 0 \pmod{p}}} \frac{1}{(Am+B)^\sigma} + \frac{1}{p} \sum_{am+b \leq x} \frac{1}{(Am+B)^\sigma},$$

where of the supplementary conditions we retain only that $Am+B$ be divisible by g^j. The second of these two sums is

$$\ll \frac{1}{p} \cdot \frac{1}{g^{j\sigma}} \sum_{r \leq (Ax+\Delta)/ag^j} \frac{1}{r^\sigma} \ll \frac{x^{1-\sigma}}{pg^j}.$$

If $Am + B = g^j r$ and $am+b \equiv 0 \pmod{p}$ then $g^j r \equiv \Delta \pmod{p}$, so that the other sum involving $Am + B$ is

$$\ll \sum_{\substack{g^j r \leq (Ax+\Delta)/a \\ g^j r \equiv \Delta \pmod{p}}} \frac{1}{(g^j r)^\sigma} \ll \frac{1}{g^{j\sigma}} + \frac{1}{g^{j\sigma}} \sum_{1 \leq u \leq (Ax+\Delta)/apg^j} \frac{1}{(up)^\sigma}$$

$$\ll \frac{1}{g^{j\sigma}} + \frac{1}{(g^j p)^\sigma} \cdot \left(\frac{x}{pg^j}\right)^{1-\sigma}.$$

Hence

$$Z_t(g^i, g^j) \ll \frac{1}{g_{ij}^\sigma} + \frac{x^{1-\sigma}}{pg_{ij}}.$$

It is readily checked that a similar bound holds when $i > j$.

Let Z_t be the sum obtained from $Z_t(g^i, g^j)$ by deleting the requirements involving g^i and g^j. Essentially we replace g by 1. Then Z_t is also bounded (uniformly in p, t).

For each integer n there is a representation

$$f(an+b) - f(An+B) = \sum_{g|a A \Delta} L(g^i, g^j) + L_t,$$

where $t = 1$ or 2, g (as before) denotes a prime, $g^i \| (an+b)$ and $g^j \| (An+B)$. By means of this representation and a change in the order of summation we have

$$Y_p = \sum_{t=1}^{2} \sum_{g|aA\Delta} \sum_{i,j} L(g^i, g^j) Z_t(g^i, g^j) + \sum_{t=1}^{2} L_t Z_t,$$

where the pairs i, j are restricted by $g_{ij} \leq Ax + |\Delta|$.

For any $V \geq 1$ the terms in the triple sum with prime-power $g_{ij} \leq V$ give towards Y_p an amount whose square is

$$\ll \left(\sum_{g|a A \Delta} \sum_{g_{ij} \leq V} |L(g^i, g^j)| \right)^2$$

$$\ll T \sum_{g|a A \Delta} \sum_{g_{ij} \leq V} g_{ij} \ll T \sum_{g|a A \Delta} V\left(1 + \frac{1}{g} + \frac{1}{g^2} + \cdots\right)$$

$$\ll TV,$$

the second step by means of the Cauchy–Schwarz inequality.

Those terms in the triple sum with $g_{ij} > V$ we estimate using the second of our bounds for $Z_t(g^i, g^j)$. Altogether they give an amount whose square is

$$\ll \left(\sum_{g|a A \Delta} \sum_{g_{ij} > V} |L(g^i, g^j)| \left\{ \frac{1}{g_{ij}^{\sigma}} + \frac{x^{1-\sigma}}{p g_{ij}} \right\} \right)^2$$

$$\ll T \sum_{g|a A \Delta} \sum_{g_{ij} > V} \left\{ \frac{1}{g_{ij}^{2\sigma-1}} + \frac{x^{2(1-\sigma)}}{p^2 g_{ij}} \right\}$$

$$\ll T \sum_{g|a A \Delta} \left(\frac{1}{V^{2\sigma-1}} + \frac{x^{2(1-\sigma)}}{p^2 V} \right) \left(1 + \frac{1}{g} + \frac{1}{g^2} + \cdots \right)$$

$$\ll T \left(\frac{1}{V^{2\sigma-1}} + \frac{x^{2(1-\sigma)}}{p^2 V} \right),$$

and we have appealed to the Cauchy–Schwarz inequality again.

Putting these results together we obtain readily the bound

$$\sum_{\substack{p \leq x^\delta \\ (p, a A \Delta) = 1}} p |Y_p|^2 \ll T \sum_{p \leq x^\delta} p \left\{ V + \frac{1}{V^{2\sigma-1}} + \frac{x^{2(1-\sigma)}}{p^2 V} \right\}$$

$$\ll T \left\{ \frac{x^\delta V}{\log x} + \frac{x^{2(1-\sigma)}}{V} \log \log x \right\}.$$

Choosing $V = x^{1-\sigma-\delta} \sqrt{\log x}$ completes the proof of theorem (8.2).

Proof of theorem (8.3). We adapt the proof of theorem (8.2) by taking advantage of the fact that

$$\log \frac{am + b}{Am + B}$$

is nearly constant for all large integers m.

Set

$$z'_m = \{\log(am+b) - \log(Am+B)\}\left(\frac{m}{Am+B}\right)^\sigma \left(1 - \frac{a(Am+B)}{Ax+\Delta}\right)$$

if $Am+B$ is positive, and $z'_m = 0$ otherwise. This definition of z'_m is modelled on that of the z_m which appears in the proof of theorems (8.1) and (8.2), the rôle of $f(n)$ now being played by $\log n$.

Applying the elementary inequality $|r-s|^2 \leq 2|r|^2 + 2|s|^2$ we see that

$$\sum_{p \leq x^\delta} p \left| \sum_{\substack{0 < am+b \leq x \\ am+b \equiv 0 \pmod p}} (z_m - Uz'_m) m^{-\sigma} - \frac{1}{p} \sum_{\substack{0 < am+b \leq x \\ (am+b, p) = 1}} (z_m - Uz'_m) m^{-\sigma} \right|^2$$

$$\leq 2 \sum_{p \leq x^\delta} p \left| \sum_{\substack{0 < am+b \leq x \\ am+b \equiv 0 \pmod p}} z_m m^{-\sigma} - \frac{1}{p} \sum_{\substack{0 < am+b \leq x \\ (am+b, p) = 1}} z_m m^{-\sigma} \right|^2$$

$$+ 2|U|^2 \sum_{p \leq x^\delta} p \left| \sum_{\substack{0 < am+b \leq x \\ am+b \equiv 0 \pmod p}} z'_m m^{-\sigma} - \frac{1}{p} \sum_{\substack{0 < am+b \leq x \\ (am+b, p) = 1}} z'_m m^{-\sigma} \right|^2.$$

As before, the first of these two majorants is

$$\ll x^{1-\sigma} \sum_{0 < am+b \leq x} |z_m|^2 m^{-\sigma}$$

$$\ll x^{1-\sigma} \sum_{an+b \leq x} \frac{|f(an+b) - f(An+B)|^2}{n^\sigma}.$$

We shall prove that the second of the two majorants is $O(x^{2\delta}|U|^2)$. The proof of theorem (8.3) may then be completed by replacing $f(n)$ by $f(n) - U \log n$ in (4) and what follows it, the rôle of R to be played by

$$\frac{x^{2(1-\sigma)}}{(\log x)^{1/4}} \sum_{q \leq Ax + |\Delta|} \frac{|f(q) - U \log q|^2}{q} + x^{2\delta}|U|^2$$

$$+ x^{1-\sigma} \sum_{an+b \leq x} \frac{|f(an+b) - f(An+B)|^2}{n^\sigma}.$$

It is also necessary to restrict the range of summation in the analogue of (4) by $p \leq x^\delta$. Since this is anyway done later in the proofs of theorem (8.1) and (8.2), it introduces no further loss of precision.

A straightforward argument shows that over the range in which we are interested

$$z'_m = \frac{1}{A^\sigma} \log \frac{a}{A} \cdot \left(1 - \frac{am+b}{x}\right) + O\left(\frac{1}{m}\right).$$

Writing λ for $A^{-\sigma} \log a/A$ we have

$$\sum_{\substack{0<am+b\leq x \\ am+b\equiv 0(\mathrm{mod}\, p)}} z'_m m^{-\sigma} = \lambda \sum_{\substack{0<am+b\leq x \\ am+b\equiv 0(\mathrm{mod}\, p)}} \frac{1}{m^\sigma}\left(1 - \frac{am+b}{x}\right)$$

$$+ O\left(\sum_{\substack{am+b\leq x \\ am+b\equiv 0(\mathrm{mod}\, p)}} \frac{1}{m^{\sigma+1}}\right)$$

$$= \lambda a^\sigma \sum_{\substack{0<am+b\leq x \\ am+b\equiv 0(\mathrm{mod}\, p)}} \frac{1}{(am+b)^\sigma}\left(1 - \frac{am+b}{x}\right)$$

$$+ O\left(\sum_{\substack{am+b>0 \\ am+b\equiv 0(\mathrm{mod}\, p)}} \frac{1}{(am+b)^{\sigma+1}}\right),$$

which by lemma (8.4) is estimated as

$$\frac{\lambda a^{\sigma-1}}{(1-\sigma)(2-\sigma)} \cdot \frac{x^{1-\sigma}}{p}\left(1 - \frac{1}{p}\right) + O(1).$$

Similarly (essentially p is replaced by 1)

$$\sum_{\substack{0<am+b\leq x \\ (am+b,p)=1}} z'_m m^{-\sigma} = \frac{\lambda a^{\sigma-1}}{(1-\sigma)(2-\sigma)} \cdot x^{1-\sigma}\left(1 - \frac{1}{p}\right) + O(1).$$

By subtraction

$$\sum_{p\leq x^\delta} p \left| \sum_{\substack{0<am+b\leq x \\ am+b\equiv 0(\mathrm{mod}\, p)}} z'_m m^{-\sigma} - \frac{1}{p} \sum_{\substack{0<am+b\leq x \\ (am+b,p)=1}} z'_m m^{-\sigma} \right|^2 \ll \sum_{p\leq x^\delta} p \ll x^{2\delta},$$

and theorem (8.3) follows directly.

Remarks. The rôles of (a, b) and (A, B) in these proofs may be interchanged.

In each of theorems (8.1) and (8.2) it is possible to replace $E(x)$ by the function
$$\alpha(x) = \sum_{\substack{l \le x \\ (l,a\Delta)=1}} \frac{f(l)}{l},$$
where l runs through prime numbers. For
$$E(x) - E\left(\frac{x}{p}\right) = \alpha(x) - \alpha\left(\frac{x}{p}\right) + \sum_{x/p < l \le x} \frac{f(l)}{l^2} + \sum_{\substack{x/p < l^k \le x \\ k \ge 2, (l,a\Delta)=1}} \frac{f(l^k)}{l^k}\left(1 - \frac{1}{l}\right),$$
and since $p \le x^\delta < x^{1/2}$ the last two sums are, by the Cauchy–Schwarz inequality,
$$\ll x^{-1/8} D(x),$$
so that the replacement introduces a negligible error.

Likewise the function $E^*(x)$ in theorem (8.3) may be replaced by
$$\sum_{\substack{l \le x \\ (l,a\Delta)=1}} \frac{f(l) - U \log l}{l},$$
l denoting a prime. Moreover, from the estimate
$$\sum_{l \le y} \frac{\log l}{l} = \log y + \text{constant} + O((\log y)^{-K-1}),$$
which is valid for each fixed K and all $y \ge 2$, and may be deduced from a sufficiently sharp version of the Prime Number Theorem, as in Davenport [1], Chapter 18, or Prachar [1], Satz (5.3), p. 77, we have
$$\log p - \sum_{x/p < l \le x} \frac{\log l}{l} = O((\log x)^{-K-1})$$
uniformly for $2 \le p \le x^\delta \ (\le x^{1/2})$.

From theorem (8.3) we therefore deduce the inequality
$$\sum_{\substack{p \le x^\delta \\ (p,a\Delta)=1}} \frac{1}{p} \left| f(p) - \alpha(x) + \alpha\left(\frac{x}{p}\right) \right|^2$$
(9) $\ll \dfrac{1}{x^{1-\sigma}} \sum_{n \le (x-b)/a} \dfrac{|f(an+b) - f(An+B)|^2}{n^\sigma} + \dfrac{1}{(\log x)^{1/4}} \sum_{\substack{q \le Ax+|\Delta| \\ (q,a\Delta)=1}} \dfrac{|f^*(q)|^2}{q}$

$+ \dfrac{1}{x^{1-\sigma-\delta}} \left(\sum_{p \mid a\Delta} \sum_{p_{ij} \le Ax+|\Delta|} \dfrac{|L^*(p^i, p^j)|^2}{p_{ij}} + \sum_{t=1}^{2} |L_t^*|^2 \right) + \dfrac{|U|^2}{(\log x)^K},$

valid uniformly for all additive functions $f(n)$, complex U, and real $x \geq 4(aA\Delta)^2$. It is in this form that we apply theorem (8.3).

EXAMPLE. Let $f(n)$ be a *completely* additive function which satisfies

(10) $$f(an+b) - f(An+B) \ll 1$$

for all large n, with $\Delta \neq 0$. Suppose, further, that

(11) $$\sum_{q \leq x} \frac{|f(q)|^2}{q} \ll (\log x)^{2h}$$

for some fixed $h > 1$ and all $x \geq 2$.

Applying theorem (8.1) gives, for each fixed ε not exceeding δ,

(12) $$\sum_{x^\varepsilon < p \leq x^\delta} \frac{1}{p} \left| f(p) - E(x) + E\left(\frac{x}{p}\right) \right|^2 \ll (\log x)^{2h-1/2}.$$

With ε replaced by $\varepsilon/2$ and x by x^2 we also obtain

$$\sum_{x^\varepsilon < p \leq x^{2\delta}} \frac{1}{p} \left| f(p) - E(x^2) + E\left(\frac{x^2}{p}\right) \right|^2 \ll (\log x)^{2h-1/2}.$$

In particular, appropriate subtraction leads to

$$\sum_{x^\varepsilon < p \leq x^\delta} \frac{1}{p} \left| E(x^2) - E(x) - \left\{ E\left(\frac{x^2}{p}\right) - E\left(\frac{x}{p}\right) \right\} \right|^2 \ll (\log x)^{2h-1/2}.$$

Since

$$\sum_{x^\varepsilon < p \leq x^\delta} \frac{1}{p} = \log \frac{\delta}{\varepsilon} + O\left(\frac{1}{\log x}\right),$$

for large enough x we can find a prime p_1, exceeding x^ε, for which

$$|E(x^2) - E(x)| \leq |E(x^2 p_1^{-1}) - E(x p_1^{-1})| + c_0 (\log x)^{h-1/4}.$$

We now repeat the argument, the rôles of x^2, x being played by x^2/p_1 and x/p_1. After k steps we reach

$$|E(x^2) - E(x)| \leq E\left(\frac{x^2}{p_1 \cdots p_k}\right) - E\left(\frac{x}{p_1 \cdots p_k}\right) + kc_0(\log x)^{h-1/4}.$$

Since at the stage $j+1$ we obtain

$$p_{j+1} > \left(\frac{x}{p_1 \cdots p_j}\right)^\varepsilon,$$

we have (inductively)

$$\frac{x}{p_1 \cdots p_k} \leq \left(\frac{x}{p_1 \cdots p_{k-1}}\right)^{1-\varepsilon} \leq \cdots \leq x^{(1-\varepsilon)^k}.$$

In particular, after $O(\log \log x)$ steps $x(p_1 \cdots p_k)^{-1}$ becomes bounded independently of x, and we obtain (suppressing details)

$$\max_{w \leq x^2} |E(w)| \leq 2 \max_{w \leq x} |E(w)| + O((\log x)^{h-1/4} \log \log x).$$

In turn, this inequality may be applied inductively, replacing x by $x^{2^{-j}}$, $j = 1, 2, \ldots$. This time $O(\log \log x)$ steps lead to a bound

$$E(x) \ll \log x + (\log x)^{h-1/4} \log \log x \sum_{2^j \leq \log x} (2^j)^{5/4-h}.$$

Here the sum over j is

$$\ll \begin{cases} (\log x)^{5/4-h} \log \log x & \text{if } h \leq \tfrac{5}{4}, \\ 1 & \text{if } h > \tfrac{5}{4}. \end{cases}$$

Returning to the inequality (12), now with $\varepsilon = \delta/2$, we see that we have secured an estimate

$$\sum_{x^{\delta/2} < p \leq x^\delta} \frac{1}{p} \cdot |f(p)|^2 \ll \{(\log x)^2 + (\log x)^{2h-1/2}\}(\log \log x)^2$$

from which we readily deduce the inequality

(13) $$\sum_{q \leq x} \frac{|f(q)|^2}{q} \ll \{(\log x)^2 + (\log x)^{2h-1/2}\}(\log \log x)^2.$$

By means of the hypothesis (10), theorem (8.1) has enabled us to improve our initial bound (11) to that of (13), essentially replacing $2h$ by $\max(2, 2h - \tfrac{1}{2})$. If $2h - \tfrac{1}{2} > 2$ then we may go around the loop again. After finitely many operations we reach

$$\sum_{q \leq x} \frac{|f(q)|^2}{q} \ll (\log x \log \log x)^2.$$

This is nearly best possible, since the function $f(n) = \log n$ fits the hypothesis (10).

However, one would like to better characterize those additive functions for which (10) holds, and for that a more careful study of the functional inequality (12) seems appropriate. This will be the subject of our next section.

In our later applications of the loop mechanism, the rôle of the *a priori* bound (11) will, of course, be played by an inequality from Chapter 3.

Third Motive

The inequality

(1) $$\sum_{x^\varepsilon < p \leq x^\delta} \frac{1}{p} \left| f(p) - \alpha(x) + \alpha\left(\frac{x}{p}\right) \right|^2 \leq h(x)^2,$$

where

(2) $$\alpha(x) = \sum_{p \leq x} \frac{f(p)}{p}$$

may be regarded as an approximate functional equation for the arithmetic function $f(p)$ which is defined on the prime numbers. We think of $h(x)$ as being in some sense "small".

Two features of this approximate equation present themselves: it is non-linear; and the example

$$f(p) = \log p \cdot (\log \log p)^{1/2}, \qquad h(x) = \log x \cdot (\log \log x)^{-1/2}$$

shows that even if $h(x) = o(\log x)$ as $x \to \infty$, $f(p)$ need not be a very good approximation to a logarithm.

To gain experience with approximate functional equations consider the following form of Cauchy's equation.

Let $\tau \geq 1$ be fixed. Let $w(t)$, possibly complex-valued, belong to the Lebesgue class $L^\tau(0, z)$ for each real $z > 0$, and satisfy

(3) $$\lim_{z \to \infty} z^{-2} \int_0^z \int_0^z |w(x+y) - w(x) - w(y)|^\tau \, dx \, dy = 0.$$

We hope to prove that in some sense $w(t)$ is nearly a linear function of t.

It is a particular consequence of (3), with $\tau = 1$, that

(4) $$\iint_{\substack{x+y \leq z \\ x, y \geq 0}} \{w(x+y) - w(x) - w(y)\} \, dx \, dy = o(z^2), \qquad z \to \infty.$$

The choice of the triangular region of integration is important. It was suggested by considerations from the probabilistic theory of numbers. (See Elliott [5], and [11], Volume II, Chapter 14, lemma (14.8).) It allows us to treat that part of the integral (4) which involves $w(x+y)$ by the change of variables $x = r(\text{Cos } \theta)^2$, $y = r(\text{Sin } \theta)^2$, with Jacobian determinant $r \text{ Sin } 2\theta$, to give

$$\iint\limits_{\substack{x+y \leq x \\ x,y \geq 0}} w(x+y) \, dx \, dy = \int_0^z rw(r) \, dr.$$

Then (4) becomes

$$\int_0^z rw(r) \, dr - 2 \int_0^z (z-x)w(x) \, dx = o(z^2).$$

Defining

$$J(z) = \int_0^z w(t) \, dt$$

integration by parts yields

(5) $\qquad zJ(z) - 3 \int_0^z J(u) \, du = o(z^2), \qquad z \to \infty.$

Let us further define

$$H(z) = \int_0^z J(u) \, du,$$

and write (5) in the form

$$z \frac{dH}{dz} - 3H = o(z^2).$$

This is an approximate form of a linear differential equation and the integrating factor z^{-4} is readily found, to give

$$\frac{d}{dz}(z^{-3}H) = o(z^{-2}), \qquad z \to \infty.$$

Assuming for the moment that $z \geq 1$, we integrate:

$$z^{-3}H(z) - H(1) = \int_1^z o(t^{-2}) \, dt = B - \int_z^\infty o(t^{-2}) \, dt$$

for some constant B. From this we readily obtain the asymptotic estimate

$$H(z) = Cz^3 + o(z^2), \qquad C = B + H(1).$$

In turn, this estimate may be substituted into the equation (5) to give

$$J(z) = 3Cz^2 + o(z), \qquad z \to \infty.$$

Returning now to our original hypothesis (3):

$$\int_0^z |zw(x) + J(z) - \{J(x+z) - J(x)\}|^\tau \, dx$$

$$= \int_0^z \left| \int_0^z (w(x) + w(y) - w(x+y)) \, dy \right|^\tau dx$$

$$\leq z^{\tau-1} \int_0^z \int_0^z |w(x) + w(y) - w(x+y)|^\tau \, dy \, dx = o(z^{\tau+1})$$

as $z \to \infty$, the third line coming from an application of Hölder's inequality. Since

$$J(z) - J(x+z) + J(x) = -6Cxz + o(z)$$

uniformly for $0 \leq x \leq z$, with $A = 6C$ we obtain

(6) $$\lim_{z \to \infty} z^{-1} \int_0^z |w(t) - At|^\tau \, dt = 0.$$

In this sense $w(t)$ is almost the linear function At. Moreover, the condition (6) implies our original hypothesis (3), so that our result is best possible. For a more detailed account of this result see Elliott [17]. For an analogous result concerning the exponential function, which, however, seems more difficult to obtain, see Elliott [15].

Note that in this treatment of the Cauchy equation it proved convenient to treat the "average" $J(t)$ rather than the original function $w(t)$, and to work in an L^1 space rather than an L^τ space. The full result (6) was then recovered by using the hypothesis (3) to relate $w(t)$ to its "average."

Returning to our number-theoretical situation (1) let us extend the definition of $f(p)$ to the continuous function $f(x)$ which is obtained by making $f(x)$ linear between successive prime values of x. Let us also introduce the measure μ for which

$$\mu((0, x]) = \sum_{p \leq x} \frac{1}{p}.$$

Then we may regard our inequality (1) as expressing $f(p)$ in terms of its "average"

$$\alpha(x) = \int_2^x f(t)\, d\mu,$$

the inequality itself taking place in an L^2 space with regard to μ. Our experience with the Cauchy equation suggests that we concentrate on the "average" $\alpha(x)$, and work in the L^1 space with respect to μ. We would then hope for an approximate differential equation.

The second feature of the approximate functional equation (1) concerns the possible forms which the function $\alpha(x)$ may take. Once again, an experience in probabilistic number theory helps.

Let

(7) $$\nu_x(n; f(n) - \alpha(x) \leq z\beta(x))$$

denote the frequency

$$[x]^{-1} N(x, z),$$

where $N(x, z)$ is the number of positive integers n not exceeding x for which the inequality

$$f(n) - \alpha(x) \leq z\beta(x)$$

is satisfied. For the moment we assume that $f(n)$ is real-valued additive, $\beta(x) > 0$. The function $\alpha(x)$ need not be that given at (2). In the author's paper [5] (see also Elliott [11], Volume II, Chapters 13 and 14) necessary and sufficient conditions were sought in order that the frequencies (7) might converge, as $x \to \infty$, to the improper distribution function with jump at the origin, $z = 0$. Of interest here is that under the condition $\beta(x) \to \infty$ as $x \to \infty$ and

$$\limsup_{x \to \infty} \frac{\beta(x^y)}{\beta(x)} < \infty$$

for each fixed $y > 0$, it was established that $\alpha(x)$ must behave somewhat like a logarithm. In fact there will be a decomposition

$$\alpha(x) = \alpha_1(x) + \alpha_2(x),$$

where the individual functions $\alpha_j(x)$ satisfy

(8) $\quad \alpha_1(x^y) = y\alpha_1(x) + o(\beta(x)), \qquad \alpha_2(x^y) = \alpha_2(x) + o(\beta(x))$

for each fixed positive value of y.

Note that this permits the choice $\alpha_1(x) = \log x (\log \log x)^{1/2}$, $\beta(x) = \log x$.

The decomposition (8) is partly achieved by (essentially) considering the behavior of the function

$$\Lambda(a, b) = \alpha(ab) - \alpha(a) - \alpha(b)$$

for a, b in intervals of the type $[x^u, x^v]$. Assuming a decomposition (8) to hold, this function would then be

$$-\alpha_2(x) + o(\beta(x)).$$

We can expect this to be small in comparison with $\alpha_1(x)$, but not to be $o(\beta(x))$. The difficulty of identifying $-\alpha_2(x)$ is overcome by considering the four-dimensional function

(9) $$|\Lambda(a_1, b_1) - \Lambda(a_2, b_2)|$$

with $x^u \leq a_i, b_j \leq x^v$ and showing that in an almost sure sense (with respect to a suitable Haar measure) it is small. Thus there exists at least one pair of values (a^*, b^*) so that

$$|\Lambda(a, b) - \Lambda(a^*, b^*)|$$

is small for almost all pairs (a, b) in $[x^u, x^v]^2$. In particular $-\alpha_2(x)$ will be within $o(\beta(x))$ of $\Lambda(a^*, b^*)$.

We shall adopt this point of view in the next chapter. In the probabilistic number theory considerations it is shown that for each $w > 0$

$$\sum_{p \leq x} \frac{1}{p} \to 0, \quad x \to \infty,$$

the sum being over primes for which

$$\left| f(p) - \alpha(x) + \alpha\left(\frac{x}{p}\right) \right| > w\beta(x).$$

This gives information concerning primes p which may be comparable with x in size (see Elliott [11], Volume II, Chapter 13). In our present situation (1) we can cover only the range $p \leq x^\delta$. This loss is offset by two gains, we are implicitly working with convergence in an L^2 sense, rather than in measure, and our function $\alpha(x)$ is specified in terms of $f(p)$, giving us a little *a priori* control on its size. Moreover, the procedure can be carried out in terms of the standard Riemann or Lebesgue integration. Once again a decomposition of the type (8) ensues.

We shall follow the two general designs sketched above, but in the opposite order.

Chapter 9

The Approximate Functional Equation

In this chapter I solve the functional equation appropriate to the loop inequality of Chapter 8. I give the solution in two forms. The first has an asymptotic sense, so as to present the result and its proof in a more intelligible shape. The second is an inequality. It will appear as theorem (9.11). The treatment of this chapter may be read without a knowledge of the previous chapters of the present volume.

Theorem (9.1). *Let $f(p)$ be a complex-valued function defined on the prime numbers $p(\geq 2)$ which satisfies, for a fixed $\delta > 0$ and every ε, $0 < \varepsilon \leq \delta$,*

(1) $$\sum_{x^\varepsilon < p \leq x^\delta} \frac{1}{p} \left| f(p) - \alpha(x) + \alpha\left(\frac{x}{p}\right) \right| = o(\rho(x)), \qquad x \to \infty,$$

with

(2) $$\alpha(x) = \sum_{p \leq x} \frac{f(p)}{p},$$

and where the measurable function $\rho(x)$ satisfies $\rho(x) > 0$ for all positive x, and

(3) $$\limsup_{x \to \infty} \frac{\rho(x^y)}{\rho(x)} < \infty$$

for each fixed $y > 0$.
   Then there is a decomposition

(4) $$\alpha(x) = \alpha_1(x) + \alpha_2(x)$$

with measurable functions $\alpha_j(x)$, such that

(5) $\quad \alpha_1(x^y) = y\alpha_1(x) + o(\rho(x)), \qquad \alpha_2(x^y) = \alpha_2(x) + o(\rho(x))$

as $x \to \infty$, holds for each fixed $y > 0$.

9. The Approximate Functional Equation

Remark. It will be seen during the course of the proof of theorem (9.1) that we only make use of the fact that the p are primes to the extent of the estimate

$$(6) \qquad \sum_{p \leq w} \frac{\log p}{p} = \log w + c_0 + O((\log w)^{-B}),$$

which holds for each fixed $B > 0$. This estimate may be obtained, using an integration by parts, from any sufficiently sharp version of the prime number theorem, as, for example, that given in Prachar [1], Satz (5.3), p. 77, or in Davenport [1], Chapter 18.

Let

$$(7) \qquad \theta(x) = \sum_{x^\varepsilon < p \leq x^\delta} \frac{1}{p} \left| f(p) - \alpha(x) + \alpha\left(\frac{x}{p}\right) \right|.$$

Lemma (9.2). *Let $0 < \varepsilon < \delta$. Then there is a constant c, depending only upon ε and δ, so that*

$$\sum_{p \leq x} \frac{|f(p)|}{p} \leq (\max_{p \leq c} |f(p)| + \sup_{w \leq x^{1/\delta}} \theta(w))(\log x)^c$$

uniformly for $x \geq 2$.

Proof. We adopt a simplified form of the argument outlined in the example at the end of Chapter 8. Let $0 < \mu < 1$. From our definition

$$\sum_{x^{\mu\varepsilon} < p \leq x^{\mu\delta}} \frac{1}{p} \left| f(p) - \alpha(x^\mu) + \alpha\left(\frac{x^\mu}{p}\right) \right| = \theta(x^\mu),$$

so that

$$\sum_{x^\varepsilon < p \leq x^{\mu\delta}} \frac{1}{p} \left| \alpha(x) - \alpha(x^\mu) - \left\{ \alpha\left(\frac{x}{p}\right) - \alpha\left(\frac{x^\mu}{p}\right) \right\} \right| \leq \theta(x) + \theta(x^\mu).$$

Let

$$\psi = \psi(x) = \sup_{w \leq x} \theta(w).$$

Let μ be fixed at a value for which $\mu\delta > \varepsilon$. Then there exists a positive d such that for all large enough values of x, say $x \geq x_0$, we have

$$\sum_{x^\varepsilon < p \leq x^{\mu\delta}} \frac{1}{p} = \log \frac{\mu\delta}{\varepsilon} + O\left(\frac{1}{\log x}\right) > \frac{2}{d},$$

The Approximate Functional Equation

and there is a prime $p \geq x^\varepsilon$ so that

$$|\alpha(x) - \alpha(x^\mu)| \leq \left|\alpha\left(\frac{x}{p}\right) - \alpha\left(\frac{x^\mu}{p}\right)\right| + d\psi.$$

If

$$\eta(z) = \sup_{w \leq z}|\alpha(w)|, \qquad \tau = \min(1-\mu, \varepsilon),$$

then we deduce that

$$\eta(x) \leq 3\eta(x^{1-\tau}) + \eta(w_0) + d\psi.$$

Arguing inductively

$$\eta(x) \leq 3^k \eta(x^{(1-\tau)^k}) + \{\eta(w_0) + d\psi\} \sum_{j=0}^{k-1} 3^j$$

so long as

$$x^{(1-\tau)^{k-1}} \geq x_0.$$

After $O(\log \log x)$ steps we reach

$$\eta(x) \ll \{\eta(w_0) + \psi(x)\}(\log x)^{c_1},$$

where w_0, c_1 and the implied constant depend only upon ε and δ.

This bound together with (7) gives

$$\sum_{x^\varepsilon < p \leq x^\delta} \frac{|f(p)|}{p} \ll \{\eta(w_0) + \psi(x)\}(\log x)^{c_1}$$

and, replacing x by $x^{1/\delta}$,

$$\sum_{x^{\varepsilon/\delta} < p \leq x} \frac{|f(p)|}{p} \ll \{\eta(w_0) + \psi(x^{1/\delta})\}(\log x)^{c_1}.$$

Replacing x by

$$x^{(\varepsilon/\delta)^j}, \qquad j = 1, 2, \ldots,$$

and summing over j gives the desired inequality of lemma (9.2).

The relevant application of this result in the circumstances of theorem (9.1) is embodied in

Lemma (9.3). *There are positive constants c_2 and c_3 so that*

$$(\log x)^{-c_2} \leq \rho(x) \leq (\log x)^{c_2},$$

$$\sum_{p \leq x} \frac{|f(p)|}{p} \leq \rho(x)(\log x)^{c_3},$$

for all sufficiently large x.

Proof. We first prove that there is a constant c_4 so that

(8) $$\sup_{x^{1/2} \leq w \leq x} \rho(w) \leq c_4 \rho(x)$$

for all large x.

Assume that this proposition is false, and suppose that there is an unbounded increasing sequence $x_1 < x_2 < \cdots < x_k < \cdots$, with associated numbers w_k, $|w_k| \leq \log 2$, so that

(9) $$\frac{\rho(x_k^{e^{w_k}})}{\rho(x_k)} \to \infty, \quad k \to \infty.$$

A simple extension of Egoroff's theorem shows that there is a subset E of the interval $[-\log 2, \log 2]$, of measure at least $(7 \log 2)/4$, and constants c_5 and c_6 so that

$$c_5 \rho(x_k) \leq \rho(x_k^{e^y}) \leq c_6 \rho(x_k)$$

uniformly for all large k, for all y in E.

Similarly, there is a subset F of this same interval, of measure at least $(7 \log 2)/4$, and positive constants c_7 and c_8 so that

$$c_7 \rho(x_k^{e^{w_k}}) \leq \rho(x_k^{e^{w_k+z}}) \leq c_8 \rho(x_k^{e^{w_k}})$$

uniformly for all large k, for all z in F.

The sets $w_k + F$ and E must intersect. Otherwise (in terms of Lebesgue measure μ)

$$(7 \log 2)/2 \leq \mu(w_k + F) + \mu(E) \leq 2 \log 2 + |w_k|,$$

since their union $(w_k + F) \cup E$ lies either in the interval $[-\log 2, \log 2 + w_k]$ (when $w_k \geq 0$), or in the interval $[-\log 2 + w_k, \log 2]$ (when $w_k < 0$). As $|w_k| \leq \log 2$ this is impossible.

Suppose, therefore, that $w_k + z_k = y_k$, where z_k belongs to F and y_k to E. Then

$$\rho(x_k^{e^{w_k}}) \leq \frac{c_8}{c_7} \cdot \rho(x_k^{e^{w_k+z_k}}) = \frac{c_8}{c_7} \cdot \rho(x_k^{e^{y_k}}) \leq \frac{c_8 \cdot c_6}{c_7} \rho(x_k)$$

for all sufficiently large values of k. This contradicts (9).

Let (8) hold for $x \geq x_1 > 1$. From hypothesis (3) in the statement of theorem (9.1), with $y = 2$ and x replaced by $x^{1/2}$,

$$\rho(x) \leq c_9 \rho(x^{1/2})$$

holds (without loss of generality) over the same range of x-values. Arguing inductively

$$\rho(x) \leq c_9^k \rho(x^{2^{-k}})$$

certainly so long as

$$x^{2^{-(k-1)}} \geq x_1^2.$$

After $O(\log \log x)$ steps and an application of the inequality (8) we obtain

$$\rho(x) \ll (\log x)^{c_{10}} \sup_{x_1 < w \leq x_1^2} \rho(w) \ll (\log x)^{c_{10}},$$

which is the first upper bound in lemma (9.3).

The lower bound for $\rho(x)$ is similarly obtained, beginning with the observation that

$$\limsup_{x \to \infty} \frac{\rho(x)}{\rho(x^y)} = \limsup_{x^y \to \infty} \frac{\rho(\{x^y\}^{1/y})}{\rho(x^y)} < \infty,$$

so that

$$\liminf_{x \to \infty} \frac{\rho(x^y)}{\rho(x)} > 0.$$

Moreover, if x is large enough then our bound for $\rho(x)$ applied with lemma (9.2) yields

$$\sum_{p \leq x} \frac{|f(p)|}{p} \ll (\max_{p \leq c}|f(p)| + \sup_{w \leq x^{1/\delta}} \rho(w))(\log x)^c$$

$$\ll \rho(x)(\log x)^{c+2c_2},$$

and the proof of lemma (9.3) is complete.

Remarks. In the proof of this theorem $w_k + F$ denotes the set of all numbers of the form $w_k + z$ where z belongs to F.

The constant c_2 in lemma (9.3) does not depend upon the function $f(p)$, and the constant c_3 only depends upon the values $|f(p)|$ with p not exceeding the c which appears in lemma (9.2).

The hypothesis of theorem (9.1) ensures that

$$\sum_{x^\varepsilon < p \leq x^\delta} \frac{1}{p} \left| f(p) - \alpha(z) + \alpha\left(\frac{z}{p}\right) \right| = o(\rho(x))$$

holds uniformly for $x \leq z \leq x^k$, where k may be given any fixed value, $k \geq 1$. It is convenient to use this estimate in the form

(10) $$\sum_{x^\varepsilon < p \leq x^\delta} \frac{\log p}{p} \left| f(p) - \alpha(z) + \alpha\left(\frac{z}{p}\right) \right| = o(\rho(x)L),$$

where L denotes $\log x$. We shall adopt this notation for the remainder of the chapter.

Our next step is to obtain a form of the equation (10) with continuous variables.

Lemma (9.4).

$$\int_{x^\varepsilon}^{x^\delta} \left| \alpha(z) - \alpha\left(\frac{z}{u}\right) - \left\{ \alpha(w) - \alpha\left(\frac{w}{u}\right) \right\} \right| \frac{du}{u} = o(\rho(x)L)$$

holds uniformly for $x \leq w$, $z \leq x^k$.

Proof. Define

$$T(u) = \left| \alpha(z) - \alpha\left(\frac{z}{u}\right) - \left\{ \alpha(w) - \alpha\left(\frac{w}{u}\right) \right\} \right|.$$

From (10) with z, and with z replaced by w, subtraction gives

$$\sum{}' \frac{\log p}{p} T(p) = o(\rho(x)L).$$

Here $'$ denotes summation over primes in the interval $x^\varepsilon < p \leq x^\delta$. If

$$R(u) = \sum_{p \leq u} \frac{\log p}{p} - \log u - c_0,$$

The Approximate Functional Equation

then we can write the left-hand side of this approximate functional equation in the form

$$\int_{x^\varepsilon}^{x^\delta} T(u)\,d\left(\sum_{p\leq u}\frac{\log p}{p}\right) = \int_{x^\varepsilon}^{x^\delta} T(u)\cdot\frac{du}{u} + \int_{x^\varepsilon}^{x^\delta} T(u)\,dR(u).$$

$R(u)$ is a continuous function except at points $u = p$ (prime). If z, w do not have integer values then $T(u)$, which has possible discontinuities only when z/u or w/u is a prime, will be continuous at these same points.

Suppose for the moment that x^ε, x^δ are not primes. Then an integration by parts shows that the Stieltjes integral which involves $dR(u)$ is

$$[T(u)R(u)]_{x^\varepsilon}^{x^\delta} - \int_{x^\varepsilon}^{x^\delta} R(u)\,dT(u).$$

By lemma (9.3)

$$\alpha(t) \ll \rho(x)(\log x)^{c_2+c_3}$$

uniformly for all $t \leq x^k$ for all sufficiently large x. In view of the estimate (6) with $B > c_2 + c_3$, the integrated portion will be $o(\rho(x))$ as $x \to \infty$.

The function $T(u)$ is constant unless u passes through a number of the form z/p or w/p for some prime p. When it does so there will be a jump in $T(u)$ of at most $p^{-1}|f(p)|$. Here we make use of the inequality

$$||\beta_1+\eta|-|\beta_2+\eta|| \leq |\beta_1+\eta-(\beta_2+\eta)| = |\beta_1-\beta_2|,$$

which is valid for all complex numbers β_1, β_2 and η. Since $u \geq x^\varepsilon > 1$ the corresponding primes p cannot exceed x^k, and by lemma (9.3)

$$\int_{x^\varepsilon}^{x^\delta} R(u)\,dT(u) \ll (\log x)^{-B} \sum_{p\leq x^k} \frac{|f(p)|}{p} = o(\rho(x)),$$

once again applying (6), this time with $B > c_3$.

This gives an estimate of the type sought in the present lemma, save that we must now remove our restrictions upon the variables.

If z has an integral value and ν, $0 < \nu \leq 1$ is sufficiently small, $z(1+\nu)$ will not be an integer and the approximate equation in the statement of the present lemma will hold with $z(1+\nu)$ in place of z. For any value of u, such a change in z will change the function $T(u)$ by an amount not more than

$$\Delta(\nu) = \left|\alpha(z(1+\nu)) - \alpha(z) - \left\{\alpha\left(\frac{z(1+\nu)}{u}\right) - \alpha\left(\frac{z}{u}\right)\right\}\right|.$$

If z is a prime p_0 then for ν sufficiently small the first of these differences is $p_0^{-1}f(p_0)$. Since p_0 belongs to the interval $[x, x^k]$, from our fundamental hypothesis (1) with x^δ replaced by p_0,

$$p_0^{-1}|f(p_0)| \leq o(\rho(x)) + p_0^{-1}|\alpha(p_0^{1/\delta}) - \alpha(p_0^{(1/\delta)-1})|,$$

which bound by lemma (9.3) is

$$\leq o(\rho(x)) + p_0^{-1}\rho(x)(\log x)^{c_2+c_3} = o(\rho(x)).$$

Moreover,

$$\int_{x^\varepsilon}^{x^\delta} \left|\alpha\left(\frac{z(1+\nu)}{u}\right) - \alpha\left(\frac{z}{u}\right)\right| \frac{du}{u} \leq \int_{x^\varepsilon}^{x^\delta} \sum_{z/u < p \leq z(1+\nu)/u} \frac{|f(p)|}{p} \frac{du}{u}$$

$$\leq \sum_{p \leq 2z} \frac{|f(p)|}{p} \int_{z/p < u \leq z(1+\nu)/p} \frac{du}{u}$$

$$= \sum_{p \leq 2z} \frac{|f(p)|}{p} \cdot \log(1+\nu).$$

Hence

$$\limsup_{\nu \to 0+} \int_{x^\varepsilon}^{x^\delta} |\Delta(\nu)| \frac{du}{u} = o(\rho(x)L)$$

as $x \to \infty$.

We can argue in a similar manner to remove the restriction that w not be an integer.

Since over the whole range of u

$$T(u) \ll \rho(x)(\log x)^{c_2+c_3},$$

changing the end-points of integration x^ε, x^δ by a small enough amount that they become not integers, will introduce an error of at most

$$O(x^{-\varepsilon}\rho(x)(\log x)^{c_2+c_3}) = o(\rho(x)).$$

Lemma (9.4) is completely proved.

It is convenient to define the function

$$\Lambda(w, u) = \alpha(w) - \alpha\left(\frac{w}{u}\right) - \alpha(u)$$

for real positive w and u. This notation differs slightly from that in the third motive. We rewrite lemma (9.4) in the form

$$(11) \qquad \int_{x^\epsilon}^{x^\delta} |\Lambda(z,u) - \Lambda(w,u)| \frac{du}{u} = o(\rho(x)L), \qquad x \to \infty,$$

uniformly for $x \leq z$, $w \leq x^k$.

We shall need the following variant of this estimate.

Lemma (9.5). *Let $0 < \tau < 1$, $a > 0$, $0 < b \leq \min(1, \delta)$, $l \geq ab$ hold. Then*

$$\iint_{\substack{x^{a(1-\tau)} \leq v \leq x^a \\ v^{b(1-\tau)} \leq uv/w \leq v^b}} |\Lambda(w,u) - \Lambda(w,v)| \frac{dw \, dv}{wv} = o(\rho(x) L^2), \qquad x \to \infty,$$

holds uniformly for all u in the interval given by

$$(12) \qquad x^{ab} \leq u^\delta \leq x^l.$$

Proof. We write the integrand in the form

$$\left| \alpha(u) - \alpha\left(\frac{u}{uv/w}\right) - \alpha\left(\frac{uv}{w}\right) - \left\{\alpha(v) - \alpha\left(\frac{v}{uv/w}\right) - \alpha\left(\frac{uv}{w}\right)\right\} \right| \frac{1}{wv}$$

and make the change of variables

$$v = \omega, \qquad w = \frac{u\omega}{\mu}.$$

This has Jacobian determinant

$$\left| \frac{\partial(v,w)}{\partial(\mu,\omega)} \right| = \left| \begin{matrix} 0 & 1 \\ -u\omega/\mu^2 & u/\mu \end{matrix} \right| = \frac{u\omega}{\mu^2},$$

and the integral becomes

$$\iint |\Lambda(u,\mu) - \Lambda(\omega,\mu)| \frac{d\mu \, d\omega}{\mu \omega}$$

over the new region of integration

$$(13) \qquad \begin{aligned} x^{a(1-\tau)} &\leq \omega \leq x^a, \\ \omega^{b(1-\tau)} &\leq \mu \leq \omega^b. \end{aligned}$$

If $x^{ab} \leq u^\delta$ then for each fixed ω not exceeding x^a,

$$\mu \leq \omega^b \leq x^{ab} \leq u^\delta,$$

so that

$$\mu \leq (\min(\omega, u))^\delta,$$

and we estimate the integral over μ by lemma (9.4) in the form (11), replacing the z, w, x of that lemma by u, ω, $\min(\omega, u)$, respectively. It is then $o(\rho(x)L)$ uniformly for

(14) $$\min(\omega, u) \leq u, \quad \omega \leq (\min(\omega, u))^k$$

for any fixed $k \geq 1$. In particular, if k is given a large enough value and u is restricted by (12), then

$$u^{1/k} \leq x^{1/(\delta k)} \leq x^{a(1-\tau)},$$

$$u^k \geq x^{kab/\delta} \geq x^a,$$

so that the range of uniformity (14) contains the interval $x^{a(1-\tau)} \leq \omega \leq x^a$ of (13).

The integral over ω contributes an extra factor $O(L)$, and lemma (9.5) is proved.

Let $0 < \delta \leq 1$ and set $a = 1 - \delta + \delta^2$, $b = \delta^2/(2a)$ so that $b \geq \min(\frac{1}{2}, \delta)$, and $h = 2ab/\delta = \delta$. For $0 < \tau < 1$ define

$$D(x) = \int_{x^{h(1-\tau)} \leq u \leq x^h} \int_{x^{a(1-\tau)} \leq v \leq x^a} \int_{v^{b(1-\tau)} \leq uv/w \leq v^b} |\Lambda(z, u) - \Lambda(w, v)| \frac{du\, dv\, dw}{uvw}.$$

Lemma (9.6). *If τ is fixed at a suitable value*

$$D(x) = o(\rho(x)L^3), \qquad x \to \infty,$$

uniformly for $x \leq z \leq x^k$.

Remark. A careful examination of the following proof shows that $\tau = \min(\frac{1}{2}, \delta^2/(2-\delta))$ may be taken.

Proof. We have

$$|\Lambda(z, u) - \Lambda(w, v)| \leq |\Lambda(z, u) - \Lambda(w, u)| + |\Lambda(w, u) - \Lambda(w, v)|,$$

$$= \Lambda_1 + \Lambda_2$$

The Approximate Functional Equation

say. Hence $D(x)$ does not exceed $D_1(x) + D_2(x)$ where, for $j = 1, 2$, $D_j(x)$ is the integral obtained by replacing the integrand in $D(x)$ by Λ_j.

We first consider

$$D_2(x) = \int_u \left(\int_v \int_w |\Lambda(w, u) - \Lambda(w, v)| \frac{dw\, dv}{wv} \right) \frac{du}{u}.$$

For each u in the outer range the condition $u^\delta \geq x^{ab}$ will certainly be satisfied if $\tau \leq \frac{1}{2}$, and we may estimate the inner double integral by lemma (9.5). Thus

$$D_2(x) = o(\rho(x) L^2) \int_{x^{ab/\delta}}^{x^h} \frac{du}{u} = o(\rho(x) L^3).$$

To estimate $D_1(x)$ we change the order of integration:

$$D_1(x) \leq \int_{x^{a(1-\tau)} \leq v \leq x^a} \int_{x^{h(1-\tau)} v^{1-b} \leq w \leq x^h v^{1-b(1-\tau)}}$$

$$\cdot \left(\int_{x^{h(1-\tau)} \leq u \leq \min(x^h, wv^{b-1})} |\Lambda(z, u) - \Lambda(w, u)| \frac{du}{u} \right) \frac{dv\, dw}{vw}.$$

In the innermost integral over u,

$$u \leq u_0 = \min(x^h, wv^{b-1}).$$

Then $u_0 \leq x^h \leq x^\delta \leq z^\delta$, and for small enough τ

$$u_0 \leq wv^{b-1} \leq w^\delta.$$

To gain this last inequality it will suffice if $w^{1-\delta} \leq v^{1-b}$. In fact

$$w^{1-\delta} \leq (x^h v^{1-b(1-\tau)})^{1-\delta}$$

and this last expression will not exceed v^{1-b} if

$$x^{h(1-\delta)} \leq v^{1-b-(1-\delta)(1-b(1-\tau))}.$$

But $v \geq x^{a(1-\tau)}$ and it will therefore suffice if

(15) $$h(1-\delta) \leq a(1-\tau)\{1 - b - (1-\delta)(1 - b(1-\tau))\}.$$

It we let τ approach zero from above, the upper bound in the inequality

(15) approaches $a\delta(1-b)$. Since

$$a\delta(1-b) - h(1-\delta) = \delta(a - ab - 1 + \delta) = \delta^3/2 > 0,$$

a small enough but fixed value of τ will guarantee the validity of (15).

We can now apply lemma (9.4) in the form (11) to majorize the u-integral in the bound for $D_1(x)$, replacing the z, w, x of (11) by $z, w, u_0^{1/\delta}$, respectively. In this way we obtain

$$D_1(x) \le o(\rho(x)L) \int_1^{x^a} \int_1^{x^{a+h}} \frac{dv\, dw}{vw} = o(\rho(x)L^3),$$

and lemma (9.6) is proved.

Let $0 < \sigma < \delta(1-\tau)/4$ and set $u_0 = x^{h(1-\sigma\tau)}$. *Define*

$$\kappa(x) = \iint \frac{dw\, dv}{wv}, \qquad \lambda(x) = \frac{1}{\kappa(x)} \iint \Lambda(w, v) \frac{dw\, dv}{wv},$$

both integrals being taken over the region

$$S: x^{a(1-\tau)} \le v \le x^a, \qquad v^{b(1-\tau)} \le \frac{u_0 v}{w} \le v^{b(1-\tau/2)}.$$

A straightforward integration shows that

$$\kappa(x) = \frac{a^2 b\tau}{4}(1 - (1-\tau)^2)(\log x)^2$$

whatever the definition of u_0, so long as it is positive.

Lemma (9.7).

$$\int_{x^{\delta(1-\sigma\tau)}}^{x^\delta} |\Lambda(x, u) - \lambda(x)| \frac{du}{u} = o(\rho(x)L)$$

as $x \to \infty$.

Proof. The integral which is to be estimated can be represented in the form

$$\int_{x^{\delta(1-\sigma\tau)}}^{x^\delta} \frac{1}{\kappa(x)} \left| \kappa(x)\Lambda(x, u) - \iint_S \Lambda(w, v) \frac{dw\, dv}{wv} \right| \frac{du}{u}$$

$$= \int_{x^{\delta(1-\sigma\tau)}}^{x^\delta} \frac{1}{\kappa(x)} \left| \iint_S (\Lambda(x, u) - \Lambda(w, v)) \frac{dw\, dv}{wv} \right| \frac{du}{u}.$$

Moreover, the region in the (u, v, w) space which is determined by

$$x^{\delta(1-\sigma\tau)} \le u \le x^{\delta}, \qquad x^{a(1-\tau)} \le v \le x^{a}, \qquad v^{b(1-\tau)} \le \frac{u_0 v}{w} \le v^{b(1-\tau/2)}$$

is contained in that restricted by

$$x^{h(1-\tau)} \le u \le x^{h}, \qquad x^{a(1-\tau)} \le v \le x^{a}, \qquad v^{b(1-\tau)} \le \frac{uv}{w} \le v^{b}.$$

For example, if (u_1, v_1, w_1) belongs to the first of these regions

$$\frac{u_1 v_1^{1-b(1-\tau)}}{w_1} \ge \frac{u_0 v_1^{1-b(1-\tau)}}{w_1} \ge 1,$$

since $u_1 \ge u_0$; and

$$\frac{u_1 v_1^{1-b}}{w_1} \le \frac{u_0 x^{\delta\sigma\tau} v_1^{1-b}}{w_1} \le \frac{x^{\delta\sigma\tau}}{v_1^{b\tau/2}} \le 1,$$

since $v_1 \ge x^{a(1-\tau)}$ and $\delta\sigma - ab(1-\tau)/2 \le 0$. The last integral above is therefore $\ll L^{-2} D(x)$ which by lemma (9.6) is $o(\rho(x)L)$.

Lemma (9.7) is proved, and with it we come to the end of the application of the second part of the third motive.

From now until the end of the proof of theorem (9.1) we shall assume τ to be a positive number fixed at a value not exceeding

(16) $$\frac{\delta}{8} \min\left(\tfrac{1}{2}, \frac{\delta^2}{2-\delta}\right).$$

With this notation the result of lemma (9.7) holds with $\sigma\tau$ formally replaced by τ.

We return now to our equation (1), and express it in the form

$$\sum_{x^{\delta(1-\tau)} < p \le t} \frac{1}{p}(f(p) - \alpha(x)) + \alpha\left(\frac{x}{p}\right) = o(\rho(x))$$

uniformly for $x^{\delta(1-\tau)} \le t \le x^{\delta}$.

This we put into a continuous form:

Lemma (9.8).

$$\alpha(t) - \int_{x^{\delta(1-\tau)}}^{t} \frac{\alpha(u)}{u \log u} du = \alpha(x^{\delta(1-\tau)}) + \int_{x^{\delta(1-\tau)}}^{t} \frac{\Lambda(x, u)}{u \log u} du + o(\rho(x))$$

holds uniformly for $x^{\delta(1-\tau)} \le t \le x^{\delta}$.

Proof. This result may be established along the lines used in the proof of lemma (9.4). We note here only that

$$\alpha(t) - \alpha(x^{\delta(1-\tau)}) = \sum_{x^{\delta(1-\tau)} < p \le t} \frac{f(p)}{p}$$

$$= \sum_{x^{\delta(1-\tau)} < p \le t} \frac{1}{p}\left(\alpha(x) - \alpha\left(\frac{x}{p}\right)\right) + o(\rho(x))$$

$$= \int_{x^{\delta(1-\tau)}}^{t} \left(\alpha(x) - \alpha\left(\frac{x}{u}\right)\right) d\left(\sum_{p \le u} \frac{1}{p}\right) + o(\rho(x));$$

and that an integration by parts allows us to deduce from (6) the asymptotic estimate

$$\sum_{p \le w} \frac{1}{p} = \log\log w + c_{11} + O((\log w)^{-B}), \qquad w \ge 3.$$

Once again B may be fixed at any positive value.

Lemma (9.9). *There is a (measurable) function $F(x)$ so that*

(17) $$\alpha(t) - \int_{x^{\delta(1-\tau)}}^{t} \frac{\alpha(u)}{u \log u} du = F(x) + \lambda(x) \log\log t + o(\rho(x))$$

holds uniformly for $x^{\delta(1-\tau)} \le t \le x^{\delta}$.

Proof. If we denote the left-hand side of this equation by $M(t)$, then by lemma (9.8), and (9.7) in accordance with the convention (16),

$$\left| M(t) - \alpha(x^{\delta(1-\tau)}) - \lambda(x) \int_{x^{\delta(1-\tau)}}^{t} \frac{du}{u \log u} \right|$$

$$\le \int_{x^{\delta(1-\tau)}}^{t} |\Lambda(x, u) - \lambda(x)| \frac{du}{u \log u} + o(\rho(x))$$

$$= O\left(\frac{1}{\log x} o(\rho(x)L)\right) + o(\rho(x)) = o(\rho(x)).$$

The result of lemma (9.9) now follows if we set

$$F(x) = \alpha(x^{\delta(1-\tau)}) - \lambda(x) \log\log x^{\delta(1-\tau)}.$$

The Approximate Functional Equation

Lemma (9.10). *There is a (measurable) function $G(x)$ so that*

$$\alpha(t) = G(x) \log t - \lambda(x) + o(\rho(x)), \qquad x \to \infty,$$

holds uniformly over the interval $x^{\delta(1-\tau)} \leq t \leq x^\delta$.

Remark. With this representation our functional equation (1) is essentially solved.

Proof. The left-hand side of equation (17) in lemma (9.9) may be expressed in the form

$$t(\log t)^2 \frac{d}{dt}\left(\frac{1}{\log t} \int_{x^{\delta(1-\tau)}}^{t} \frac{\alpha(u)}{u \log u} du\right), \qquad x^{\delta(1-\tau)} \leq t \leq x^\delta,$$

save possibly at the finitely many points when $\alpha(u)$ is not continuous, or at the end-points $t = x^{\delta(1-\tau)}$, $t = x^\delta$. Using this fact with t replaced by z, and integrating over the range $x^{\delta(1-\tau)} \leq z \leq t$, gives

$$\frac{1}{\log t} \int_{x^{\delta(1-\tau)}}^{t} \frac{\alpha(u)}{u \log u} du = \int_{x^{\delta(1-\tau)}}^{t} \frac{F(x) + \lambda(x) \log \log z + o(\rho(x))}{z (\log z)^2} dz$$

$$= F(x)\left(\frac{1}{\delta(1-\tau) \log x} - \frac{1}{\log t}\right)$$

$$+ \lambda(x) \int_{x^{\delta(1-\tau)}}^{t} \frac{\log \log z}{z (\log z)^2} dz + o\left(\frac{\rho(x)}{\log x}\right).$$

We can now eliminate between this equation and that of lemma (9.9), to reach

$$\alpha(t) = \frac{F(x)}{\delta(1-\tau) \log x} \log t$$

$$+ \lambda(x)\left\{\log \log t + \log t \cdot \int_{x^{\delta(1-\tau)}}^{t} \frac{\log \log z}{z (\log z)^2} dz\right\} + o(\rho(x)).$$

An integration by parts gives

$$\int_{2}^{t} \frac{\log \log z}{z (\log z)^2} dz = \text{constant} - \frac{1}{\log t} - \frac{\log \log t}{\log t}$$

uniformly for $t \geq 2$, and we obtain the estimate of lemma (9.10) with

$$G(x) = \frac{F(x)}{\delta(1-\tau)\log x} + \frac{\lambda(x)\{1+\log\log x^{\delta(1-\tau)}\}}{\delta(1-\tau)\log x}.$$

Proof of theorem (9.1). Define

$$\alpha_1(x) = G(x)\log x, \qquad \alpha_2(x) = \alpha(x) - \alpha_1(x).$$

Let d be a positive number so small that the intervals $[y\delta(1-\tau), y\delta]$ and $[\delta(1-\tau), \delta]$ overlap in at least an interval I for each value of y in the range $1-d \leq y \leq 1+d$.

It follows from lemma (9.10) that for each fixed y, $|y-1| \leq d$,

$$\alpha(t) = G(x)\log t - \lambda(x) + o(\rho(x)),$$

$$\alpha(t) = G(x^y)\log t - \lambda(x^y) + o(\rho(x)),$$

holds as $x \to \infty$, uniformly for all $t = x^\beta$ with β in I. For these same values of t, therefore

(18) $$\{G(x) - G(x^y)\}\log t = \lambda(x) - \lambda(x^y) + o(\rho(x)).$$

We apply this estimate with two distinct values of β, say β_1 and β_2 and eliminate between the resulting estimates to obtain

$$\{G(x) - G(x^y)\}(\beta_1 - \beta_2)\log x = o(\rho(x)).$$

For such values of y,

$$\alpha_1(x^y) = G(x^y) y \log x = y\left\{G(x) + o\left(\frac{\rho(x)}{L}\right)\right\}\log x$$

(19)
$$= y\alpha_1(x) + o(\rho(x)).$$

That this relation holds for every $y > 0$ may be obtained in the following manner. If $0 < y < 1$ then there is a non-negative integer k so that $y = (1-d)^k r$ with $1 - d \leq r \leq 1$. Then arguing inductively

$$\alpha_1(x^y) = \alpha_1(x^{(1-d)^k r}) = r\alpha_1(x^{(1-d)^k}) + o(\rho(x))$$

by (19) with x replaced by $x^{(1-d)^k}$,

$$= r(1-d)\alpha_1(x^{(1-d)^{k-1}}) + o(\rho(x))$$

...

$$= r(1-d)^k \alpha_1(x) + o(\rho(x))$$

$$= y\alpha_1(x) + o(\rho(x)).$$

For $y > 1$ we may argue similarly, using $1+d$ in place of $1-d$.
From (18)

$$\lambda(x^y) = \lambda(x) + o(\rho(x)).$$

However, by lemma (9.10) with $t = x^\delta$, using what we have already proved concerning the function $G(x)$,

$$\alpha_2(x^\delta) = \alpha(x^\delta) - G(x^\delta) \log x^\delta$$

$$= \{G(x) - G(x^\delta)\} \log x^\delta - \lambda(x) + o(\rho(x))$$

$$= -\lambda(x) + o(\rho(x)).$$

Hence, applying this with x replaced by $x^{y/\delta}$,

$$\alpha_2(x^y) = -\lambda(x^{y/\delta}) + o(\rho(x))$$

$$= -\lambda(x^{1/\delta}) + o(\rho(x))$$

$$= \alpha_2(x) + o(\rho(x)), \qquad x \to \infty,$$

for $|y-1| \le d$.

This last result is easily extended to hold for every positive y, and the proof of theorem (9.1) is complete.

Remarks. Let $0 < d_1 < d_2$ hold. From what we have just proved, the representation

$$\alpha(x) = G(x) \log t - \lambda(x) + o(\rho(x)), \qquad x \to \infty,$$

of lemma (9.10) is now seen to hold uniformly for $x^{d_1} \le t \le x^{d_2}$. This may be applied to the basic hypothesis (1) to give

$$\sum_{x^\varepsilon < p \leq x^\delta} \frac{1}{p} |f(p) - G(x) \log p| = o(\rho(x)),$$

showing that in an L^1 sense, $f(p)$ is a logarithm.
This ends the remarks.

We now formulate our above argument as an inequality.

Theorem (9.11). *Let $0 < \varepsilon < \delta$, $0 < d_1 < d_2$ be given. Let $f(p)$ be a complex-valued function defined on the prime numbers $p (\geq 2)$, and define*

$$\theta(x) = \sum_{x^\varepsilon < p \leq x^\delta} \frac{1}{p} \left| f(p) - \alpha(x) + \alpha\left(\frac{x}{p}\right) \right|,$$

where

$$\alpha(x) = \sum_{p \leq x} \frac{f(p)}{p}$$

for $x \geq 2$.
Then there is a representation

$$\alpha(t) = G(x) \log t - \lambda(x) + O(\Delta(x))$$

valid for $x^{d_1} \leq t \leq x^{d_2}$, with

$$\sup_{x^{1/2} \leq w \leq x^2} |G(x) - G(w)| \ll \frac{\Delta(x)}{\log x},$$

$$\sup_{x^{1/2} \leq w \leq x^2} |\lambda(x) - \lambda(w)| \ll \Delta(x),$$

and

$$\Delta(x) = \sup_{x^{c_{12}} \leq w \leq x^{c_{13}}} \theta(w) + (\log x)^{-B} (\max_{p \leq c} |f(p)| + \sup_{w \leq x^{c_{12}}} \theta(w))$$

for all $x \geq 2$.
The constant B may be given any (fixed) positive value. The constants c_{12}, c_{13}, c, and those which are implied, then depend at most upon ε, δ, d_1, d_2 and B.

The Approximate Functional Equation

Remarks. If $f(p)$ is only defined for $2 \leq p \leq N$, then for fixed ε, δ, d_1, d_2 and B, theorem (9.11) makes sense for those values of x which satisfy $x \geq 2$ and $x^\beta \leq N$, where $\beta = \max(d_2, 2, c_{13})$.

If $\theta(x) = o(\rho(x))$, where $\rho(x)$ satisfies the hypothesis which it does in theorem (9.1), then from (8) in lemma (9.3)

$$\sup_{x^{c_{12}} \leq w \leq x^{c_{13}}} \theta(w) = o(\rho(x)).$$

If $B > \max(c_3, c_2)$ then by lemma (9.3), $\Delta(x) = o(\rho(x))$ and we recover the result of theroem (9.1). Here we have only used the hypothesis (1) for a *fixed* pair ε, δ.

It is a corollary of theorem (9.11) that if $0 < \delta < 1$, $d_1 \leq \min(1, 1-\delta)$ and $d_2 > 1$, then

$$\sum_{x^\varepsilon < p \leq x} \frac{1}{p} |f(p) - G(x) \log p| \ll \Delta(x).$$

This ends our remarks.

Proof of theorem (9.11). This follows the proof of theorem (9.1), so only the salient features will be given.

Let

$$J(x) = \max_{p \leq c} |f(p)| + \sup_{w \leq x} \theta(w).$$

This is a non-decreasing function of x. Lemma (9.2) then asserts that for $x \geq 2$

$$\sum_{p \leq x} \frac{|f(p)|}{p} \leq J(x^{1/\delta})(\log x)^{c_2}.$$

We do not need an analogue of lemma (9.3).

If B denotes the constant which appears in (6) then the format of lemma (9.4) is the same, save that the error term $o(\rho(x)L)$ is now

(20) $$\ll \sup_{x \leq w \leq x^k} \theta(w) L + L^{-B/2} J(x^{k/\delta})$$

for large enough B. Note that

$$\sup_{2 \leq t \leq x^k} |\alpha(t)| \ll \sup_{t \leq x^k} J(t^{1/\delta})(\log t)^{c_2} \ll J(x^{k/\delta})(\log x)^{c_2},$$

and that for p_0 in the interval $[x, x^k]$

$$p_0^{-1}|f(p_0)| \leq \sup_{x \leq w \leq x^k} \theta(w) + p_0^{-1}|\alpha(p_0^{1/\delta}) - \alpha(p_0^{(1/\delta)-1})|$$

$$\ll \sup_{x \leq w \leq x^k} \theta(w) + x^{-1}J(x^{k/\delta})(\log x)^{c_2}.$$

Let us denote the upper bound at (20) by $R(x)$.

Concerning the analogue of lemma (9.5) we need an upper bound for the integral

$$\int_{\omega^{b(1-\tau)} \leq \mu \leq (\min(\omega,u))^\delta} |\Lambda(u,\mu) - \Lambda(\omega,\mu)| \frac{d\mu}{\mu}.$$

If we denote $\min(\omega, \mu)$ by β then the range of integration is contained in the union of the intervals

$$(\beta^{\delta^j})^\delta \leq \mu \leq \beta^{\delta^j}, \quad j = 1, 2, \ldots, m,$$

for some positive integer m depending only on δ, b and τ. If the constant k implicit in the definition of $J(x)$ is sufficiently large the above integral is, by lemma (9.4) as (11),

$$\ll \sum_{j=0}^{m-1} R(\beta^{\delta^j})$$

$$\ll \sup_{x^{c_{14}} \leq w \leq x^k} \theta(w) + L^{-B/2}J(x^{k/\delta}).$$

The estimate of lemma (9.5) remains valid with

$$\ll \sup_{x^{c_{14}} \leq w \leq x^k} \theta(w)L^2 + L^{-B/3}J(x^{k/\delta})$$

in place of $= o(\rho(x)L^2)$.

It is now straightforward to prove that the function $D(x)$ is

$$\ll \sup_{x^{c_{15}} \leq w \leq x^{c_{16}}} \theta(w)L^3 + L^{-B/4}J(x^{c_{16}})$$

for suitably chosen positive constants c_{15} and c_{16}, and all large enough (but fixed) B.

A similar bound but with L in place of L^3 also serves in the analogue of lemma (9.7).

The Approximate Functional Equation

The modifications required to establish theorem (9.11) are now straightforward.

The following result will be useful in connection with the application of theorem (9.11).

Theorem (9.12). *The function $G(x)$ which occurs in the statement of theorem (9.11) satisfies*

$$G(x) \ll \sum_{p \leq x^{c_{14}}} \frac{|f(p)|}{p}, \qquad x \geq 2,$$

for a certain positive constant c_{14}.

Proof. Since

$$\theta(w) \ll \sum_{p \leq \max(w, w^\delta)} \frac{|f(p)|}{p}$$

it is clear from its definition that for a suitably large but fixed value of c_{14}

$$\Delta(x) \ll \sum_{p \leq x^{c_{14}}} \frac{|f(p)|}{p}.$$

Since

$$G(x) - G(x^{1/2}) \ll \frac{\Delta(x)}{\log x}$$

the desired bound may be obtained with an argument by induction, provided we note that

$$\sum_{2^j \leq \log x / \log 2} \frac{1}{2^{-j} \log x} \ll 1.$$

This completes our consideration of the approximate functional equation.

Chapter 10

Additive Arithmetic Functions on Differences

The Basic Inequality

In this chapter I establish inequalities which measure the control the differences of an additive function have over its values on the powers of primes.

For the duration of the chapter $a > 0$, b, $A > 0$, B will be integers for which the determinant

$$\Delta = \begin{vmatrix} a & b \\ A & B \end{vmatrix} = aB - Ab$$

is non-zero.

Theorem (10.1). *The inequality*

$$\underset{\substack{q \leq x \\ (q,aA\Delta)=1}}{\Sigma} \frac{1}{q} |f(q) - F(x)\log q|^2 \ll \sup_{x \leq w \leq x^c} \frac{1}{w} \sum_{x < n \leq w} |f(an+b) - f(An+B)|^2,$$

with

$$F(x) = \underset{\substack{x^{1/2} < q \leq x \\ (q,aA\Delta)=1}}{\Sigma} \frac{f(q)}{q} \Big/ \underset{\substack{x^{1/2} < q \leq x \\ (q,aA\Delta)=1}}{\Sigma} \frac{\log q}{q}$$

holds uniformly for all additive functions $f(n)$ for all $x \geq x_0$. Here x_0, c and the implied constant depend at most upon a, b, A and B; q denotes a prime-power.

Remark. We shall use Σ' to denote summation over an integer variable which is prime to $aA\Delta$.

In order to state the supplementary results concerning the values of $f(n)$

The Basic Inequality

on the powers of primes which divide $aA\Delta$, it is necessary to recall some notation from Chapter 3.

Let $\mu = (a, b)$, $\nu = (A, B)$. For every integer n the conditions

$$an + b \equiv 0 \pmod{\mu}, \qquad An + B \equiv 0 \pmod{\nu}$$

are simultaneously satisfied. When $aA\Delta$ is odd it is possible to find an integer n so that

$$(an+b)\mu^{-1} \quad \text{and} \quad (An+B)\nu^{-1}$$

have no further prime factor in common with $aA\Delta$. The pair $(\mu; \nu)$ is called the *fixed-divisor pair* associated with the sequences $\{an+b\}$, $\{An+B\}$, and we define

$$L_1 = f(\mu) - f(\nu).$$

For even values of $aA\Delta$ the same procedure may be followed unless $a\mu^{-1}$, $A\nu^{-1}$ are both odd, and $b\mu^{-1} + B\nu^{-1}$ is odd. In this case we can find an n such that 2μ divides $an+b$ and

$$((an+b)/2\mu, aA\Delta) = 1 \quad \text{with} \quad ((An+B)/\nu, aA\Delta) = 1;$$

and also an(other) n so that 2ν divides $An+B$ and

$$((an+b)/\mu, aA\Delta) = 1 \quad \text{with} \quad ((An+B)/2\nu, aA\Delta) = 1.$$

We call $(2\mu; \nu)$ and $(\mu; 2\nu)$ the *fixed-divisor pairs* associated with the sequences $\{an+b\}$, $\{An+B\}$. We then define

$$L_1 = f(2\mu) - f(\nu); \qquad L_2 = f(\mu) - f(2\nu).$$

Suppose that p is a prime which divides $aA\Delta$, and that for some integer n, $p^i \| (an+b)$, $p^j \| (An+B)$. Then (in the notation of Chapter 3) it is useful to define

$$L(p^i, p^j) = f(p^i) - f(p^j) - \{f(p^r) - f(p^s)\},$$

where $p^r \| u$, $p^s \| v$ and $(u; v)$ is an appropriate fixed-divisor pair. For simplicity of exposition, when $aA\Delta$ is even and there is a choice of fixed-divisor pair we shall take $(u; v)$ to be $(2\mu; \nu)$.

In this notation there is a representation

(1) $\quad f(an+b) - f(An+B) = \sum\limits_{\substack{q \| (an+b) \\ (q, aA\Delta) = 1}} - \sum\limits_{\substack{q \| (An+B) \\ (q, aA\Delta) = 1}} f(q) + \sum\limits_{p | aA\Delta} L(p^i, p^j) + L_t,$

where $t = 1$ or 2. Of course the values of i, j depend upon p and n. As we mentioned already in Chapter 3, it is convenient to treat $L(p^i, p^j)$ as a single entity. Note that if $p^m \| \Delta$ then since

$$a(An + B) - A(an + b) = \Delta,$$

we must have $\min(i, j) \leq m$. As in Chapter 8, p_{ij} will denote $\max(p^i, p^j)$.

For complex numbers U we now define

$$L(p^i, p^j, U) \quad \text{and} \quad L_t(U)$$

to be the functions $L(p^i, p^j)$ and L_t, respectively, but with $f(q)$ everywhere replaced by $f(q) - U \log q$. In Chapter 8 these same functions were indicated by *, but we shall presently need to draw attention to differing choices for U.

We can now state our supplementary result.

Theorem (10.2). *For all primes p which divide $aA\Delta$, and all fixed-divisor pairs (α, β), the expressions*

$$\sum_{p_{ij} \leq x} \frac{|L(p^i, p^j, F(x))|^2}{p_{ij}},$$

$$\left| f(\alpha) - f(\beta) - F(x) \log \frac{\alpha}{\beta} \right|^2,$$

are

$$\ll \sup_{x \leq w \leq x^c} \frac{1}{w} \sum_{x < n \leq w} |f(an + b) - f(An + B)|^2.$$

Here $F(x)$ is the function which appears in the statement of theorem (10.1).

If $a \neq A$ then $|F(x)|^2$ has this same upper bound.

If $a = A$ let n_0 be a positive integer so that $an + b$ and $An + B$ are positive for all $n > n_0$, and let

$$D = \sum_{n_0 < n \leq x} \left(\log \left(\frac{an + b}{An + B} \right) \right)^2 > 0.$$

Then

$$x^{-1/2} |F(x)| = D^{-1/2} \left(x^{-1} \sum_{n_0 < n \leq x} |f(an + b) - f(An + B)|^2 \right)^{1/2}$$

$$+ O\left(\sup_{x \leq w \leq x^c} \left\{ \frac{1}{w} \sum_{x < n \leq w} |f(an + b) - f(An + B)|^2 \right\}^{1/2} \right)$$

for all large x.

The Basic Inequality

Remarks. Since $a = A$ the series

$$\sum_{n=n_0+1}^{\infty} \left(\log\left(\frac{an+b}{An+B}\right)\right)^2$$

converges.

It follows from theorem (10.2) that when $a = A$

$$x^{-1}|F(x)|^2 \ll \sup_{x \le w \le x^c} \frac{1}{W} \sum_{n \le w} |f(an+b) - f(An+B)|^2.$$

It is tempting to believe that as in the other inequalities of these two theorems, one may replace the summation condition $n \le w$ here by $x < n \le w$. This is in fact not possible. For if $f(n) = \log n$ then it would lead to a bound

$$|F(x)|^2 \ll x \sup_{x \le w \le x^c} \frac{1}{W} \sum_{x < n \le w} \frac{1}{n^2} \ll x^{-1},$$

in contrast to the actual value $F(x) = 1$. All that theorem (10.2) will predict in these circumstances is the consistent estimate

$$F(x) = 1 + O(x^{-1/2}).$$

As we shall show later, theorems (10.1) and (10.2) are in some sense best possible.

In what follows it will sometimes be convenient to collect together expressions of the type considered in theorem (10.2) in the form

$$\sum_{p|a A \Delta} \sum_{p_{ij} \le x} \frac{|L(p^i, p^j, F(x))|^2}{p_{ij}} + \sum_{t=1}^{2} |L_t(F(x))|^2.$$

Here, and on similar occasions, it is to be understood that if there is only one fixed-divisor pair, then the summand(s) involving $L_2(F(x))$, and so on, are to be omitted.

Before proving these theorems it will be convenient to establish a simple result which illustrates a significance of the function $F(x)$.

Lemma (10.3). *The inequality*

$$\sum_{q \le x}' \frac{1}{q} |f(q) - F(x) \log q|^2 \ll \sum_{q \le x}' \frac{1}{q} |f(q) - \lambda \log q|^2$$

holds uniformly for all complex values of λ, additive functions $f(n)$, and real

$x \geq 2$. Moreover, if the upper bound is denoted by $H(\lambda)$ then

$$\sum_{p_{ij} \leq x} \frac{|L(p^i, p^j, F(x))|^2}{p_{ij}} \ll \sum_{p_{ij} \leq x} \frac{|L(p^i, p^j, \lambda)|^2}{p_{ij}} + H(\lambda)(\log x)^{-2}$$

$$|F(x) - \lambda|^2 \ll H(\lambda)(\log x)^{-2},$$

$$|L_t(F(x))|^2 \ll |L_t(\lambda)|^2 + H(\lambda)(\log x)^{-2},$$

hold with the same uniformities, and for $t = 1$ and 2.

Remark. This result shows that in theorems (10.1) and (10.2) it will be enough to find any function of x, in place of $F(x)$, for which the asserted inequalities hold.

Proof. From the definition of $F(x)$

$$F(x) - \lambda = \sum_{x^{1/2} < q \leq x}' \frac{1}{q}(f(q) - \lambda \log q) \Big/ \sum_{x^{1/2} < q \leq x}' \frac{\log q}{q}.$$

An application of the Cauchy–Schwarz inequality shows that the numerator of this fraction is not more than

$$\left(\sum_{q \leq x}' \frac{1}{q} |f(q) - \lambda \log q|^2 \cdot \sum_{x^{1/2} < q \leq x} \frac{1}{q} \right)^{1/2}.$$

From lemma (1.2)

$$\sum_{x^{1/2} < q \leq x} \frac{1}{q} \ll 1$$

and

$$\sum_{x^{1/2} < q \leq x}' \frac{\log q}{q} = \tfrac{1}{2} \log x + O(1),$$

so that

$$|F(x) - \lambda|^2 \ll H(\lambda)(\log x)^{-2}.$$

By the Cauchy–Schwarz inequality the first of the sums to be majorized in lemma (10.3) does not exceed

$$2 \sum_{q \leq x}' \frac{1}{q} |f(q) - \lambda \log q|^2 + 2 \sum_{q \leq x}' \frac{1}{q} |\lambda - F(x)|^2 (\log q)^2.$$

The Basic Inequality

What we have so far proved and a further application of lemma (1.2) show that this expressions is $\ll H(\lambda)$.

The proof of lemma (10.3) is now readily completed.

The proof of theorems (10.1) and (10.2) consists of finitely many improving steps.

Let $K > 0$ be given and define

$$T(x) = \frac{1}{(\log x)^{1/4}} \sum_{q \leq x^{c_1}}' \frac{|f(q) - U \log q|^2}{q}$$

$$+ \frac{1}{(\log x)^K} \left(|U|^2 + \sum_{p | a A \Delta} \sum_{p_{ij} \leq x^{c_1}} \frac{|L(p^i, p^j, U)|^2}{p_{ij}} + \sum_{t=1}^{2} |L_t(U)|^2 \right)$$

$$+ \sup_{x^{1/c_1} \leq w \leq x^{c_1}} \frac{1}{w^{1-\sigma}} \sum_{n \leq w} \frac{|f(an+b) - f(An+B)|^2}{n^\sigma},$$

where, as usual, the term involving $L_2(U)$ is to be omitted if it cannot occur.

It is convenient to note here that applying the Cauchy–Schwarz inequality to the representation (1) gives

$$|f(an+b) - f(An+B)|^2 \ll \log 2n \left\{ \sum_{q \| (an+b)}' |f(q)|^2 + \sum_{q \| (An+B)}' |f(q)|^2 \right.$$

$$\left. + \sum_{p | a A \Delta} |L(p^i, p^j)|^2 + \sum_{t=1}^{2} |L_t|^2 \right\}.$$

We multiply by $n^{-\sigma}$ and sum over the integers not exceeding w, inverting the order of summation in the upper bound, to obtain

$$w^{-1+\sigma} \sum_{n \leq w} n^{-\sigma} |f(an+b) - f(An+B)|^2$$

$$\ll \log 2w \left\{ \sum_{q \leq w_1}' \frac{|f(q)|^2}{q} + \sum_{p | a A \Delta} \sum_{p_{ij} \leq w_1} \frac{|L(p^i, p^j)|^2}{p_{ij}} + \sum_{t=1}^{2} |L_t|^2 \right\},$$

where $w_1 = \max(aw + |b|, Aw + |B|)$.

Since, for example,

$$|L(p^i, p^j)|^2 \ll |L(p^i, p^j, U)|^2 + (|U| \log p_{ij})^2,$$

it is straightforward to prove that for all large x

(2) $$\sup_{y\leq x} T(y) \ll T(x^2)(\log x)^{\max(5/4, K+1)}.$$

Lemma (10.4). *Let $0 < \psi < 1$. Then there is a (measurable) function $G(x)$ so that for suitably chosen positive constants σ, $\tfrac{1}{2} < \sigma < 1$, x_0 and c_1,*

$$\sum_{\substack{x^\psi < p \leq x \\ (p, aA\Delta) = 1}} \frac{1}{p} |f(p) - G(x) \log p|^2 \ll T(x)$$

holds uniformly for all $x \geq x_0$. Here $G(x)$ depends upon the function $f(\ )$ but not upon U. The constants x_0, σ, c_1 and the implied constant depend only upon a, b, A, B and ψ.

Proof. Define

$$\alpha(x) = \sum_{\substack{p \leq x \\ (p, aA\Delta) = 1}} \frac{f(p)}{p}.$$

Then from theorem (8.3) in the form (9) of Chapter 8, with $\delta = \tfrac{1}{8}$, $\sigma = (25 - \sqrt{97})/24$,

(3) $$\sum_{\substack{x^\varepsilon < p \leq x^{1/8} \\ (p, aA\Delta) = 1}} \frac{1}{p} \left| f(p) - \alpha(x) + \alpha\left(\frac{x}{p}\right) \right|^2 \ll T(x)$$

holds for every fixed ε, $0 < \varepsilon < \tfrac{1}{8}$, with suitable values for the constants.

An application of the Cauchy–Schwarz inequality together with lemma (1.2) now gives

$$\sum_{\substack{x^\varepsilon < p \leq x^{1/8} \\ (p, aA\Delta) = 1}} \frac{1}{p} \left| f(p) - \alpha(x) + \alpha\left(\frac{x}{p}\right) \right| \ll T(x)^{1/2}.$$

We apply theorem (9.11), and obtain a representation

(4) $$\alpha(t) = G(x) \log t - \lambda(x) + O(D(x))$$

valid for $x^{d_1} \leq t \leq x^{d_2}$, where $0 < d_1 < d_2$ but otherwise the d_j may be freely chosen. Here (in the notation of theorem (9.11)) $D(x) \ll$

$$\sup_{x^{c_{12}} \leq w \leq x^{c_{13}}} T(w)^{1/2} + (\log x)^{-2K-1} \Big(\max_{p \leq c, (p, aA\Delta) = 1} |f(p)| + \sup_{w \leq x^{c_{12}}} T(x)^{1/2} \Big)$$

for a certain c_1. In view of the bound (2), we may find a possibly larger value of c_1 so that $D(x) \ll T(x)^{1/2}$. According to theorem (9.11) we shall then have

$$(5) \qquad \sup_{x^{1/2} < w \leq x^2} |G(x) - G(w)| \ll \frac{T(x)^{1/2}}{\log x}.$$

With d_1, d_2 fixed at suitable values, we deduce from (3) and (4) that

$$\sum_{\substack{x^\varepsilon < p \leq x^\delta \\ (p, aA\Delta) = 1}} \frac{1}{p} |f(p) - G(x) \log p|^2 \ll T(x),$$

certainly if $x^\varepsilon > aA|\Delta|$, $x \geq 2$.

We replace x by x^8, and set $\varepsilon = \psi/8$. With c_1 sufficiently large, applying (5), the inequality of lemma (10.4) is established.

Lemma (10.5). *If c_1 is fixed at a large enough value then*

$$\sum_{q \leq x}' \frac{1}{q} |f(q) - F(x) \log q|^2 \ll T(x),$$

$$\sum_{p | aA\Delta} \sum_{p_{ij} \leq x} \frac{|L(p^i, p^j, F(x))|^2}{p_{ij}} \ll T(x),$$

$$|L_t(F(x)) + F(x) \log a/A|^2 \ll T(x)$$

hold for $x \geq 2$. Here $F(x)$ is the function which appears in theorem (10.1), and our inequality holds uniformly in all complex U.

Remark. The sums which are here majorized do not involve the parameter U although, of course, $T(x)$ does.

Proof. For ease of notation set

$$r(n) = f(n) - G(x) \log n, \qquad J = G(x) \log a/A,$$

where $G(x)$ is the function which appears in lemma (10.4).

Let n_0 be a positive integer so that $an + b$ and $An + B$ are positive for all $n > n_0$.

We begin with the inequality

$$(6) \qquad \sum_{n_0 < n \leq x} |r(an + b) - r(An + B) + J|^2 \ll xT(x).$$

This follows from the definition of $r(\ )$ and $T(\ )$ provided we note that over the range $n_0 < n \leq x$

$$G(x)\{\log(an+b) - \log(An+B)\} = J + O(|G(x)|n^{-1}),$$

and that, applying lemma (9.12),

$$\sum_{n \leq x} |G(x)n^{-1}|^2 \ll |G(x)|^2 \ll \left(\sum_{p \leq x^{c_{14}}}' \frac{|f(p)|}{p}\right)^2 \ll \sum_{p \leq x^{c_{14}}}' \frac{|f(p)|^2}{p} \log\log x$$

$$\ll \log\log x \left\{\sum_{p \leq x^{c_{14}}}' \frac{|f(p) - U\log p|^2}{p} + |U|^2(\log x)^2\right\} \ll xT(x)$$

if c_1 is sufficiently large.

Let m run through the integers n, in the range $n_0 < n \leq x$, for which neither $an+b$ nor $An+B$ is exactly divisible by a prime-power p^k which satisfies both $k \geq 2$ and $p^k > x^{1/4}$. From the set of all integers in the interval $(n_0, x]$ this removes

$$\sum_{\substack{k \geq 2 \\ p^k > x^{1/4}}} \sum_{\substack{n \ll x \\ n \equiv 0 \pmod{p^k}}} 1 \ll x \sum_{\substack{k \geq 2 \\ p^k > x^{1/4}}} \frac{1}{p^k} \ll x^{7/8}$$

members.

Let (α, β) be a fixed-divisor pair associated with the sequences $\{an+b\}$, $\{An+B\}$.

We specialize the inequality (6) to

$$\sum_m^\dagger |r(am+b) - r(Am+B) + J|^2 \ll xT(x),$$

where † indicates that $am+b \equiv 0 \pmod{\alpha}$ and $Am+B \equiv 0 \pmod{\beta}$ are to hold.

Let

$$s_2(n) = \sum_{\substack{p \| (an+b) \\ p > x^{1/4}}} r(p) - \sum_{\substack{p \| (An+B) \\ p > x^{1/4}}} r(p),$$

$$s_1(n) = r(an+b) - r(An+B) - s_2(n).$$

Then by the Cauchy–Schwarz inequality

$$\sum_m^\dagger |s_1(m)|^2 \leq 2 \sum_m^\dagger |s_2(m)|^2 + O(xT(x)).$$

To estimate the sum in this upper bound we define

$$E_2(x) = \sum_{x^{1/4} < p \leq x_1} \frac{r(p)}{p}\left(1 - \frac{1}{p}\right) = \sum_{x^{1/4} < p \leq x_1} \frac{(f(p) - G(x) \log p)}{p}\left(1 - \frac{1}{p}\right),$$

where $x_1 = \max(ax + |b|, Ax + |B|)$, and apply the Cauchy–Schwarz inequality, the Turán–Kubilius inequality, and then lemma (10.4) with $\psi = \frac{1}{4}$ and a suitable value of c_1:

$$\sum_m{}^\dagger |s_2(m)|^2 \leq 2 \sum_m \left| \sum_{\substack{p \| (am+b) \\ p > x^{1/4}}} r(p) - E_2(x) \right|^2 + 2 \sum_m \left| \sum_{\substack{p \| (Am+B) \\ p > x^{1/4}}} r(p) - E_2(x) \right|^2$$

(7)

$$\ll x \sum_{x^{1/2} < p \leq x_1} \frac{|f(p) - G(x) \log p|^2}{p} \ll xT(x).$$

This last argument may be compared with that used to obtain the inequality (4) in the remarks which follow the proof of lemma (1.5).

We next show that the range of summation in the bound

$$\sum_m{}^\dagger |s_1(m)|^2 \ll xT(x)$$

can be extended to all the integers in the interval $(n_0, x]$ for which † holds. We define

$$\sigma_1(n) = \sum_{\substack{q \| (an+b) \\ q \leq x^{1/4}}}{}' r(q) - \sum_{\substack{q \| (An+B) \\ q \leq x^{1/4}}}{}' r(q),$$

$$\sigma_2(n) = \sum_{\substack{p | aA\Delta \\ p_{ij} \leq x^{1/4}}} L(p^i, p^j, G(x)),$$

so that

$$s_1(m) = \sigma_1(m) + \sigma_2(m) + L_t(G(x)),$$

where the value $t = 1$ or 2 is to be chosen consistent with the choice of fixed-divisor pair (α, β).

Let $''$ denote summation over integers n for which $An + B$ is exactly divisible by some prime-power p^k exceeding $x^{1/4}$, with $k \geq 2$. Then

$$\sigma_1(n)^2 \ll \log x_1 \left\{ \sum_{\substack{q \| (an+b) \\ q \leq x^{1/4}}}{}' |f(q)|^2 + \sum_{\substack{q \| (An+B) \\ q \leq x^{1/4}}}{}' |f(q)|^2 \right\}$$

since $an+b$ and $An+B$ have at most $\{\log(an+b)\}/\log 2$ and $\{\log(An+B)\}/\log 2$ exact prime-power factors, respectively. Hence

(8)
$$\sum_{n \leq x}{}'' \sigma_1(n)^2 \ll \log x_1 \Bigg\{ \sum_{q \leq x^{1/4}}{}' |f(q)|^2 \sum_{\substack{n \leq x \\ an+b \equiv 0 (\bmod q)}}{}'' 1$$
$$+ \sum_{q \leq x^{1/4}}{}' |f(q)|^2 \sum_{\substack{n \leq x \\ An+B \equiv 0 (\bmod q)}}{}'' 1 \Bigg\}.$$

Consider the innermost sum in the first double-sum which appears in this upper bound. It does not exceed

$$\sum_{\substack{x^{1/4} < p^k \leq x_1 \\ k \geq 2}} \sum_{\substack{An+B \leq x_1 \\ An+B \equiv 0 (\bmod p^k) \\ an+b \equiv 0 (\bmod q)}} 1 \leq \sum_{\substack{x^{1/4} < p^k \leq x_1 \\ k \geq 2}} \sum_{\substack{p^k t \leq x_1 \\ ap^k t \equiv \Delta (\bmod q)}} 1$$

$$\ll \sum_{\substack{x^{1/4} < p^k \leq x_1 \\ k \geq 2}} \left\{ 1 + \frac{x}{p^k q} \right\} \ll \frac{x^{1/2}}{\log x} + \frac{x^{7/8}}{q} \ll \frac{x^{7/8}}{q},$$

where we have made use of the fact that the highest common factors (p^k, q) all divide Δ, and so are uniformly bounded.

The second double sum in the upper bound at (8) is similar but simpler to treat, and we reach

$$\sum_{n \leq x}{}'' |\sigma_1(n)|^2 \ll xT(x).$$

Arguments in this manner applied to $\sigma_2(n)$ and $L_t(G(x))$ enable us to safely assert that

(9)
$$\sum_{n_0 < n \leq x}^{\dagger} |\sigma_1(n) + \sigma_2(n) + L_t(G) + J|^2 \ll xT(x),$$

where G denotes $G(x)$.

It has been necessary to take a slightly indirect route to establish this inequality since we have, for the moment, no information concerning the size of $f(p^k)$ when $p^k > x^{1/4}$ and $k \geq 2$. These prime-powers are too few in number for the arguments of Chapter 9 to be directly applicable. To pay for this ignorance we have traded in the non-negativity of the summands at (6).

The Basic Inequality

Expanding the square in the expression at (9) we write the sum in the form

(10)
$$\sum_{n_0<n\leq x}^{\dagger} |\sigma_1(n)+\sigma_2(n)|^2 + 2\operatorname{Re}(\bar{L}_t(G)+\bar{J}) \sum_{n_0<n\leq x}^{\dagger} (\sigma_1(n)+\sigma_2(n))$$

$$+|L_t(G)+J|^2 \sum_{n_0<n\leq x}^{\dagger} 1 = Z_1 - 2\operatorname{Re} Z_2 + Z_3,$$

say.

Before we consider these sums it is convenient to make a few remarks concerning the estimation of numbers of integers in residue classes.

A sequence of positive integers $a_1 < a_2 < \cdots$ is said to have the (asymptotic) density d if

$$d = \lim_{x\to\infty} x^{-1} \sum_{a_i \leq x} 1$$

exists.

Consider the sequence of integers which belong to a certain collection of k_1 residue classes (mod D_1). It is easy to see that they have the density k_1/D_1. Moreover, the number of such integers which do not exceed a given x is

(11)
$$\frac{k_1 x}{D_1} + O(k_1).$$

Suppose now the further requirement that these integers belong to (another) collection of k_2 residue classes (mod D_2) is added, where D_1 and D_2 are coprime. Then it follows from the Chinese Remainder Theorem that the new density is $k_1 k_2 / D_1 D_2$, the product of the old densities, and the estimate corresponding to (11) will be

$$\frac{k_1}{D_1} \cdot \frac{k_2}{D_2} \cdot x + O(k_1 k_2).$$

In our present applications it will be arranged that the k_i are uniformly bounded (independent of the appropriate moduli) and the moduli $D_1 D_2$, and so on, are essentially $\ll x^{1/2}$. The estimates which are analogous to (11) may all then be given the form

$$\{1 + O(x^{-1/2})\} x \cdot \text{(density)}.$$

The sum Z_1. The sum Z_1 may itself be expanded as

$$\sum_{n_0<n\leq x}^{\dagger} |\sigma_1(n)|^2 - 2\operatorname{Re} \sum_{n_0<n\leq x}^{\dagger} \sigma_1(n)\overline{\sigma_2(n)} + \sum_{n_0<n\leq x}^{\dagger} |\sigma_2(n)|^2 = Y_1 - 2Y_2 + Y_3,$$

say. Then

$$Y_1 = \sum_{n_0 < n \leq x}^{\dagger} \left| \sum_{\substack{q \| (an+b) \\ q \leq x^{1/4}}}' r(q) \right|^2 - 2 \operatorname{Re} \sum_{n_0 < n \leq x}^{\dagger} \left\{ \sum_{\substack{q \| (an+b) \\ q \leq x^{1/4}}}' r(q) \right\} \left\{ \sum_{\substack{q \| (An+B) \\ q \leq x^{1/4}}}' \bar{r}(q) \right\}$$

$$+ \sum_{n_0 < n \leq x}^{\dagger} \left| \sum_{\substack{q \| (An+B) \\ q \leq x^{1/4}}}' r(q) \right|^2$$

$$= W_1 - 2W_2 + W_3.$$

Expanding the square and inverting the order of summation gives

$$W_1 = \sum_{q \leq x^{1/4}}' |r(q)|^2 \sum_{\substack{n_0 < n \leq x \\ an+b \equiv 0 (\operatorname{mod} q)}}^{\dagger} 1 + \sum_{\substack{q_1,q_2 \leq x^{1/4} \\ (q_1,q_2)=1}}' r(q_1)\bar{r}(q_2) \sum_{\substack{n_0 < n \leq x \\ an+b \equiv 0 (\operatorname{mod} q_i), i=1,2}}^{\dagger}$$

Let D be the density of those integers n for which $an + b \equiv 0 \pmod{\alpha}$ and $An + B \equiv 0 \pmod{\beta}$. Note that α and β are made up of primes which divide $aA\Delta$, and so are coprime to the q which appears in this representation of W_1. The density associated with the integers counted in the last of the above sums is therefore

$$\frac{D}{q_1 q_2}\left(1 - \frac{1}{q_1}\right)\left(1 - \frac{1}{q_2}\right).$$

Bearing in mind the remarks on density we readily obtain the estimate

$$(Dx)^{-1} W_1 = \{1 + O(x^{-1/2})\} \sum_{q \leq x^{1/4}}' \frac{|r(q)|^2}{q}\left(1 - \frac{1}{q_0}\right)$$

$$+ \left| \sum_{q \leq x^{1/4}}' \frac{r(q)}{q}\left(1 - \frac{1}{q_0}\right) \right|^2 - \sum_{q_0 \leq x^{1/4}}' \left| \sum_{q = q_0^f \leq x^{1/4}} \frac{r(q)}{q}\left(1 - \frac{1}{q_0}\right) \right|^2,$$

where, as in Chapter 8, q_0 denotes the prime of which q is a power.

A similar estimate is obtained for W_3.

When estimating W_2 we note that if $q_1 \| (an+b)$ and $q_2 \| (An+B)$ then $(q_1, q_2) = 1$. For if q_1 and q_2 were powers of a common prime p, then p would divide Δ, a possibility that is ruled out by the condition '. Thus

$$(Dx)^{-1} W_2 = \left| \sum_{q \leq x^{1/4}}' \frac{r(q)}{q}\left(1 - \frac{1}{q_0}\right) \right|^2$$

$$- \sum_{q \leq x^{1/4}}' \left| \sum_{q = q_0^f \leq x^{1/4}} \frac{r(q)}{q}\left(1 - \frac{1}{q_0}\right) \right|^2 + O\left(x^{-1/2} \sum_{q \leq x^{1/4}}' \frac{|r(q)|^2}{q}\right).$$

Putting together these estimates for the W_j the squares cancel and

$$(Dx)^{-1} Y_1 = 2 \sum_{q \leq x^{1/4}}' \frac{|r(q)|^2}{q} \left(1 - \frac{1}{q_0}\right) + O(T(x)).$$

We next consider the sum Y_2. It may be written as the real part of the difference

$$\sum_{\substack{q \| (an+b) \\ q \leq x^{1/4}}}' r(q) \bar{\sigma}_2(n) - \sum_{\substack{q \| (An+B) \\ q \leq x^{1/4}}}' r(q) \bar{\sigma}_2(n) = V_1 - V_2,$$

say. The treatment of these two sums is similar. There is a representation

$$\sigma_2(n) = \sum_{p | aA\Delta} L(p^i, p^j, G), \qquad G = G(x),$$

for certain integers i, j restricted by $p_{ij} \leq x^{1/4}$. Thus $\bar{\sigma}_2(n)$ involves only powers of primes which divide $aA\Delta$, whilst the $r(q)$ involves only powers of primes which do not. The Chinese Remainder Theorem may be invoked to give products of densities, and we readily obtain the estimate

$$V_1 = Dx \sum_{q \leq x^{1/4}}' \frac{|r(q)|^2}{q} \left\{ \sum_{p | aA\Delta} \sum_{p_{ij} \leq x^{1/4}} \frac{\bar{L}(p^i, p^j, G)}{p_{ij}} \delta_{ij}(p) \right\}$$

$$+ O\left(\left(x \sum_{q \leq x^{1/4}}' \frac{|r(q)|^2}{q} \left\{ \sum_{p | aA\Delta} \sum_{p_{ij} \leq x^{1/4}} \frac{|L(p^i, p^j, G)|^2}{p_{ij}} \right\}\right)^{1/2}\right).$$

Here $D\delta_{ij}(p)/p_{ij}$ is the density of the integers n which satisfy

$$an + b \equiv 0 \pmod{\alpha}, \qquad An + B \equiv 0 \pmod{\beta},$$

$$an + b \cong 0 \pmod{p^i}, \qquad An + B \cong 0 \pmod{p^j}.$$

This density can be built in the following way. For each prime g let α_g, β_g denote the exact powers of g which divide α and β, respectively. Let d_g be the density of those integers n which satisfy

$$an + b \equiv 0 \pmod{\alpha_g}, \qquad An + B \equiv 0 \pmod{\beta_g}.$$

Then

$$D = \prod_{g | aAB} d_g.$$

If $e_{ij}(p)$ is the density of those integers which satisfy

$$an + b \equiv 0 \pmod{p^i}, \quad An + B \equiv 0 \pmod{p^j},$$

then $i \geq \alpha_p, j \geq \beta_p$ and

$$\frac{D\delta_{ij}(p)}{p_{ij}} = \frac{De_{ij}(p)}{d_p}.$$

In particular there is an h_0, depending only upon a, A, b and B, so that

$$0 < h_0 < \delta_{ij}(p) \leq 1$$

holds uniformly for all permissible i, j and p.

Since the main terms in the estimates for the V_i cancel we have

$$(Dx)^{-1} Y_2 \ll x^{-1/2} \left(\sum_{q \leq x^{1/4}}' \frac{|r(q)|^2}{q} \left\{ \sum_{p | aA\Delta} \sum_{p_{ij} \leq x^{1/4}} \frac{|L(p^i, p^j, G)|^2}{p_{ij}} \right\} \right)^{1/2}.$$

Apart from the factor $x^{-1/2}$ this error term is of the form $(uv)^{1/2}$, which does not exceed $u + v$. It is then easy to check that

$$(Dx)^{-1} Y_2 \ll T(x).$$

We come now to Y_3. Using g to denote a prime and temporarily writing G for $G(x)$ we have

$$Y_3 = \sum_{p | aA\Delta} \sum_{p_{ij} \leq x^{1/4}} \sum_{g | aA\Delta} \sum_{g_{uv} \leq x^{1/4}} L(p^i, p^j, G) \bar{L}(g^u, g^v, G) \sum_{n_0 < n \leq x}^{\dagger} 1,$$

where the integers n in the innersum are restricted by the requirements

$$p^i \| (an + b), \quad p^j \| (An + B),$$

$$g^u \| (an + b), \quad g^v \| (An + B).$$

Those terms with $g = p$, so that $u = i$ and $v = j$, contribute

$$\{(1 + O(x^{-1/2})\} Dx \sum_{p | aA\Delta} \sum_{p_{ij} \leq x^{1/4}} \frac{|L(p^i, p^j, G)|^2}{p_{ij}} \delta_{ij}(p).$$

If $g \neq p$ we can again employ the Chinese Remainder Theorem, to get

The Basic Inequality

a contribution of

$$Dx \sum_{p \mid aA\Delta} \sum_{p_{ij} \leq x^{1/4}} \sum_{g \mid aA\Delta} \sum_{g_{uv} \leq x^{1/4}} \frac{L(p^i, p^j, G)}{p_{ij}} \delta_{ij}(p) \cdot \frac{\bar{L}(g^u, g^v, G)}{g_{uv}} \delta_{uv}(g)$$
$$p \neq g$$

$$+ O\left(x^{1/2} \sum_{p \mid aA\Delta} \sum_{p_{ij} \leq x^{1/4}} \frac{|L(p^i, p^j, G)|^2}{p_{ij}} \right).$$

Hence

$$(Dx)^{-1} Y_3 = \{1 + O(x^{-1/2})\} \sum_{p \mid aA\Delta} \sum_{p_{ij} \leq x^{1/4}} \frac{|L(p^i, p^j, G)|^2}{p_{ij}} \delta_{ij}(p)$$

$$+ |\Lambda|^2 - \sum_{p \mid aA\Delta} \left| \sum_{\substack{i,j \\ p_{ij} \leq x^{1/4}}} \frac{L(p^i, p^j, G)}{p_{ij}} \delta_{ij}(p) \right|^2,$$

with

$$\Lambda = \sum_{p \mid aA\Delta} \sum_{p_{ij} \leq x^{1/4}} \frac{L(p^i, p^j, G)}{p_{ij}} \delta_{ij}(p).$$

Let

$$\rho_1 = \sum_{q \leq x^{1/4}}' \frac{|r(q)|^2}{q} \left(1 - \frac{1}{q_0}\right),$$

$$\rho_2 = \sum_{p \mid aA\Delta} \sum_{p_{ij} \leq x^{1/4}} \frac{|L(p^i, p^j, G)|^2}{p_{ij}} \delta_{ij}(p).$$

Then the estimates for the various sums Y_j together with lemma (10.3) show

(12)
$$(Dx)^{-1} Z_1 = 2\rho_1 + \rho_2 + |\Lambda|^2 - \sum_{p \mid aA\Delta} \left| \sum_{\substack{i,j \\ p_{ij} \leq x^{1/4}}} \frac{L(p^i, p^j, G)}{p_{ij}} \delta_{ij}(p) \right|^2 + O(T(x)).$$

It is useful to note here that for each fixed p dividing $aA\Delta$

$$\sum_{i,j} \frac{\delta_{ij}(p)}{p_{ij}} = 1.$$

In fact for all $P \geq 2$

$$D \cdot \sum_{p_{ij} \leq P} \frac{\delta_{ij}(p)}{p_{ij}} = \text{(density of those integers } n \text{ for which } an+b \equiv 0 \pmod{p^i}$$
$$\text{and } An+B \equiv 0 \pmod{p^j} \text{ for some pair } i,j \text{ with } i \geq \alpha_p,$$
$$j \geq \beta_p \text{ and } p_{ij} \leq P)$$

$$= \text{(density of those integers } n \text{ for which } an+b \equiv 0 \pmod{\alpha}$$
$$\text{and } An+B \equiv 0 \pmod{\beta})) + O(P^{-1})$$

$$= D + O(P^{-1}).$$

Letting $P \to \infty$ we justify our assertion.

Since (α, β) is a fixed-divisor pair there is a positive c_0 so that

$$\frac{\delta_{i_p j_p}(p)}{p_{i_p j_p}} > c_0$$

holds uniformly for the prime divisors of $aA\Delta$. In particular, if $\hat{\ }$ denotes that the pair (i_p, j_p) is to be deleted from the summation, then

$$\hat{\sum_{i,j}} \frac{\delta_{ij}(p)}{p_{ij}} < 1 - c_0.$$

Noting that for $i = i_p, j = j_p$ we have $L(p^i, p^j, G) = 0$, an application of the Cauchy–Schwarz inequality shows that

(13)
$$\sum_{p \mid aA\Delta} \left| \sum_{\substack{i,j \\ p_{ij} \leq x^{1/4}}} \frac{L(p^i, p^j, G)}{p_{ij}} \delta_{ij}(p) \right|^2$$

$$\leq \sum_{p \mid aA\Delta} \sum_{\substack{i,j \\ p_{ij} \leq x^{1/4}}} \frac{|L(p^i, p^j, G)|^2}{p_{ij}} \delta_{ij}(p) \left\{ \hat{\sum_{i,j}} \frac{\delta_{ij}(p)}{p_{ij}} \right\}$$

$$\leq (1 - c_0)\rho_2.$$

The estimates (12) and (13) together constitute a lower bound for Z_1 which is sufficient for our needs.

The sum Z_2. This is much simpler to treat than Z_1. Apart from the factor $\bar{L}_t(G) + \bar{J}$ we have to consider the sum(s)

$$\sum_{n_0 < n \leq x}^{\dagger} \sigma_1(n) + \sum_{n_0 < n \leq x}^{\dagger} \sigma_2(n).$$

The first of these is readily seen to be

$$\ll \sum_{q \leq x^{1/4}}' \frac{|r(q)|}{q} \ll \left\{ \sum_{q \leq x^{1/4}} \frac{|r(q)|^2}{q} \log \log x \right\}^{1/2} \ll x^{1/2} T(x)^{1/2}.$$

The second of them has the estimate

$$Dx \sum_{p \mid aA\Delta} \sum_{p_{ij} \leq x^{1/4}} \frac{L(p^i, p^j, G)}{p_{ij}} \delta_{ij}(p) + O(x^{1/2} \rho_2^{1/2}).$$

Altogether (we apply lemma (9.12))

$$(Dx)^{-1} Z_2 = \text{Re}\{\bar{L}_t(G) + \bar{J}\} \Lambda + O(T(x)).$$

The sum Z_3. This is clearly

$$|L_t(G) + J|^2 \{Dx + O(1)\},$$

so that

$$(Dx)^{-1} Z_3 = |L_t(G) + J|^2 + O(T(x)).$$

Collecting together our estimates for the Z_j, $j = 1, 2, 3$, we see from (9) and (10) that

$$2\rho_1 + \rho_2 - \sum_{p \mid aA\Delta} \left| \sum_{\substack{i,j \\ p_{ij} \leq x^{1/4}}} \frac{L(p^i, p^j, G)}{p_{ij}} \delta_{ij}(p) \right|^2 + |\Lambda + L_t(G) + J|^2 \ll T(x),$$

and in view of (13)

$$\rho_1 + \rho_2 + |\Lambda + L_t(G) + J|^2 \ll T(x).$$

In particular the upper bound for ρ_1 given here shows that

$$\sum_{q \leq x^{1/4}}' \frac{1}{q} |f(q) - G(x) \log q|^2 \ll T(x).$$

Replacing x by x^4 we see that for a large enough c_1 (in the definition of $T(x)$)

$$\sum_{q \leq x}' \frac{1}{q} |f(q) - G(x^4) \log q|^2 \ll T(x).$$

The first estimate of the present lemma (10.5) now follows from the first inequality of lemma (10.3) with $\lambda = G(x^4)$. Note that with this choice of λ the $H(\lambda)$ of lemma (10.3) is $\ll T(x)$.

The second inequality of lemma (10.5) is likewise obtained, and the third follows from the fact that by the Cauchy–Schwarz inequality $|\Lambda|^2 \ll \rho_2$. The proof of lemma (10.5) is complete.

It is convenient to introduce some further notation. Once again $K(\geq \frac{1}{4})$ will be assumed given. We write σ for $(25 - \sqrt{97})/24$ and

$$M(y) = \frac{1}{(\log y)^K} \sum_{q \leq y^{c_1}}' \frac{|f(q)|^2}{q}$$

$$+ \sup_{y^{1/c_1} \leq w \leq y^{c_1}} \frac{1}{w^{1-\sigma}} \sum_{n \leq w} \frac{|f(an+b) - f(An+B)|^2}{n^\sigma},$$

$$S(y, U) = \sum_{q \leq y}' \frac{|f(q) - U \log q|^2}{q} + \sum_{p | a A \Delta} \sum_{p_{ij} \leq y} \frac{|L(p^i, p^j, U)|^2}{p_{ij}}$$

$$+ \sum_{t=1}^{2} |L_t(U) + U \log a/A|^2.$$

Lemma (10.6). *The inequality*

$$S(x, F(x)) \ll \frac{1}{(\log x)^{1/4}} S(x^{c_1}, F(x^{c_1})) + M(x)$$

holds for all $x \geq 2$.

Proof. We apply lemma (10.5) with $U = F(x^{c_1})$. Since

$$F(x^{c_1})^2 \ll \frac{1}{(\log x)^2} \sum_{q \leq x^{c_1}}' \frac{|f(q)|^2}{q},$$

the desired inequality is readily obtained.

We are now in a position to given an argument by induction, which goes up rather than down, as has been the case so far in this volume. In fact with $\theta = c_1 > 1$ the inequality of lemma (10.6) may be applied (iteratively) k times, with $k \geq 1$, to give

(14) $$S(x, F(x)) \ll \frac{1}{(\log x)^{k/4}} S(x^{\theta^k}, F(x^{\theta^k})) + \sum_{j=0}^{k-1} \frac{M(x^{\theta^j})}{(\log x)^{j/4}}.$$

From this we shall readily deduce

The Basic Inequality

Lemma (10.7). *Let $K \geq \frac{1}{4}$ be given. Then there is a constant $d > 1$ so that*

$$S(x, F(x)) \ll (\log x)^{-K} S(x^d, 0)$$
$$+ \sup_{x^{1/d} \leq w \leq x^d} \frac{1}{w} \sum_{x^{1/d} < n \leq w} |f(an+b) - f(An+B)|^2$$

holds for all $x \geq 2$.

Proof. It will be enough to establish this inequality for all values of x which are sufficiently large in terms of K.

If in (14) we fix k at a value which exceeds $4K$, and if $d \geq c_1^k$ then we obtain

$$S(x, F(x)) \ll (\log x)^{-K} S(x^d, 0)$$
$$+ \sup_{x^{c_1} \leq w \leq x^d} \frac{1}{w^{1-\sigma}} \sum_{n \leq w} \frac{|f(an+b) - f(An+B)|^2}{n^\sigma}.$$

From the inequalities which immediately precede the estimate (2) we see that for $x^{c_1} \leq w \leq x^d$, $d \geq c_1 \geq 2$,

$$\frac{1}{w^{1-\sigma}} \sum_{n \leq x^{1/d}} \frac{|f(an+b) - f(An+B)|^2}{n^\sigma}$$

$$\leq \frac{1}{w^{1-\sigma}} \sum_{n \leq w^{1/2}} \frac{|f(an+b) - f(An+B)|^2}{n^\sigma}$$

$$\ll w^{-(1-\sigma)/3} S(w_2, 0) \ll (\log x)^{-K} S(x^d, 0),$$

where $w_2 = \max(aw^{1/2} + |b|, Aw^{1/2} + |B|)$.

Moreover, an integration by parts shows that over this same range for w

$$\frac{1}{w^{1-\sigma}} \sum_{x^{1/d} < n \leq w} \frac{|f(an+b) - f(An+B)|^2}{n^\sigma}$$

$$\ll \sup_{x^{1/d} < y \leq x^d} \frac{1}{y} \sum_{x^{1/d} < n \leq y} |f(an+b) - f(An+B)|^2.$$

This completes the proof.

Lemma (10.7) represents as far as we wish to go on our loop mechanism, and we now step off it.

In principle we could apply the inequality (14) with a value of k depending upon x. This would require an estimation of the dependence of the implied constant upon k.

It is convenient to include here

Lemma (10.8). *The inequality*

$$\sum_{n \leq x} |f(an+b) - f(An+B)|^2 \ll xS(x_1, 0)$$

with $x_1 = \max(ax + |b|, Ax + |B|)$ holds uniformly for all additive functions $f(\ )$, for all real $x \geq 2$.

Proof (Compare with the derivation of (2).) By means of the Cauchy–Schwarz inequality applied to the representation (1) it will suffice to obtain an upper bound of the above type for each of the sums

$$\sum_{n \leq x} \left| \sum_{\substack{q \| (an+b) \\ (q, aA\Delta) = 1}} f(q) - \sum_{\substack{q \| (An+B) \\ (q, aA\Delta) = 1}} f(q) \right|^2$$

$$\sum_{n \leq x} \left| \sum_{p | aA\Delta} L(p^i, p^j) \right|^2$$

and

$$\sum_{n \leq x} |L_t|^2.$$

The third of these sums is clearly $\ll xS(x, 0)$.

The second sum, after another application of the Cauchy–Schwarz inequality, is

$$\ll \sum_{n \leq x} \sum_{p | aA\Delta} |L(p^i, p^j)|^2$$

$$\ll \sum_{p | aA\Delta} \sum_{p_{ij} \leq x_1} |L(p^i, p^j)|^2 \sum_{\substack{n \leq x \\ an+b \equiv 0 \pmod{p^i} \\ An+B \equiv 0 \pmod{p^j}}} 1 \ll xS(x_1, 0).$$

To estimate the first of the sums we interpose the function

$$E(x_1) = \sum_{q \leq x_1}{}' \frac{f(q)}{q}\left(1 - \frac{1}{q_0}\right)$$

into the summand and apply the Cauchy–Schwarz inequality and then the

The Basic Inequality

Turán–Kubilius inequality in the manner used to treat the function $s_2(m)$ at (7):

$$\sum_{n\leq x}\left|\sum_{\substack{q\|(an+b)\\(q,aA\Delta)=1}}f(q)-\sum_{\substack{q\|(An+B)\\(q,aA\Delta)=1}}f(q)\right|^2$$

$$\leq 2\sum_{n\leq x}\left|{\sum_{q\|(an+b)}}'f(q)-E(x_1)\right|^2+2\sum_{n\leq x}\left|{\sum_{q\|(An+B)}}'f(q)-E(x_1)\right|^2$$

$$\ll {\sum_{q\leq x_1}}'\frac{|f(q)|^2}{q}\ll xS(x_1,0).$$

This completes the proof of lemma (10.8).

Lemma (10.9). *Let $K>0$ be given. There is a constant $d>1$ so that*

$$S(x,F(x))\ll\frac{|F(x)|^2}{(\log x)^K}+\sup_{x^{1/d}\leq w\leq x^d}\frac{1}{W}\sum_{x^{1/d}<n\leq w}|f(an+b)-f(An+B)|^2$$

for all large x. Moreover, if $0<\beta\leq 1$ then

$$|F(x)|^2\ll |F(x^\beta)|^2+\frac{1}{(\log x)^2}\sup_{x^{\beta/d}\leq w\leq x^d}\frac{1}{W}\sum_{x^{\beta/d}<n\leq w}|f(an+b)-f(An+B)|^2$$

for large x.

Remark. Here x needs to be large enough only in terms of a, A, b, B and K. The inequalities are uniform in all additive functions $f(\ )$, and in β, $\beta_0\leq\beta\leq 1$ for any fixed $\beta_0>0$.

For each $c\geq 1$ define

$$N(c,x)=\sup_{x^{1/c}\leq w\leq x^c}\frac{1}{W}\sum_{x^{1/c}<n\leq w}|f(an+b)-f(An+B)|^2.$$

Proof of lemma (10.9). By theorem (3.1) there are certain positive constants c_0, c_1 and c_2 so that

$$S(y,0)\ll(\log y)^{c_1}\left\{S(c_2,0)+\sum_{n\leq y^{c_0}}n^{-1}|f(an+b)-f(An+B)|^2\right\}$$

holds for all $y\geq 2$.

From lemma (10.8), with an integration by parts, we obtain the bound

$$\sum_{n \leq x^{1/2d}} n^{-1} |f(an+b) - f(An+B)|^2 \ll S(x_2, 0) \log x,$$

where $x_2 = \max(ax^{1/2d} + |b|, Ax^{1/2d} + |B|)$.

Moreover, another integration by parts shows that

$$\sum_{x^{1/2d} < n \leq x^{c_0}} n^{-1} |f(an+b) - f(An+B)|^2 \ll N(d_1, x) \log x$$

for a certain positive constant d_1.

If K is large enough and we apply lemma (10.7) with K there replaced by $2K$, then these results show that for a suitably large constant d

$$S(x, F(x)) \ll (\log x)^{-2K+c_1+1} S(x, 0) + N(d, x).$$

Since

$$S(x, 0) \ll S(x, F(x)) + |F(x)|^2 (\log x)^2,$$

we have

$$S(x, F(x)) \ll (\log x)^{-2K+c_1+1} S(x, F(x)) + (\log x)^{-2K+c_1+3} |F(x)|^2 + N(d, x).$$

With $K \geq c_1 + 3$ this gives the first asserted inequality of lemma (10.9). As for the second, we have (using x^β in place of x)

$$\sum_{q \leq x}' \frac{|f(q) - F(x) \log q|^2}{q} \ll \frac{|F(x)|^2}{(\log x)^K} + N(d, x),$$

$$\sum_{q \leq x^\beta}' \frac{|f(q) - F(x^\beta) \log q|^2}{q} \ll \frac{|F(x^\beta)|^2}{(\log x)^K} + N(d, x^\beta).$$

Between these inequalities we can eliminate the $f(q)$, $x^{\beta/2} < q \leq x^\beta$:

$$\sum_{x^{\beta/2} < q \leq x^\beta}' \frac{|F(x) - F(x^\beta)|^2 (\log q)^2}{q}$$

$$\leq 2 \sum_{x^{\beta/2} < q \leq x^\beta}' \frac{|F(x) \log q - f(q)|^2}{q} + 2 \sum_{x^{\beta/2} < q \leq x^\beta}' \frac{|f(q) - F(x^\beta) \log q|^2}{q}$$

$$\ll \frac{|F(x)|^2 + |F(x^\beta)|^2}{(\log x)^K} + N(d, x) + N(d, x^\beta).$$

The Basic Inequality

Since

$$\sum_{x^{\beta/2}<q\leq x^{\beta}}' \frac{(\log q)^2}{q} \sim \frac{3\beta^2}{8}(\log x)^2$$

as $x \to \infty$, we readily deduce that

$$|F(x)|^2 \ll |F(x^{\beta})|^2 + \frac{1}{(\log x)^2}\{N(d, x) + N(d, x^{\beta})\},$$

which gives the desired result.
Lemma (10.9) is proved.

We can now establish our main results. The proofs differ slightly, depending upon whether or not $a = A$.

Proof of theorems (10.1) *and* (10.2) *when* $a \neq A$. Define the additive function

$$\psi(m) = f(m) - F(x)\log m.$$

Let $an + b$ and $An + B$ be positive for all $n > n_0$. For each such n we apply the Cauchy–Schwarz inequality in the form

$$\left|F(x)\log\left(\frac{an+b}{An+B}\right)\right|^2 = |F(x)\log(an+b) - F(x)\log(An+B)|^2$$

$$\leq 2|\psi(an+b) - \psi(An+B)|^2 + 2|f(an+b) - f(An+B)|^2.$$

Let y be chosen so that $\max(ay+|b|, Ay+|B|) = x$. Then from lemma (10.8) applied to the function $\psi(n)$

$$x^{-1} \sum_{x^{1/2}<n\leq y} |\psi(an+b) - \psi(An+B)|^2 \ll S(x, F(x)),$$

an upper bound which by lemma (10.9) is

$$\ll \frac{|F(x)|^2}{(\log x)^K} + N(d, x).$$

Moreover, if $d \geq 2$ then by definition

$$x^{-1} \sum_{x^{1/2}<n\leq y} |f(an+b) - f(An+B)|^2 \ll N(d, x).$$

Hence

$$(15) \quad |F(x)|^2 x^{-1} \sum_{x^{1/2} < n \le y} \left(\log\left(\frac{an+b}{An+B}\right)\right)^2 \ll \frac{|F(x)|^2}{(\log x)^K} + N(d, x).$$

For the first time we make essential use of the condition $a \ne A$. It allows us to assert that for all large n

$$\left|\log\left(\frac{an+b}{An+B}\right)\right| \ge \tfrac{1}{2}\left|\log\frac{a}{A}\right| > 0,$$

so that (15) gives

$$|F(x)|^2 \ll \frac{|F(x)|^2}{(\log x)^K} + N(d, x).$$

In particular

$$|F(x)|^2 \ll N(d, x).$$

This last bound, used in conjunction with the first inequality of lemma (10.9), gives

$$S(x, F(x)) \ll N(d, x).$$

We replace x by x^{2d}:

$$(16) \quad S(x^{2d}, F(x^{2d})) \ll N(d, x^{2d}),$$

and omit from the sums implicit in the left-hand expression those terms corresponding to $q > x$, to reach

$$(17) \quad S(x, F(x^{2d})) \ll N(d, x^{2d}).$$

For example, it follows from (16) that

$$\sum_{q \le x^{2d}} \frac{|f(q) - F(x^{2d}) \log q|^2}{q} \ll N(d, x^{2d}).$$

Omitting the terms with $q > x$ gives

$$\sum_{q \le x} \frac{|f(q) - F(x^{2d}) \log q|^2}{q} \ll N(d, x^{2d}),$$

which is one of the bounds implicit in (17).

Note that
$$|F(x^{2d})|^2 \ll N(d, x^{2d})$$
also holds.

It follows from lemma (10.3) that

(18) $$S(x, F(x)) + |F(x)|^2 \ll S(x, \lambda) + |\lambda|^2$$

holds for all complex λ. With $\lambda = F(x^{2d})$ we obtain

$$S(x, F(x)) + |F(x)|^2 \ll N(d, x^{2d}),$$

and since

$$N(d, x^{2d}) = \sup_{x^2 < w \le x^{2d^2}} \frac{1}{W} \sum_{x^2 < n \le w} |f(an+b) - f(An+b)|^2$$

we shall have established those parts of theorems (10.1) and (10.2) which deal with the cases $a \ne A$ provided that we fix c at a value as large as $2d^2$.

Proof of theorems (10.1) *and* (10.2) *when* $a = A$. Consider the first inequality which appears in the statement of lemma (10.9). If we replace the values $f(q)$ everywhere by $f(q) - F(x) \log q$ then $F(x)$ becomes zero, $S(x, F(x))$ remains the same, and $N(d, x)$ becomes

(19) $$\sup_{x^{1/d} \le w \le x^d} \frac{1}{W} \sum_{x^{1/d} < n \le w} \left| f(an+b) - f(An+B) - F(x) \log\left(\frac{an+b}{An+B}\right) \right|^2.$$

In our present circumstances $a = A$, so that

$$\log\left(\frac{an+b}{An+B}\right) \ll \frac{1}{n}$$

and

$$\sum_{x^{1/d} < n \le w} \left| F(x) \log\left(\frac{an+b}{An+B}\right) \right|^2 \ll x^{-1/d} |F(x)|^2.$$

The expression (19) is now easily seen to be

$$\ll N(d, x) + x^{-2/d} |F(x)|^2.$$

In turn, we apply the second inequality of lemma (10.9) with $\beta = 1/d$:

$$|F(x)|^2 \ll F(x^{1/d})|^2 + N(d^2, x).$$

Altogether this gives

$$S(x, F(x)) \ll N(d^2, x) + x^{-2/d}|F(x^{1/d})|^2.$$

We replace x by x^d

$$S(x^d, F(x^d)) \ll N(d^2, x^d) + x^{-2}|F(x)|^2,$$

truncate the sums implicit in the left-hand expression to run over only those prime-powers not exceeding x, and apply lemma (10.3) with $\lambda = F(x^d)$, to obtain

(20) $$S(x, F(x)) \ll N(d^3, x) + x^{-2}|F(x)|^2.$$

In certain respects this is stronger than the corresponding inequality of lemma (10.9). At the expense of replacing the constant d which appears there by d^3, we have replaced the factor $(\log x)^{-K}$ by the much stronger x^{-2}. Note that if $a \neq A$, then an argument of this type leads to a weaker result.

We can now follow the proof used in the case $a \neq A$, but applying (20) in place of the first inequality of lemma (10.9), until we reach the following analogue of (15):

$$|F(x)|^2 x^{-1} \sum_{x^{1/2} < n \leq y} \left(\log\left(\frac{an+b}{An+B}\right)\right)^2 \ll x^{-2}|F(x)| + N(d^3, x).$$

Since $a = A$,

$$\log\left(\frac{an+b}{An+B}\right) = -\frac{\Delta}{aAn} + O\left(\frac{1}{n^2}\right),$$

so that for all large x (and so y)

$$x^{1/2} \sum_{x^{1/2} < n \leq y} \left(\log\left(\frac{an+b}{An+B}\right)\right)^2 > c_3 > 0$$

for some constant c_3. This gives us

$$|F(x)|^2 \ll x^{3/2} N(d^3, x).$$

Applying this bound together with that of (20) we obtain

$$S(x, F(x)) \ll N(d^3, x).$$

Employing a now familiar argument, we replace x by x^{2d^3} and restrict our range of summation:

$$S(x, F(x^{2d^3})) \ll N(d^3, x^{2d^3}).$$

Replacing $F(x^{2d^3})$ by $F(x)$ is justified by lemma (10.3) and we reach

(21) $$S(x, F(x)) \ll N(d^3, x^{2d^3}).$$

Provided we ultimately choose $c \geq 2d^6$ this inequality gives everything remaining in theorems (10.1) and (10.2) save for the estimate of $|F(x)|^2$, in theorem (10.2), when $a = A$.

Let $x_1 = \max(ax+|b|, Ax+|B|)$, as before. Then we may deduce from (21) that

(22) $$S(x_1, F(x_1)) \ll N(d^3, x_1^{2d^3}) \ll N(2d^3, x^{2d^3}).$$

In particular

$$\sum_{q \leq x_1}' \frac{|f(q) - F(x_1) \log q|^2}{q} \ll N(2d^3, x^{2d^3}).$$

By eliminating the $f(q)$, for q in the interval $x^{1/2} < q \leq x$, between this inequality and the inequality

$$\sum_{q \leq x}' \frac{|f(q) - F(x) \log q|^2}{q} \ll N(2d^3, x^{2d^3}),$$

which is implied by (21), we obtain

$$(|F(x) - F(x_1)| \log x)^2 \ll N(2d^3, x^{2d^3}).$$

This result may be applied in conjunction with that of (22) to give

$$S(x_1, F(x)) \ll N(2d^3, x^{2d^3}).$$

For the third time we employ the argument at the beginning of the proof of the cases $a \neq A$. In the notation used there

$$x^{-1} \sum_{n_0 < n \leq x} |\psi(an+b) - \psi(An+B)|^2 \ll S(x_1, F(x)) \ll N(2d^3, x^{2d^3}),$$

this last step by what we have just shown.

Let

$$\Lambda_1 = x^{-1} \sum_{n_0 < n \leq x} |f(an+b) - f(An+B)|^2,$$

$$\Lambda_2 = x^{-1} |F(x)|^2 \sum_{n_0 < n \leq x} \left(\log\left(\frac{an+b}{An+B}\right)\right)^2.$$

By Minkowski's inequality

$$|\Lambda_2^{1/2} - \Lambda_1^{1/2}| \leq \left(x^{-1} \sum_{n_0 < n \leq x} |\psi(an+b) - \psi(An+B)|^2\right)^{1/2}$$

$$\ll N(2d^3, x^{2d^3})^{1/2},$$

and with $c = 4d^6$ the proof of theorem (10.2) is complete.

Our main results, theorem (10.1) and (10.2) are now proved.

The Decomposition of the Mean

In this section we round out the results of theorems (10.1) and (10.2) by studying the asymptotic behaviour of the function

$$E(x) = \sum_{\substack{p^k \leq x \\ (p, aA\Delta) = 1}} \frac{f(p^k)}{p^k}\left(1 - \frac{1}{p}\right).$$

Let m be a positive integer for which $am + b$, $Am + B$ are positive and distinct.

Define

$$\tau = \begin{cases} \dfrac{f(am+b) - f(Am+B)}{\log(am+b) - \log(Am+B)} & \text{if } a = A, \\ 0 & \text{otherwise.} \end{cases}$$

Theorem (10.10).

$$F(x) = \tau + O\left(\sup_{x \leq w \leq x^c} \left\{\frac{1}{w} \sum_{x < n \leq w} |f(an+b) - f(An+B)|^2\right\}^{1/2}\right)$$

for all sufficiently large x.

The Decomposition of the Mean 233

Proof. We need only consider the case $a = A$, since when $a \neq A$ our result is already guaranteed by theorem (10.2).

According to the representation (1) from the beginning of this chapter, for $t = 1$ or 2

$$f(am+b) - f(Am+B) - F(x)\{\log(am+b) - \log(Am+B)\}$$

$$= \sum_{\substack{q\|(am+b) \\ (q,aA\Delta)=1}} \{f(q) - F(x)\log q\} - \sum_{\substack{q\|(Am+B) \\ (q,aA\Delta)=1}} \{f(q) - F(x)\log q\}$$

$$+ \sum_{p|aA\Delta} L(p^i, p^j, F(x)) + L_t(F(x)).$$

Each of these sums, together with the last term, may be estimated by theorems (10.1) and (10.2) to lie within the error given in the statement of the present lemma. Dividing by $\{\log(am+b)/(Am+B)\}$ we obtain the asserted result, with an implicit constant which depends upon m.

We now recall the function

$$N(c, x) = \sup_{x^{1/c} \leq w \leq x^c} \frac{1}{W} \sum_{x^{1/c} < n \leq w} |f(an+b) - f(An+B)|^2,$$

which appeared after the statement of lemma (10.9).

Theorem (10.11). *Let $K > 0$ and $0 < d_1 < d_2 < \infty$ be given. Then there is a measurable function $F_2(x)$ and a constant c such that*

(23) $\qquad E(t) = F(x)\log t + F_2(x) + O(|\tau|(\log x)^{-K}) + O(N(c, x)^{1/2})$

holds uniformly for $x^{d_1} \leq t \leq x^{d_2}$. Moreover,

(24) $\qquad \left.\begin{array}{c} \sup_{x^{1/2} \leq z \leq x^2} |F(z) - F(x)| \log x \\ \\ \sup_{x^{1/2} \leq z \leq x^2} |F_2(z) - F_2(x)| \end{array}\right\} \ll |\tau|(\log x)^{-K} + N(2c, x)^{1/2}$

uniformly for all sufficiently large x.

Remarks. In theorem (10.10) the constant c is the same as that which appears in the corresponding theorems (10.1) and (10.2). In theorem (10.11) it may be larger, depending upon the values of d_1 and d_2.

The error term which involves τ may be improved by appealing to a sufficiently sharp version of the Prime Number Theorem.

Proof. As before, let

$$\psi(n) = f(n) - F(x) \log n,$$

and let q_0 denote the prime of which q is a power. Define

$$\tilde{E}(y) = \sum_{p^k \leq y}{}' \frac{\psi(p^k)}{p^k}\left(1 - \frac{1}{p}\right),$$

where $'$ indicates that the primes p do not divide $aA\Delta$.
Then uniformly for $x^{1/4} < q$ (a prime-power) $\leq x^{1/2}$

$$\tilde{E}(x) - \tilde{E}\left(\frac{x}{q}\right) = \sum_{xq^{-1} < p^k \leq x}{}' \frac{\psi(p^k)}{p^k}\left(1 - \frac{1}{p}\right) \ll N(c, x)^{1/2},$$

the last step by the Cauchy–Schwarz inequality and theorem (10.1). Hence

$$\sum_{x^{1/4} < q \leq x^{1/2}}{}' \frac{1}{q}\left|\psi(q)\left(1 - \frac{1}{q_0}\right) - \left\{\tilde{E}(x) - \tilde{E}\left(\frac{x}{q}\right)\right\}\right| \ll N(c, x)^{1/2}.$$

However, by the Prime Number Theorem

$$\sum_{x^{1/4} < q \leq x^{1/2}}{}' \frac{1}{q}\left|\log q\left(1 - \frac{1}{q_0}\right) - \sum_{xq^{-1} < p^k \leq x}{}' \frac{\log p^k}{p^k}\left(1 - \frac{1}{p}\right)\right|$$

$$\ll \sum_{x^{1/4} < q \leq x} \frac{1}{q} O((\log x)^{-K}) \ll (\log x)^{-K}.$$

Note that

$$\sum_{p^k > x^{1/2}} p^{-(k+1)} \ll \sum_{p > x^{1/2}} p^{-2} + x^{-1/4} \sum_{\substack{p^k > x^{1/2} \\ k \geq 2}} p^{-(k/2+1)} \ll x^{-1/4}.$$

Therefore, applying theorem (10.10),

$$\sum_{x^{1/4} < q \leq x^{1/2}}{}' \frac{1}{q}\left(1 - \frac{1}{q_0}\right)\left|f(q)\left(1 - \frac{1}{q_0}\right) - E(x) + E\left(\frac{x}{q}\right)\right| \ll |\tau|(\log x)^{-K}$$

$$+ N(c, x)^{1/2}.$$

Since

$$\sum_{q \leq w}{}' \frac{1}{q}\left(1 - \frac{1}{q_0}\right) = \log w + \text{constant} + O((\log w)^{-D}), \qquad w \geq 2,$$

for every fixed D, we may apply theorem (9.11); here the remark (see (6)) which follows the statement of theorem (9.1) is relevant. We obtain a representation

(25) $$E(t) = H(x) \log t - R(x) + O(\Delta(x))$$

uniformly for $x^{d_1} \leq t \leq x^{d_2}$, with

$$\Delta(x) = \sup_{x^{c_{12}} \leq w \leq x^{c_{13}}} (|\tau|(\log x)^{-K} + N(c, w)^{1/2})$$

$$+ (\log x)^{-K} \{ \max_{\substack{q \leq c_{14} \\ (q, aA\Delta) = 1}} |f(q)| + \sup_{w \leq x^{c_{12}}} (|\tau| (\log x)^{-K} + N(c, w)^{1/2}) \}.$$

For all large enough x

$$\max_{\substack{q \leq c_{14} \\ (q, aA\Delta) = 1}} |f(q)| \leq \sum'_{q \leq c_{14}} |f(q) - F(x) \log q| + |F(x)| \sum'_{q \leq c_{14}} \log q$$

$$\ll N(c, w)^{1/2} + |\tau|.$$

Moreover,

$$\sup_{x^{c_{12}} \leq w \leq x^{c_{13}}} N(c, w)^{1/2} \ll N(c', x)^{1/2},$$

provided that c' is fixed at a sufficiently large value in terms of c, c_{12}, c_{13}. Here we may safely assume that $c_{12} \leq 1/(2c)$.

Arguing as at the beginning of the proof of theorem (10.1) we have

$$\frac{1}{z} \sum_{n \leq z} |f(an + b) - f(An + B)|^2$$

$$\ll \frac{1}{z} \sum_{n \leq z} |\psi(an + b) - \psi(An + B)|^2 + \frac{1}{z} \sum_{n \leq z} \left| F(x) \log \left(\frac{an + b}{An + B} \right) \right|^2$$

$$\ll S(z, F(x)) + |F(x)|^2$$

$$\ll S(x, F(x)) + |F(x)|^2$$

$$\ll N(c, x) + |\tau|^2,$$

uniformly for $z \le x^{1/2}$. Therefore

$$\sup_{w \le 1/(2c)} N(c, w)^{1/2} \ll N(c, x)^{1/2} + |\tau|.$$

These remarks together show that

(26) $$\Delta(x) \ll |\tau|(\log x)^{-K} + N(c', x)^{1/2}$$

for a suitable $c' > 0$.

In order to relate the functions $H(x)$ and $F(x)$ to each other we apply the estimate (25) with $t = x$, $x^{1/2}$, and subtract, to obtain

$$E(x) - E(x^{1/2}) = \tfrac{1}{2} H(x) \log x + O(\Delta(x)).$$

Here we have assumed, as we may, that $d_1 \le \tfrac{1}{2}$.

However, by the definition of $F(x)$,

$$E(x) - E(x^{1/2}) = F(x) \sum_{x^{1/2} < q \le x}' \frac{\log q}{q} - \sum_{x^{1/2} < p^k \le x}' \frac{f(p^k)}{p^{k+1}}$$

$$= F(x)\{\tfrac{1}{2} \log x + O((\log x)^{-K})\}$$

$$- \sum_{x^{1/2} < p^k \le x}' \frac{(f(p^k) - F(x) \log p^k)}{p^{k+1}} - F(x) \sum_{x^{1/2} < p^k \le x}' \frac{\log p^k}{p^{k+1}}$$

$$= \frac{1}{2} F(x) \log x + O(|\tau|(\log x)^{-K} + N(c, x)^{1/2}),$$

where we have made use of the Prime Number Theorem and theorem (10.1). Hence

$$|H(x) - F(x)| \log x \ll |\tau|(\log x)^{-K} + N(c, x)^{1/2} + \Delta(x).$$

This result together with (25) and (26) gives the (first) desired asymptotic estimate of theorem (10.11) with $F_2(x) = -R(x)$, and c' in place of c.

Suppose now that $x^{1/2} \le z \le x^2$. Then if $d_2 > 4d_1$ by (23) with x, and x replaced by z, we have

$$E(t) = F(x) \log t + F_2(x) + O(R),$$

$$E(t) = F(z) \log t + F_2(z) + O(R),$$

where

$$R = |\tau|(\log x)^{-K} + N(2c, x)^{1/2},$$

and both estimates hold uniformly for $x^{2d_1} \le t \le x^{d_2/2}$.

By subtraction

(27) $$\{F(z)-F(x)\}\log t + \{F_2(z)-F_2(x)\} = O(R).$$

We set $t = t_1, t_2$ in turn, where

$$(\log t_1)/\log x = (d_2/2 - 2d_1)/3,$$

$$(\log t_2)/\log x = (d_2/2 - 2d_1)2/3.$$

If we write (27) in the form

$$M \log t + N = O(R)$$

then this argument will give, by elimination,

$$M \log t_2/t_1 = O(R), \qquad N \log t_2/t_1 = O(R \log t_1 t_2),$$

from which the remaining assertions of theorem (10.11) follow readily.

EXAMPLE. The following example explains the title of this section. Let $a = A = b = 1$, $B = 0$. Then

$$\sum_{n \leq x} \psi(n) = \sum_{p^k \leq x} \psi(p^k) \left\{ \left[\frac{x}{p^k}\right] - \left[\frac{x}{p^{k+1}}\right] \right\}$$

$$= x \sum_{p^k \leq x} \frac{\psi(p^k)}{p^k} \left(1 - \frac{1}{p}\right) + O\left(\sum_{p^k \leq x} |\psi(p^k)|\right).$$

Applying the Cauchy–Schwarz inequality and then theorem (10.1)

$$\sum_{p^k \leq x} |\psi(p^k)| p^{-k/2} p^{k/2} \ll \left(\sum_{q \leq x} \frac{1}{q} |f(q) - F(x) \log q|^2 \sum_{q \leq x} q\right)^{1/2}$$

$$\ll \{N(c, x) x^2 (\log x)^{-1}\}^{1/2}.$$

An integration by parts readily gives the estimate

$$\sum_{n \leq x} \log n = x \log x - x + O(\log x),$$

whilst by the Prime Number Theorem

$$\sum_{p^k \leq x} \frac{\log p^k}{p^k} \left(1 - \frac{1}{p}\right) = \log x + c_0 + O((\log x)^{-K})$$

for some constant c_0.

By subtraction, therefore.

$$\frac{1}{x}\sum_{n\leq x} f(n) = E(x) + (1-c_0)F(x) + O(|\tau|(\log x)^{-K}$$

$$+ \{N(c, x)(\log x)^{-1}\}^{1/2}).$$

Putting this together with the first assertion of theorem (10.11) gives

$$\frac{1}{x}\sum_{n\leq x} f(n) = F(x)\{\log x + 1 - c_0\} - F_2(x) + O(|\tau|\log x)^{-K} + N(c, x)^{1/2}),$$

which may be regarded as an asymptotic formula for the mean-value of $f(n)$.

Suppose, further, that

$$x^{-1} \sum_{n\leq x} |f(n+1) - f(n)|^2 \to 0$$

as $x \to \infty$. Then $N(c, x) \to 0$ also, and

$$\frac{1}{x}\sum_{n\leq x} f(n) = F(x)\{\log x + 1 - c_0\} - F_2(x) + o(1),$$

as $x \to \infty$.

Let us write

(28) $$\frac{1}{x}\sum_{n\leq x} f(n) = U(x) + V(x),$$

where

$$U(x) = F(x) \log x.$$

We shall prove that for each fixed positive value of y,

(29) $U(x^y) - yU(x) = o(1), \qquad V(x^y) - V(x) = o(1)$

as $x \to \infty$.

From part (24) of theorem (10.11)

$$|U(x^y) - yU(x)| = |F(x^y) - F(x)|y \log x = o(1)$$

uniformly for $\frac{1}{2} \leq y \leq 2$. It is now easy to extend this result to hold o. any finite interval of positive y-values which are bounded away from zero. For example, if $2 \le y \le 4$ then

$$U(x^y) = U(\{x^2\}^{y/2}) = (y/2)U(x^2) + o(1) = yU(x) + o(1)$$

having applied the above partial result twice.
Since

$$V(x) = (1 - c_0)F(x) - F_2(x) + o(1),$$

part (24) of theorem (10.11) will also yield the second asymptotic relation of (29).

We can now regard our mean-value at (28) as being decomposed into two pieces, the first of which "behaves like a logarithm", and the second of which is "in some sense small".

A more general case goes in much the same way.

Concluding Remarks

Suppose that $a \ne A$ and that the notation of the proof of theorems (10.1), (10.2) is in force. Then

$$w^{-1} \sum_{n \le w} |\psi(an+b) - \psi(An+B)|^2 \ll S(w_1, F(x)), \qquad w \ge 2,$$

with $w_1 = \max(aw + |b|, Aw + |B|)$. Since

$$w^{-1} \sum_{n \le w} \left| F(x) \log\left(\frac{an+b}{An+B}\right) \right|^2 \ll |F(x)|^2,$$

we see that for all sufficiently large values of x

$$\sup_{x \le w \le x^c} \frac{1}{w} \sum_{x < n \le w} |f(an+b) - f(An+B)|^2 \ll S(aAx^c, F(x)) + |F(x)|^2.$$

This shows that when $a \ne A$ theorems (10.1) and (10.2) are in some sense best possible. What is not clear is whether in these theorems one can take a value of c which is near to 1. To obtain a value which is independent of the parameters a, A, b, B is probably feasible.

Suppose now that $a = A$. It follows from theorems (10.1) and (10.2) that

$$x^{-1}|F(x)|^2 + S(x, F(x)) \ll \sup_{x \le w \le x^c} \frac{1}{w} \sum_{n \le w} |f(an+b) - f(An+B)|^2.$$

In this case

$$w^{-1} \sum_{n \leq w} \left| F(x) \log\left(\frac{an+b}{An+B}\right) \right|^2 \ll w^{-1} |F(x)|^2 \sum_{n \leq w} n^{-2},$$

so that the analogue of the above argument gives

$$\sup_{x \leq w \leq x^c} \frac{1}{w} \sum_{n \leq w} |f(an+b) - f(An+B)|^2 \ll S(aAx^c, F(x)) + x^{-1}|F(x)|^2.$$

We see that our results cannot be much improved in the case $a = A$ either. As before, there remains the possibility of reducing the size of c.

These results may be compared with certain others from the circle of interest of the Turán–Kubilius inequality. For a background discussion of this inequality and of the associated argument of Turán see Elliott [11] Chapter 4 and Chapter 15, respectively. The most general inequalities involving the Turán–Kubilius inequality are presently due to Ruzsa [3], and are obtained by applying the full force of Halász' treatment of Dirichlet series, together with results derived by sieve methods. Typically, if

$$E_0 = \sum_{p \leq x} \frac{f(p)}{p},$$

then one readily derives from the standard Turán–Kubilius inequality the bound

$$x^{-1} \sum_{n \leq x} |f(n) - E_0|^2 \ll \sum_{q \leq x} \frac{1}{q} |f(q) - \lambda \log q|^2 + |\lambda|^2$$

valid for all complex λ and real $x \geq 2$. Here, as elsewhere in this chapter, q denotes a typical prime-power. In his paper Ruzsa shows that if λ is given the value

$$\lambda_0 = \left(\sum_{q \leq x} q^{-1} f(q) \log q \right) \bigg/ \left(1 + \sum_{q \leq x} q^{-1} (\log q)^2 \right),$$

then an inequality of this type goes in the other direction:

(30) $$\sum_{q \leq x} \frac{1}{q} |f(q) - \lambda_0 \log q|^2 + |\lambda_0|^2 \ll x^{-1} \sum_{n \leq x} |f(n) - E_0|^2.$$

If, for example, $a \neq A$, then arguing as in the proof of lemma (10.3)

$$|\lambda_0 - F(x)|^2 (\log x)^2 \ll S(x, F(x)) + |F(x)|^2,$$

Concluding Remarks

and an easy modification of the inequality of theorem (10.1) yields

(31) $\quad \sum'_{q \le x} \dfrac{|f(q) - \lambda_0 \log q|^2}{q} + |\lambda_0|^2 \ll \sup_{x \le w \le x^c} \dfrac{1}{w} \sum_{n \le w} |f(an+b) - f(An+B)|^2.$

In view of the similarity of these results it is natural to ask how far towards an inequality of the type (30) we can apply the present method, and so avoid the use of arguments of the Halász type.

Indeed, beginning with a weak form of the Dual of the Turán–Kubilius inequality, say

$$\sum_{x^{1/4} < p \le x^{1/3}} p \left| \sum_{\substack{n \le x \\ p \| n}} a_n - \dfrac{1}{p}\left(1 - \dfrac{1}{p}\right) \sum_{n \le x} a_n \right|^2 \ll x \sum_{n \le x} |a_n|^2,$$

in place of the inequality (3) in the proof of theorem (8.1), we may follow the lines of the proof of theorem (8.3), but presently set

$$z_m = f(m) - E_0, \qquad z'_m = \log m - \sum_{p \le x} \dfrac{\log p}{p}.$$

We then reach the following analogue of the inequality (9) of Chapter 8:

(32) $\quad \sum_{x^{1/4} < p \le x^{1/3}} \dfrac{1}{p} \left| f(p) - \alpha(x) + \alpha\left(\dfrac{x}{p}\right) \right|^2 \ll x^{-1/3} \sum_{p \le x} \dfrac{|f(p) - U \log p|^2}{p}$

$\qquad + x^{-1/3} |U|^2 \log x + \sup_{x^{2/3} \le w \le x} \dfrac{1}{w} \sum_{n \le w} |f(n) - \alpha(w)|^2,$

where

$$\alpha(w) = \sum_{p \le w} \dfrac{f(p)}{p}.$$

Moreoever, this inequality is obtained without the intervention of any of the results, from Chapter 7, which concerned the distribution of additive functions in residue classes.

An a priori bound on the $f(p)$, needed to replace the bounds obtained in Chapter 3, may be obtained from (32) by setting $U = 0$. Then

$$\sum_{x^{1/4} < p \le x^{1/3}} \dfrac{1}{p} \left| f(p) - \alpha(x) + \alpha\left(\dfrac{x}{p}\right) \right|^2 \ll x^{-1/3} \sum_{p \le x} \dfrac{|f(p)|^2}{p} + \gamma(x),$$

where

$$\gamma(x) = \sup_{w \leq x} \frac{1}{w} \sum_{n \leq w} |f(n) - \alpha(w)|^2.$$

Several applications of the Cauchy–Schwarz inequality give

$$\left| \alpha\left(\frac{x}{p}\right) - \alpha(x^{2/3}) \right|^2 \ll \sum_{p \leq x^{3/4}} \frac{|f(p)|^2}{p}$$

uniformly for $x^{1/4} < p \leq x^{1/3}$, and

$$|\alpha(x)|^2 \sum_{x^{1/4} < p \leq x^{1/3}} \frac{1}{p} \ll |\alpha(x^{2/3})|^2 + \sum_{p \leq x^{3/4}} \frac{|f(p)|^2}{p} + x^{-1/3} \sum_{p \leq x} \frac{|f(p)|^2}{p} + \gamma(x).$$

Here

$$\sum_{x^{1/4} < p \leq x^{1/3}} \frac{1}{p} = \log \tfrac{4}{3} + O\left(\frac{1}{\log x}\right) > \tfrac{1}{2} \log \tfrac{4}{3}$$

for all large x. In particular, therefore

$$|\alpha(x)|^2 \ll x^{-1/3} \sum_{p \leq x} \frac{|f(p)|^2}{p} + \log \log 2x \sum_{p \leq x^{3/4}} \frac{|f(p)|^2}{p} + \gamma(x).$$

By means of this last inequality, and the obvious inequalities

$$\sum_{x/2 < p \leq x} \frac{1}{p} |f(p) - \alpha(x)|^2 \ll x^{-1} \sum_{x/2 < n \leq x} |f(n) - \alpha(x)|^2 \ll \gamma(x),$$

we see that

$$\sum_{x/2 < p \leq x} \frac{|f(p)|^2}{p} \ll \frac{|\alpha(x)|^2}{\log x} + \gamma(x),$$

and that the sum on the left-hand side here does not exceed

$$\frac{c_0}{x^{1/3} \log x} \sum_{p \leq x} \frac{|f(p)|^2}{p} + c_0 \sum_{p \leq x^{3/4}} \frac{|f(p)|^2}{p} + c_0 \gamma(x)$$

for some constant c_0.

For all large values of x

$$\sum_{p \leq x} \frac{|f(p)|^2}{p} \left(1 - \frac{c_0}{x^{1/3} \log x}\right) \leq \sum_{p \leq x/2} \frac{|f(p)|^2}{p} + c_0 \sum_{p \leq x^{3/4}} \frac{|f(p)|^2}{p} + c_0 \gamma(x),$$

Concluding Remarks

and an argument by induction now shows that a bound

$$\sum_{p \leq z} \frac{|f(p)|^2}{p} \leq c_1 \gamma(z)(\log z)^d$$

is valid, with certain constants c_1 and d, for all $z \geq 2$. This bound serves as the analogue of that of theorem (3.1).

In this manner one obtains an inequality

(33) $$\sum_{q \leq x} \frac{1}{q}|f(q) - \lambda_0 \log q|^2 + |\lambda_0|^2 \ll \sup_{x \leq w \leq x^c} \frac{1}{w} \sum_{n \leq w} |f(n) - \alpha(w)|^2,$$

with a value of c not necessarily the same as that in (31). Perhaps this constant c may be brought down to 2 or even to any fixed value $c > 1$. It seems unlikely that it will get right down to $c = 1$ without a refinement in the treatment of the approximate functional equation.

To compensate for a certain loss of precision in the inequality (33) compared to that of (30), the method of the present monograph is flexible enough to approach problems which at present are beyond the reach of the more traditional methods, including that of Halász in the theory of Dirichlet series.

Chapter 11

Some Historical Remarks

In the next three chapters I solve completely a problem of Kátai; and improve and generalize a theorem of Wirsing. These problems involve the characterization of additive arithmetic functions in terms of their differences.

The first such characterization is apparently that of Erdös [2]. He proved that if a real-valued additive function is non-decreasing, or satisfies

(1) $$f(n+1)-f(n) \to 0$$

as $n \to \infty$, then it must have the form $A \log n$ for some constant A. He had a separate argument for each case.

Let n be an odd positive integer. For a non-decreasing additive function the integers m in the interval $n \leq m \leq 2n$ satisfy

$$f(n) \leq f(m) \leq f(2n) = f(n) + f(2),$$

and therefore

$$|f(m) - f(n)| \leq f(2).$$

In Erdös' terminology the function was *finitely distributed*. His argument came at the end of a complicated paper in which he characterized such functions, and he was able to assert the existence of a constant A so that the series

$$\sum_{|f(p)-A\log p|>1} \frac{1}{p}, \quad \sum_{|f(p)-A\log p|\leq 1} \frac{(f(p) - A \log p)^2}{p},$$

taken over the rational prime numbers, converged. This gave him a good beginning for a proof. For background notions see the author's book [11]; for a ramification of this argument and result see Erdös and Ryavec [1].

Other proofs of this result were given by Sós [1], Csaszsár [1], Lambek and Moser [1] and Schönberg [1].

By adding to it a multiple of $\log n$, if necessary, we may assume that $f(n)$ is non-negative. Let a be a (temporarily fixed) positive integer, and n

a (large) integer which lies in the interval $a^k < n \le a^{k+1}$ for some integer k. Then

$$f(n) \ge f(a^k - a)$$
$$= f(a) + f(a^{k-1} - 1) \quad \text{since } (a, a^{k-1} - 1) = 1$$
$$\ge f(a) + f(a^{k-1} - a)$$
$$= 2f(a) + f(a^{k-2} - 1),$$

and so on by induction, until we reach

$$f(n) \ge (k-1)f(a), \quad \frac{f(n)}{\log n} \ge \frac{k-1}{k+1} \cdot \frac{f(a)}{\log a}.$$

Similarly

$$f(n) \le f(a^{k+1} + a) = f(a) + f(a^k + 1) \le f(a) + f(a^k + a).$$

Altogether

$$\left| \frac{f(n)}{\log n} - \frac{f(a)}{\log a} \right| \le \frac{2f(a)}{(k+1)\log a} + \frac{f(a+1)}{k \log a}.$$

Letting n (and so k) $\to \infty$ we see that

$$A = \lim_{n \to \infty} \frac{f(n)}{\log n}$$

exists and is finite, and that for every positive integer a, $f(a) = A \log a$.

When (1) was assumed Erdös gave an independent proof. Rényi [1] and Besicovitch [1] gave alternative proofs.

Besides many theorems Erdös' paper contained several interesting questions. In particular he conjectured that:

The above conclusion remains valid if the hypothesis (1) is replaced by

(2) $$\lim_{x \to \infty} x^{-1} \sum_{n \le x} |f(n+1) - f(n)| = 0,$$

and,
  If for some constant c

$$f(n+1) - f(n) \ge c$$

holds for all positive n, then there is a further constant A so that the additive function

$$f(n) - A \log n$$

is uniformly bounded.

The first of these two conjectures was established by Kátai [5], and Wirsing [2]. In fact Wirsing needed only that

$$\liminf_{x \to \infty} x^{-1} \sum_{x < n \leq \gamma x} |f(n+1) - f(n)| = 0$$

should hold for some constant $\gamma > 1$. The following modification of Wirsing's argument gives the general flavour of his proof.

Let

$$S(x) = \sum_{n \leq x} f(n).$$

Since $f(1) = 0$

$$f(n) = \sum_{r=1}^{n-1} \{f(r+1) - f(r)\} = o(n)$$

as $n \to \infty$.

Let q be a positive integer. For each integer n let $t = t(n)$ be a further integer, $0 \leq t \leq q-1$, which satisfies $(n+t, q) = 1$. Then

$$f(qn) - f(q) - f(n) = f(qn) - f(q\{n+t\}) + f(q\{n+t\}) - f(q) - f(n+t)$$
$$+ f(n+t) - f(n),$$

so that

$$\sum_{m \leq y} |f(qm) - f(q) - f(m)| \leq q \sum_{2 \leq r \leq q(y+q)} |f(r) - f(r-1)| = o(y)$$

as $y \to \infty$. With $y = x/q$ this allows us to assert that

$$\sum_{\substack{n \leq x \\ n \equiv 0 \pmod{q}}} f(n) = xf(q)/q + S(x/q) + o(x).$$

Further applications of the hypothesis (2) now give

$$\sum_{\substack{n \leq x \\ n \equiv r \pmod{q}}} f(n) = xf(q)/q + S(x/q) + o(x)$$

for each fixed value of $r \geq 0$. Setting $r = 0, \ldots, q-1$ in turn, and adding, we obtain

$$S(x) = xf(q) + qS(x/q) + o(x)$$

as $x \to \infty$. This asymptotic relation lends itself to an argument by induction, giving

$$\frac{f(q)}{\log q} = \lim_{x \to \infty} \frac{S(x)}{x \log x}.$$

This establishes Erdös' first conjecture, and gives a proof of his earlier result which assumed (1). Notice that in all of these arguments the form $f(n+1) - f(n)$ of the differences was important. If this were replaced by something more general, $f(3n+1) - f(5n+2)$ for example, then the arguments would no longer work.

Erdös' second conjecture lay deeper. It was established by Wirsing [1] with an argument by contradiction which employed, amongst other things, the construction of a finite probability space. Once again only the form $f(n+1) - f(n)$ was considered.

The scope of these questions was widened, in different directions, by Kátai, and by Wirsing.

Kátai [3], [4] asked for a characterization of those additive functions which satisfied

(3) $\qquad f(an+b) - f(An+B) \to C \quad \text{as } n \to \infty,$

for some integers $a > 0$, b, $A > 0$, B, and constant C.

Moreover, if additive functions $f_i(\ )$, $i = 1, \ldots, k$, satisfy

(4) $\qquad f_1(n+1) + f_2(n+2) + \cdots + f_k(n+k) \to 0 \quad \text{as } n \to \infty,$

must there be constants D_i so that the functions $f_i(n) - D_i \log n$ have finite support; that is to say, vanish except possibly on the powers of finitely many primes and on the integers which they generate?

The earliest studies of these questions confined themselves to particular cases.

With $B = 0$ and small values of a and b, Kátai considered the situation (3) in the papers indicated. With general a and b but still with $B = 0$, satisfactory results, in an L^1 rather than L^∞ sense, were announced by Mauclaire [1].

Assuming that each $f_i(n)$ had the form $\lambda_i f_1(n)$ for some constant λ_i, a solution of Kátai's second problem was obtained by Elliott [12], and by Kátai.

In all of these treatments the argument was a more or less sophisticated reduction of the problem to an earlier one involving $f(n+1)-f(n)$, and hence unlikely to adapt to the general situation.

Adopting a quite different point of view I proved (Elliott [9]) that a strongly additive arithmetic function satisfies

$$\limsup_{x\to\infty} x^{-1} \sum_{n\leq x} |f(n+1)-f(n)|^2 < \infty$$

if and only if there is a constant D such that the series

$$\sum_p \frac{(f(p)-D\log p)^2}{p},$$

taken over the prime numbers, converges.

The method given there I generalized (Elliott [10]) to characterize those additive functions $f_i(n)$ which satisfied *both*

(5) $$\limsup_{x\to\infty} x^{-1} \sum_{n\leq x} \left| \sum_{i=1}^k f_i(a_i n + b_i) \right|^2 < \infty$$

and

(6) $$\limsup_{x\to\infty} x^{-1} (\log x)^{-2} \sum_{n\leq x} |f_i(n)|^2 < \infty \quad \text{for } i=1,\ldots,k.$$

Here $a_i > 0$, b_i were integers which satisfy

$$\det\begin{pmatrix} a_i & b_i \\ a_j & b_j \end{pmatrix} \neq 0, \quad 1 \leq i < j \leq k.$$

Thus, to obtain a solution of both of Kátai's questions it would suffice to prove that the hypothesis (6) follows from that of (5). However, this question seems to be of independent difficulty.

In another direction it is natural to enquire how weak an hypothesis of the form

(7) $$f(n+1)-f(n) = o(\beta(n))$$

with a non-decreasing function $\beta(x) > 0$, will still suffice to force a completely additive function to have the form $D \log n$—in fact to quantize Erdös' early theorems.

The first result in this direction was due to Suck [1], who showed that $\beta(n) = (\log \log n)^{1-\delta}$, with $\delta > 0$, will suffice. The best possible $\beta(n) = \log n$

Some Historical Remarks   249

was obtained by Wirsing [3], whose characteristically ingenious proof applies Markov chains to elaborate his earlier method [1].

Although, as remarked by Erdös, the same conclusion cannot be drawn from (7) for functions which are additive rather than completely additive, it is reasonable to ask what forms an additive function which satisfies the hypothesis (7) may take, even if the function $\beta(n)$ is larger than log n. As presently constituted the method of Wirsing leads to satisfactory upper bounds for $f(n)$ if

(8) $$\beta(x^2) \leq 2^{6/5} \beta(x)$$

for all $x \geq 1$.

In these investigations only $f(n+1) - f(n)$ was considered.

We can embrace the two problems of Kátai with a natural generalization of the investigations of Wirsing by asking for a characterization, in terms of their behaviour on the primes, of those additive functions which satisfy

$$\sum_{n \leq x} \left| \sum_{i=1}^{k} f_i(a_i n + b_i) \right|^2 = o(x\beta(x)^2)$$

as $x \to \infty$, for functions $\beta(x)$ of a suitable growth.

When $k = 2$, the functions $f_1(\ )$ and $f_2(\ )$ were the same, and

(9) $$\beta(x^2) \leq C\beta(x), \qquad x \geq 1,$$

for some positive C, I had by the beginning of 1980 reduced the problem to a study of the approximate functional equation which appears in Chapter 9 of the present volume. A satisfactory treatment of that equation I obtained in the early Spring of that year, whilst supported by a fellowship from the John Simon Guggenheim Foundation.

The method, which was conceived and carried out quite independently of that of Wirsing [3], and runs along entirely different lines, I have expounded, along with some necessary introductory material, in the first ten chapters of the present volume.

As I shall indicate later; it was indeed already clear in 1980; this method may readily be extended to deal with the case of two arbitrary functions $f_i(n)$, $i = 1, 2$, and offers an approach to the general problem.

In the next three chapters I shall apply the results of Chapter 10 to give a complete solution of the first problem of Kátai, and to obtain a natural generalization of Wirsing's result. Moreover, I shall obtain the latter under an hypothesis which allows the weaker (9), rather than condition (8).

Related results and possibilities will be considered in the final chapter, on problems.

Chapter 12

From L^2 to L^∞

In this chapter a number of auxiliary results are obtained which enable us to deduce L^∞-estimates from the L^2 theory of Chapter 10.

Let $a>0$, b, $A>0$, B be integers for which

$$\Delta = \det\begin{pmatrix} a & b \\ A & B \end{pmatrix} \neq 0.$$

In the notation of Chapter 3, let $(\alpha; \beta)$ be a fixed-divisor pair associated with the sequences $\{an+b;\ n=1, 2, \ldots\}$, $\{An+B;\ n=1, 2, \ldots\}$. Thus (α, β) will be one of the pairs $(\mu; \nu)$, $(2\mu; \nu)$ or $(\mu; 2\nu)$, where

$$\mu = (a, b) \quad \text{and} \quad \nu = (A, B).$$

Let D be a positive integer, and c, x positive real numbers. Let $N(x, D)$ denote the number of integers in the interval $x < n \leq 2x$ for which the following conditions are satisfied:

(i) $an + b \equiv 0 \pmod{D}$,
(ii) $((an+b)/D, D) = 1$,
(iii) We have

$$an + b = \alpha D r_1, \qquad An + B = \beta r_2,$$

where each r_j is squarefree, and if a prime p divides r_j then $p > x^c$.

Lemma (12.1). *Let $0 < 105c \leq 1$, $\delta > 2$ hold. Then there are positive constants c_1 and c_2 so that*

$$N(x, D) \geq \frac{c_1 x}{c^2 \varphi(D)(\log x)^2}$$

holds for all $x \geq \max(D^\delta, c_2)$; and integers D which are prime to $aA\Delta$, and which are even when $aA\Delta$ is odd.

Remark. The constant c_1 depends at most upon the four integers a, b, A and B.

Proof. Choose a residue class $n_0 \pmod{\lambda}$ so that if $n \equiv n_0 \pmod{\lambda}$ then the four conditions

$$an + b \equiv 0 \pmod{\alpha}, \qquad ((an+b)\alpha^{-1}, aA\Delta) = 1,$$

$$An + B \equiv 0 \pmod{\beta}, \qquad ((An+B)\beta^{-1}, aA\Delta) = 1,$$

are simultaneously satisfied. According to the remarks made at the beginning of Chapter 3 this may be effected with a λ comprised of powers of only the primes which divide $aA\Delta$. We may also arrange that both α and β divide λ.

Let n_1 be the least positive solution to the congruences

$$an + b \equiv 0 \pmod{D}, \qquad n \equiv n_0 \pmod{\lambda}.$$

Since $(D, \lambda) = 1$ the existence of such an n_1, not exceeding $D\lambda$, is guaranteed by the Chinese Remainder Theorem.

We consider the sequence of integers

$$n = \lambda Dm + n_1$$

which lie in the interval $(x, 2x]$. For such values of n our required condition (i) will be satisfied, and we may write

$$an + b = \alpha D r_1, \qquad An + B = \beta r_2,$$

where for each value of $j = 1, 2$, $(r_j, aA\Delta) = 1$.

Let Q be the product of those primes p, not exceeding x^c, which do not divide $aA\Delta$. In order to satisfy conditions (ii) and (iii) we begin with an application of the Selberg sieve, to arrange that $(r_1 r_2, Q) = 1$.

We apply lemma (1.6) with $k(n)$ identically one, and with the a_n of that lemma played by the sequence

$$(\alpha D)^{-1}(a\{\lambda Dm + n_1\} + b)\beta^{-1}(A\{\lambda Dm + n_1\} + B)$$

for

$$(x - n_1)/\lambda D < m \leq (2x - n_1)/\lambda D.$$

Then

$$N = \left[\frac{2x - n_1}{\lambda D}\right] - \left[\frac{x - n_1}{\lambda D}\right] = \frac{x}{\lambda D} + \theta, \qquad |\theta| \leq 1.$$

With $X = (\lambda D)^{-1}x$, and for each divisor d of Q, we have the estimate

$$\sum_{a_n \equiv 0 (\bmod d)} 1 = \eta(d)X + R_d,$$

where

$$R_d \ll 2^{\omega(d)}$$

and the function $\eta(d)$, which is multiplicative on Q, is determined by

$$p\eta(p) = \begin{cases} 2 & \text{if } (p, D) = 1, \\ 1 & \text{if } p|D. \end{cases}$$

The value of $p\eta(p)$ is the number of solutions to the congruence

$$(\alpha^{-1}\lambda am + \psi_1)(\beta^{-1}\lambda ADm + \psi_2) \equiv 0 \ (\bmod \ p),$$

where

$$\psi_1 = (an_1 + b)/\alpha D, \qquad \psi_2 = (An_1 + B)/\beta.$$

Note that if p divides D (and so each of our $an + b$), then it cannot divide ψ_2. Otherwise for some n it will divide both $an + b$ and $An + B$, and therefore Δ.

In the notation of lemma (1.6)

$$S = \sum_{p|Q} \frac{\eta(p)}{1 - \eta(p)} \log p \le \sum_{p \le x^c} \frac{2 \log p}{p} + O(1) \le 2c \log x + O(1).$$

Thus if $z = x^{17c} = r^{17}$, and x is large enough (depending only upon c), $\max(\log r, S) \le \frac{1}{8} \log z$, and

(1) $$I(N, Q) \ge \frac{c_3 x}{D} \prod_{p \le x^c} (1 - \eta(p)) + O\left(\sum_{\substack{d|Q \\ d \le z^3}} 3^{\omega(d)} 2^{\omega(d)} \right).$$

Here the product in the "main" term is at least as large as

$$\prod_{\substack{p \le x^c \\ (p, a A \Delta D) = 1}} \left(1 - \frac{2}{p}\right) \prod_{\substack{p \le x^c \\ p|D}} \left(1 - \frac{1}{p}\right) \ge c_4 \prod_{p \le x^c} \left(1 - \frac{1}{p}\right)^2 \prod_{\substack{p \le x^c \\ p|D}} \left(1 - \frac{1}{p}\right)^{-1}$$

$$\ge \frac{c_5 D}{c^2 \varphi(D)(\log x)^2}$$

since our restrictions upon D ensure that $aA\Delta D$ is always even. We have applied lemma (1.2) and the fact that

$$\prod_{\substack{p|D \\ p \leq x^c}} \left(1 - \frac{1}{p}\right) - \frac{\varphi(D)}{D} \ll \sum_{\substack{p|D \\ p > x^c}} \frac{1}{p} \ll x^{-c} \log D.$$

As to our "error" term in the lower bound (1), it is

$$\ll z^3 \sum_{\substack{d|Q \\ d \leq z^3}} 6^{\omega(d)} d^{-1} \ll z^3 \prod_{p \leq x^c} (1 + 6/p) \ll x^{52c}.$$

The restrictions $105c \leq 1$ and $x \geq D^2$ now ensure that

$$(2) \qquad I(N, Q) > \frac{c_6 x}{c^2 \varphi(D)(\log x)^2}.$$

The integers counted in $I(N, Q)$ will fulfil the requirements of conditions (ii) and (iii) up to prime divisors p, of D or the r_j, which do not exceed x^c.

The number of integers $n \ (= \lambda Dm + n_1)$ for which $(an + b)/D$ is divisible by some prime exceeding x^c which also divides D, does not exceed

$$\sum_{\substack{p|D \\ p > x^c}} \sum_{\lambda am + \alpha \psi_1 \equiv 0 \,(\mathrm{mod}\, p)} 1 \ll \sum_{\substack{p|D \\ p > x^c}} \left(\frac{x}{Dp} + 1\right)$$

$$\ll \frac{x}{D} \frac{\log D}{x^c} + \log D;$$

an amount which is negligible in comparison with the lower bound at (2).

The number of integers $n \ (= \lambda Dm + n_1)$ for which $(an + b)/D$ (and so ar_1) or $An + B$ (and so βr_2) is divisible by the square of some prime p greater than x^c which is prime to D, is

$$\ll \sum_{x^c < p \ll x^{1/2}} \sum_{\lambda am + \alpha \psi_1 \equiv 0 \,(\mathrm{mod}\, p^2)} 1 + \sum_{\substack{x^c < p \ll x^{1/2} \\ (p, D) = 1}} \sum_{\lambda ADm + \beta \psi_2 \equiv 0 \,(\mathrm{mod}\, p^2)} 1$$

$$\ll \sum_{x^c < p \ll x^{1/2}} \left(\frac{x}{Dp^2} + 1\right) \ll \frac{x^{1-c}}{D} + x^{1/2},$$

an amount which is also negligible in comparison with the lower bound at (2). Here essential use has been made of the condition $\delta > 2$.

For all large enough values of x we have

$$N(x, D) > \tfrac{1}{3} I(N, Q) > \frac{c_1 x}{c^2 \varphi(D)(\log x)^2},$$

and the proof of the lemma is complete.

We now need an upper bound for functions slightly more general than $N(x, D)$. This will allow us to eliminate from our considerations a small number of primes for which we have poor information.

Lemma (12.2). *Let $0 < \eta < 1$. Let p be a prime number. Let $N_1(x, D, p)$ and $N_2(x, D, p)$ denote the number of integers counted in the sum $N(x, D)$ which also satisfy $r_1 \equiv 0 \pmod{p}$ and $r_2 \equiv 0 \pmod{p}$, respectively.*
Then

$$N_j(x, D, p) \ll \frac{N(x, D)}{\varphi(p)}$$

holds uniformly for $pD \leq x^{1-\eta}$, $j = 1, 2$ and all large x.

Here the implied constants depend only upon a, b, A, B, η and the constant c which appears in the statement of lemma (12.1).

Proof. Let $0 < \nu \leq c$. It will be enough to obtain an upper bound for the number of integers n in the interval $(x, 2x]$ which satisfy

$$an + b \equiv 0 \pmod{Dp},$$

or

$$an + b \equiv 0 \pmod{D} \quad \text{with} \quad An + B \equiv 0 \pmod{p},$$

as the case may be, and for which

$$D^{-1}(an + b)(An + B)$$

is not divisible by any prime q in the range $q \leq x^\nu$, $(q, aA\Delta p) = 1$.
For example, if

$$an_2 + b \equiv 0 \pmod{Dp} \quad \text{with} \quad 0 \leq n_2 < Dp,$$

then we can set $n = Dpm + n_2$ and proceed as in lemma (12.1), with Dp now playing the rôle of D. For integers of this kind the argument will lead to an upper bound

$$\ll \frac{x}{\nu^2 \varphi(Dp)(\log x)^2} + x^{52\nu}.$$

With ν fixed at a value which satisfied $52\nu < \eta$, this bound will suffice.

If $An + B \equiv 0 \pmod{p}$ is to be considered, then by the Chinese Remainder Theorem we can find n_3, $0 \le n_3 < pD$, so that

$$an_3 + b \equiv 0 \pmod{D}, \qquad An_3 + B \equiv 0 \pmod{p}.$$

We set

$$n = Dpm + n_3$$

and proceed as before.

This completes our proof of lemma (12.2).

We supplement this upper bound with the following result which is to take care of the "large" prime divisors of the integers $an + b$ and $An + B$.

Lemma (12.3). *Let $0 < \eta < c \le \frac{1}{250}$, $\tau > 0$. Then there is a positive constant γ, depending at most upon a, b, A, B, η and τ, so that for all sufficiently large values of x*

$$\sum_{x^{1-\eta}D^{-1} < p \le \tau x D^{-1}} N_1(x, D, p) \le \gamma c N(x, D)$$

uniformly for $D \le x^{1/2}$, whilst

$$\sum_{x^{1-\eta}D^{-1} < p \le \tau x} N_2(x, D, p) \le \gamma c N(x, D)$$

uniformly for $D \le x^{(c-\eta)/2}$.

Remark. The constant γ depends upon c only through the restriction $\eta < c$.

Proof. Consider first the case $j = 1$.

If for some integer n in the interval $(x, 2x]$ we have $an + b = \alpha p D w$ where $p > x^{1-\eta}D^{-1}$, then $w \ll x^\eta$. Since pw $(=r_1)$ has no prime factor less than x^c whenever n is counted in $N(x, D)$, the restriction $\eta < c$ will for all large values of x force $w = 1$.

We now wish to estimate the number of primes p, not exceeding $\tau x D^{-1}$, for which

$$\beta^{-1}(Aa^{-1}(\alpha pD - b) + B) \quad (=r_2)$$

has no prime factors which do not exceed x^c. In particular

$$A\alpha pD + \Delta$$

will have no prime factors in the interval $(a\beta, x^c]$.

This number may be estimated by Selberg's sieve; lemma (1.6) will suffice. One may either employ the Bombieri–Vinogradov theorem on primes in arithmetic progression (see, for example, Davenport [1], Chapter 28), or estimate the number of integers t, not exceeding $\tau x D^{-1}$, for which

$$t(A\alpha Dt + \Delta)$$

has no prime factor in the interval $(\alpha\beta, x^c]$, in addition to which t also has no prime factor in the interval $(x^c, x^{1/250}]$. An argument similar to that given in the proof of lemma (12.1) then leads to a bound

$$\ll \frac{x}{c\varphi(D)(\log x)^2}.$$

Note that c^{-1} appears here, and not c^{-2} as in the lower bound of lemma (12.1). This arises from the analogue of the product which appears in the bound (1), which in the present circumstance will be

$$\prod_{\substack{a\beta < p \leq x^c \\ (p, aA\Delta D) = 1}} \left(1 - \frac{2}{p}\right) \prod_{\substack{a\beta < p \leq x^c \\ p \mid D}} \left(1 - \frac{1}{p}\right) \prod_{x^c < p \leq x^{1/250}} \left(1 - \frac{1}{p}\right).$$

This concludes our treatment of the case $j = 1$.

The case of $j = 2$ runs along similar lines with, however, an interesting aspect to the application of the sieve.

Thus we now have $An + B = \beta pw$ with $w \ll xp^{-1} \ll Dx^\eta$. Over the range of uniformity given in the statement of the lemma we shall once again have $w = 1$.

This reduces us to estimating the number of primes p up to τx for which

$$aA^{-1}(\beta p - B) + n \equiv 0 \pmod{D},$$

and such that

$$(\alpha D)^{-1}\{aA^{-1}(\beta p - B) + b\} \quad (= r_1)$$

has no prime factors in the interval $(\alpha\beta, x^c]$. In particular

$$(a\beta p - \Delta)/D$$

has no prime factors in this same interval.

We may again apply Selberg's sieve in the form of lemma (1.6). This time it is not possible to apply a so-called one-dimensional sieve, and to use the Bombieri–Vinogradov theorem; for D may be too large in comparison with the length of the interval in which the primes lie. For a discussion of what would be needed see my paper, Elliott [7], in particular lemma 3 and what precedes it.

However, it is still possible to apply a two-dimensional sieve, and to count the number of integers t, not exceeding τx, for which

$$\alpha\beta t \equiv \Delta \pmod{D},$$

and

$$t(\alpha\beta t - \Delta)/D$$

has no prime factor in $(\alpha\beta, x^c]$, whilst t is to have no prime factor in the further interval $(x^c, x^{1/250}]$. No new difficulty appears, and the proof of lemma (12.3) is readily completed.

We can summarize the results of lemmas (12.1)–(12.3), as far as we shall need them, in the following result.

Lemma (12.4). *Let σ be a (possibly infinite) sequence of prime numbers. If the constant c in the definition of $N(x, D)$ is fixed at a small enough value, depending upon a, b, A and B, then there will be a further constant $\check{c}$ so that the number of integers counted in $N(x, D)$ which have the property that r_1 or r_2 is divisible by some prime p in σ, will not exceed*

$$\left(\tfrac{1}{4} + \check{c} \sum_{\substack{x^c < p \leq x^{1-c/4} \\ p \in \sigma}} \frac{1}{p}\right) N(x, D)$$

uniformly for $D \leq x^{c/3}$, and all sufficiently large x.

Proof. We set $\eta = c/4$ and fix c at a value so small that the combined upper bounds in lemma (12.3) contribute an amount which is less than $\tfrac{1}{4}N(x, D)$. To each of the primes p in σ which lie in the interval $(x^c, x^{1-c/4}]$ we then apply lemma (12.2).

Remark. If the primes in σ are few, in the sense that

$$\check{c} \sum_{\substack{x^c < p \leq x \\ p \in \sigma}} \frac{1}{p} < \tfrac{1}{4}$$

for large x, then we may conclude that at least one half of the integers counted in $N(x, D)$ have the extra property that neither r_1 nor r_2 is divisible by a prime in σ, no matter what its size in comparison with n.

The following variant of lemma (12.1) will be useful in Chapter 14.

Let D_1, D_2 be positive integers constructed from powers of primes which divide $2aA\Delta$. Suppose that there is a residue class $n \equiv n_0 \pmod{L}$ for which

$$an + b \equiv 0 \pmod{D_1}, \qquad ((an+b)/D_1, 2aA\Delta) = 1,$$

$$An + B \equiv 0 \pmod{D_2}, \qquad ((An+B)/D_2, 2aA\Delta) = 1.$$

Let $M(x, D_1, D_2)$ denote the numbers of integers n in the range $x < n \leq 2x$ for which these four conditions are satisfied, and for which neither of $an+b$ or $An+B$ is divisible by a prime $p \leq x^c$, $(p, 2aA\Delta) = 1$.

Lemma (12.5). *If $0 < 105c \leq 1$ then there is a positive constant c_0 so that*

$$M(x, D_1, D_2) \geq \frac{c_0 x}{D_1 D_2 (\log x)^2}$$

uniformly for $D_1 D_2 \leq x^{1/2}$, and all sufficiently large values of x.

Remark. As before, the constants (explicit or implicit) depend at most upon a, b, A, B and c.

Proof. If there is a residue class $n \equiv n_0 \pmod{L}$ which satisfies the hypothesis of the lemma, then appropriate applications of the Chinese Remainder Theorem show that one can find such an L which is a multiple of each D_i, and which is bounded by $1 \leq L \leq 2aA|\Delta|D_1 D_2$.

If now we set $n = n_0 + Lm$ then it will be enough to count the integers m, in the interval

$$(x - n_0)/L < m \leq (2x - n_0)/L,$$

for which

$$\{amLD_1^{-1} + (an_0 + b)D_1^{-1}\}\{AmLD_2^{-1} + (An_0 + B)D_2^{-1}\}$$

has no prime factor satisfying $p \leq x^c$ and $(p, 2aA\Delta) = 1$.

An application of Selberg's sieve, similar but simpler than that in the proof of lemma (12.1), now completes the proof of the present result.

Chapter 13

A Problem of Kátai

In the present chapter $a > 0$, b, $A > 0$ and B continue to be integers for which $\Delta = aB - Ab$ is not zero.

Theorem (13). *Let $f(n)$ be a real additive arithmetic function which for some constant C satisfies*

$$f(an+b) - f(An+B) \to C$$

as $n \to \infty$.

Then there is a further constant F such that

$$f(D) = F \log D$$

holds for all positive integers which are prime to $aA\Delta$.

Moreover, in the notation of theorem (10.2),

$$L(p^i, p^j, F) = 0, \qquad L_t(F) + F \log(a/A) = C$$

for all permitted powers of the primes p which divide $aA\Delta$, and for $t = 1$ and/or 2, as is appropriate.

Remark. This theorem solves completely the first of the two problems of Kátai which was discussed in Chapter 11.

Proof. Assume first that $a \neq A$, and let

$$\mu = C(\log a/A)^{-1}.$$

Then the additive function

$$h(n) = f(n) - \mu \log n$$

satisfies

$$h(an+b) - h(An+B) \to 0, \qquad n \to \infty.$$

In fact we can work with the weaker assumption that

$$\lim_{x \to \infty} x^{-1} \sum_{n \leq x} |h(an+b) - h(An+B)|^2 = 0.$$

From theorem (10.1) there is a function $H(x)$ so that

$$\lim_{x \to \infty} \sum_{\substack{q \leq x \\ (q, aA\Delta) = 1}} \frac{1}{q} |h(q) - H(x) \log q|^2 = 0.$$

Clearly

$$H = \lim_{x \to \infty} H(x)$$

exists and is finite. Then for prime-powers q which are prime to $aA\Delta$

$$f(q) - \mu \log q = H \log q,$$

which justifies the first assertion of the present theorem with $F = \mu + H$.
For $a = A$ this result seems to lie a little deeper. In this case we have

$$\sum_{\substack{q \leq x \\ (q, aA\Delta) = 1}} \frac{1}{q} |f(q) - F(x) \log q|^2 \ll 1$$

for all sufficiently large x. Since $F(x)$ is bounded we can find an unbounded sequence $\{x_\nu\}$ so that

$$F = \lim_{\nu \to \infty} F(x_\nu)$$

exists and is finite. Setting $x = x_\nu$ and letting $\nu \to \infty$ we obtain the convergence of the series

$$\sum_{(p, aA\Delta) = 1} \frac{1}{p} |f(p) - F \log p|^2.$$

Let $0 < \varepsilon < 1$. Then the series

$$\sum_{|f(p) - F \log p| > \varepsilon} \frac{1}{p}$$

converges. Let σ denote the sequence of primes which are counted in this

A Problem of Kátai 261

sum. For fixed positive c, $č$ and all large values of x

$$\check{c} \sum_{\substack{x^c < p \leq x \\ p \in \sigma}} \frac{1}{p} < \tfrac{1}{4},$$

and we may apply lemmas (12.1) and (12.4).

Suppose now that (α, β) is a fixed-divisor pair associated with the sequences $\{an+b\}$ and $\{An+B\}$, and that D is a positive integer, prime to $aA\Delta$, which is even when $aA\Delta$ is odd. Then we can find infinitely many integers n corresponding to which there are representations

$$an+b = \alpha D r_1, \qquad An+B = \beta r_2, \qquad (r_1, D) = 1,$$

where the r_i have at most c^{-1} prime factors, none of which occurs more than once, and none of which belongs to σ. In particular

$$|f(r_i) - F \log r_i| \leq \varepsilon/c$$

for each value of i.

For these particular n, and an appropriate choice of t,

$$\left| f(D) - F \log D + L_t(F) + F \log\left(\frac{an+b}{An+B}\right) - \{f(an+b) - f(An+B)\} \right|$$

$$\leq 2\varepsilon/c.$$

By taking a large enough value of n and appealing to the hypothesis of our theorem we see that

$$\left| f(D) - D \log D + L_t(F) + F \log\left(\frac{a}{A}\right) - C \right| \leq 2\varepsilon/c,$$

and since ε may be chosen arbitrarily small, that

$$f(D) - F \log D = -L_t(F) - F \log\left(\frac{a}{A}\right) + C.$$

For even values of $aA\Delta$ we may set $D=1$ in this equation. The left-hand (and so the right-hand) side then becomes zero. Hence

$$f(D) = F \log D$$

for all D prime to $aA\Delta$.

For odd values of $aA\Delta$ we first set $D = 2$, to obtain

$$C - L_t(F) - F\log\left(\frac{a}{A}\right) - f(2) + F\log 2 = 0.$$

By considering $2D$ in place of D we then obtain the representation

$$f(D) = F\log D$$

for all D which are prime to $2aA\Delta$.

Whether $a = A$ or not, the additive function

$$f(n) - F\log n$$

now satisfies the hypothesis of the theorem (with an adjusted value of C if $a \neq A$), and is zero on the powers of primes which do not divide $2aA\Delta$.

The proof of the theorem is now completed by appropriate applications of the Chinese Remainder Theorem.

EXAMPLES. (i) Let

$$f(3n+1) - f(n) \to C, \qquad n \to \infty.$$

This is the example, related to the Syracuse/Kakutani map, which was featured in the first motive. With $a = 3$, $b = 1$, $A = 1$, $B = 0$ we have $\Delta = -1$, and theorem 13 gives at first

$$f(n) = F\log n$$

for all n not divisible by 3. However, there is only the fixed divisor pair $(1; 1)$, and the remaining assertions of theorem 13 show that this relation holds for all positive integers n. Clearly $F\log 3 = C$.

(ii) Let

$$f(3n+1) - f(3n+2) \to C, \qquad n \to \infty.$$

In this case $aA\Delta = 27$, and once again

$$f(n) = F\log n$$

for all n prime to 3; with $F\log 3 = C$. This time, however, we can say nothing about the behaviour of $f(\ )$ on the powers of 3.

(iii) Now let

$$f(8n+4) - f(8n+2) \to C, \qquad n \to \infty.$$

Here $aA\Delta = -2^{10}$, so that the first part of theorem 13 guarantees a representation

$$f(n) = F \log n$$

for odd integers n. In the notation used during the preliminary discussion of Chapter 3, $a_1 = 2$, $b_1 = 1$, $A_1 = 4$, $B_1 = 2$, and we have only the single fixed-divisor pair $(4; 2)$ to consider. The second part of theorem 13 now yields

$$f(4) - F \log 4 = f(2) - F \log 2,$$

and $C = 0$.

Chapter 14

Inequalities in L^∞

I begin this chapter with a general result. Later I specialize to obtain an extension of a theorem of Wirsing.

For integers $a>0$, b, $A>0$, B which satisfy $\Delta = aB - Ab \neq 0$, and a positive constant c, let

$$\psi(x) = \max_{x<n\leq x^c} |f(an+b) - f(An+B)|.$$

If F is a further number define

$$\theta(x, F) = \max_{x<n\leq x^c} \left| f(an+b) - f(An+B) - F \log\left(\frac{an+b}{An+B}\right) \right|.$$

Theorem (14.1). *There are positive constants c_0 and c and a function $F(x)$, so that*

$$|f(D) - F(x) \log D| \leq c_0 \theta(x, F(x))$$

holds uniformly for all integers D, not exceeding x, which are prime to $aA\Delta$, for all $x \geq 2$.

The same upper bound serves for the expressions

$$|L(p^i, p^j, F(x))|, \quad |L_t(F(x))|,$$

which are defined at the beginning of Chapter 10; uniformly for all permissible prime-powers which satisfy $\max(p^i, p^j) \leq x$, $p|aA\Delta$ and for all possible choices of t.

Moreover

$$|F(x)| \leq c_0 \psi(x)$$

if $a \neq A$, whilst

$$x^{-1/2}|F(x)| \leq c_0 \max_{n \leq x^c} |f(an+b) - f(An+B)|$$

if $a = A$.

Remarks. The asymptotic behavior of the function $F(x)$, and so of $f(n)$, for large values of the variable, may be deduced from the results in the section of Chapter 10 devoted to the decomposition of the mean. We give an example later in this chapter.

When $a = A$ a sharper estimate for $x^{-1/2}|F(x)|$ may be deduced from the final assertion of theorem (10.2).

A permissible choice for $F(x)$ is the function

$$\sum_{\substack{x^{1/2} < q \leq x \\ (q, aA\Delta) = 1}} \frac{f(q)}{q} \bigg/ \sum_{\substack{x^{1/2} < q \leq x \\ (q, aA\Delta) = 1}} \frac{\log q}{q},$$

which appears in the statements of theorems (10.1) and (10.2).

If the constants c, c_0 are fixed at suitably large values then we may further assert that

$$|F(z) - F(x)| \log x \leq c_0 \psi(x)$$

holds uniformly for $x^{1/2} \leq z \leq x$.

Proof of the theorem. We start with a careful treatment of the argument used to settle the problem of Kátai.

In view of theorem (10.1) we may choose constants c_1 and c (in the definition of $\psi(x)$) so that

$$\check{c} \sum_{\substack{q \leq x \\ |f(q) - F(x) \log q| > c_1 \psi(x)}}' \frac{1}{q} < \tfrac{1}{4},$$

certainly for all x large enough. Here q denotes a prime-power, and $'$ that q is prime to $aA\Delta$.

Let D be an integer not exceeding $x^{1/350}$, which is prime to $aA\Delta$ and is even if that number is odd. Applying lemmas (12.1) and (12.4) we obtain a representation

$$an + b = \alpha D r_1, \qquad An + B = \beta r_2,$$

where $(\alpha; \beta)$ is a fixed-divisor pair, $x < n \leq 2x$, and the r_k have no prime factor not exceeding $x^{1/350}$, and no squared factor. Thus

$$|f(r_k) - F(x) \log r_k| \leq 350 \, c_1 \psi(x)$$

for each k.

We write F for $F(x)$ and, as before, deduce that for these particular n, and with a suitable value of t,

$$\left| f(D) - F \log D + L_t(F) + F \log\left(\frac{an+b}{An+B}\right) - \{f(an+b) - f(An+B)\} \right|$$

does not exceed $700 c_1 \psi(x)$.

However, a direct application of theorem (10.2) gives

$$|L_t(F)| \le c_2 \psi(x),$$

so that

$$|f(D) - F \log D| \le c_3(\psi(x) + \theta(x, F))$$

for all $D \le x^{1/400}$ which satisfy $(D, aA\Delta) = 1$. Note that if $aA\Delta$ is odd we first obtain this inequality (with some value of c_3) for $D = 2$, and then deduce the full inequality from a similar one with D replaced by $2D$.

Replacing $f(n)$ by $f(n) - F \log n$ and then x by x^{400} gives an inequality of the first type stated in the theorem.

Suppose now that p is a prime which divides $aA\Delta$, and that $\max(p^i, p^j)$ does not exceed $x^{1/1800}$. If we can find an integer n so that simultaneously

$$p^i \| (an + b), \qquad p^j \| (An + B),$$

then by lemma (12.5) we can find such an integer in the interval $x^{1/401} < n \le 2x^{1/401}$ for which in addition

$$an + b = p^i \alpha_1 r_1, \qquad An + B = p^j \beta_1 r_2,$$

where α_1 and β_1 are the appropriate factors of an appropriate fixed-divisor pair $(\alpha; \beta)$, and where the r_k have no prime factor not exceeding $x^{1/43000}$. Since a typical prime factor of these r_k does not exceed $x^{1/400}$ we may assert that for each value of k

$$|f(r_k) - F \log r_k| \le c_4(\psi(x^{1/500}) + \theta(x^{1/500}, F(x))).$$

Hence

$$\left| f(p^i \alpha_1) - f(p^j \beta_1) - F \log\left(\frac{p^i \alpha_1}{p^j \beta_1}\right) \right| \le c_4(\psi(x^{1/500}) + \theta(x^{1/500}, F(x))),$$

where the constant c upon which the functions $\psi(x)$ and $\theta(x, F)$ depend is to be increased by the factor 500.

If p^r and p^s are the exact powers of the prime p which divide α and β, respectively, then

$$\left| f(p^r) - F \log p^r - \{f(p^s) - F \log p^s\} + f(\alpha_1) - f(\beta_1) - F \log \frac{\alpha_1}{\beta_1} \right|$$

$$= |L_t(F)| \le c_3(\psi(x) + \theta(x, F)).$$

With what we have already established

$$|L(p^i, p^j, F)| \le c(\psi(x^{1/500}) + \theta(x^{1/500}, F(x))).$$

Replacing $f(n)$ by $f(n) - F \log n$ and then x by x^{1800} gives bounds of the second and third types stated in the theorem. In order that the same function $F(x)$ should feature in the various bounds it will suffice to prove that

$$|F(x^{1800}) - F(x^{400})| \le c_6 \psi(x).$$

In fact if $\delta \ge 1$, and the c in $\psi(x)$ is chosen suitably large, then two applications of theorem (10.1) give

$$\sum_{q \le x}' \frac{1}{q} |f(q) - F(x) \log q|^2 \ll \psi(x)^2,$$

and

$$\sum_{q \le x^\delta}' \frac{1}{q} |f(q) - F(x^\delta) \log q|^2 \ll \psi(x)^2.$$

Eliminating the $f(q)$ with q not exceeding x^δ, an argument which we used several times in Chapter 10, gives

$$|F(x^\delta) - F(x)|^2 \sum_{q \le x}' \frac{(\log q)^2}{q} \ll \psi(x)^2,$$

and after an appeal to lemma (1.2)

$$|F(x^\delta) - F(x)| \log x \ll \psi(x).$$

This justifies our replacements, and the last two of the remarks which follow the statement of the theorem.

The inequalities of the third part of theorem (13.1) follow directly from the corresponding ones of theorem (10.2).

Theorem (14.1) is proved.

EXAMPLE. Let

$$f(an+b) - f(An+B) = o((\log n)^\tau)$$

as $n \to \infty$, for some fixed $\tau > 0$. The case $\tau = 0$ is already covered by our treatment of Kátai's problem.

Theorem (14.1) gives

(1) $\qquad f(n) = F(x) \log n + o((\log x)^\tau), \qquad x \to \infty,$

uniformly for $n \le x$, $(n, aA\Delta) = 1$. Moreover

(2) $\qquad F(z) - F(x) = o((\log x)^{\tau-1}), \qquad x \to \infty,$

uniformly for $x^{1/2} \le z \le x^2$.

We now need a characterization of functions which satisfy this last asymptotic estimate. Such asymptotic relations are considered in Volume II, Chapter 11 of my book [11]. Our present argument will be from first principles.

The cases $0 < \tau < 1$. Let

$$u_k = F(2^{2^k}), \qquad k = 1, 2, \ldots.$$

Then our temporary hypothesis (2) gives

$$u_{k+1} - u_k = o(2^{-k(1-\tau)})$$

as $k \to \infty$. It is straightforward to establish the finite existence of

$$U = \lim_{k \to \infty} u_k,$$

and so to obtain the estimate

$$u_k = U + o(2^{-k(1-\tau)}), \qquad k \to \infty.$$

Since every $x \ge 4$ lies in an interval

$$2^{2^k} \le x < 2^{2^{k+1}} \quad (\le x^2),$$

a further application of the asymptotic relation (2) gives

$$F(x) = U + o((\log x)^{\tau-1}), \qquad x \to \infty.$$

Altogether, therefore, for those integers which are prime to $aA\Delta$

$$f(n) = U \log n + o((\log n)^\tau), \qquad n \to \infty,$$

a result which is clearly best possible.

The cases $\tau > 1$. Defining the sequence $\{u_k\}$ as before, we induct downwards rather than up, to obtain the bounds

$$u_k = o(2^{k(\tau-1)}), \qquad k \to \infty,$$

and then

$$F(x) = o((\log x)^{\tau-1}), \qquad x \to \infty.$$

This leads to the estimate

$$f(n) = o((\log n)^\tau), \qquad (n, aA\Delta) = 1,$$

which is once again best possible.

The case $\tau = 1$. In this case (2) becomes

$$F(z) - F(x) = o(1), \qquad x \to \infty,$$

uniformly for $x^{1/2} \le z \le x^2$. The example $F(x) = (\log \log x)^{1/2}$ shows that $F(x)$ need not approach a finite limit as x increases. The exact description of those additive functions $f(n)$ which possess an approximation of the form (1) which satisfies (2), but with $\tau = 1$, we do not pursue here. A related problem, due to Hardy and Ramanujan, I solve in Chapter 15, Volume II, of my book [11], within the format of the weak law of large numbers for additive functions.

Here we note that a necessary and sufficient condition for $F(x)$ to converge to a limit as $x \to \infty$ is that for some real number $\mu > 1$,

$$\lim_{k \to \infty} F(\mu^{2^k})$$

is to exist and be finite.

Thus we may conclude from

$$f(an + b) - f(An + B) = o(\log n), \qquad n \to \infty,$$

that for some constant U

$$f(n) = U \log n + o(\log n), \qquad n \to \infty,$$

for all $(n, aA\Delta) = 1$, if and only if there is a positive integer $d \geq 2$, prime to $aA\Delta$, so that

$$\lim_{k \to \infty} 2^{-k} f(d^{2^k})$$

exists and is finite. $U \log d$ will then have the value of this limit.

Supposing that $f(\ )$ is *completely* additive, that is to say, satisfies

$$f(rs) = f(r) + f(s)$$

for all positive integers r and s, this last requirement is clearly met by every $d \geq 2$, which is prime to $aA\Delta$. We deduce that for such integers

$$f(d) = U \log d.$$

In order to relate this example to Wirsing's treatment of the difference $f(n+1) - f(n)$, which we mentioned in Chapter 11, we need the following result of Ruzsa. This shows that a two-sided bound for $f(n+1) - f(n)$ can be derived from a one-sided bound.

Let

$$E(x) = \max_{1 \leq n \leq x} (f(n+1) - f(n)).$$

Lemma (14.2). *The inequality*

$$|f(n+1) - f(n)| \leq 59 \max(E(5x^2), 0) - 4f(2)$$

holds uniformly for $1 \leq n \leq x$.

Remark. Our treatment is based upon that of Ruzsa [2]. We have made precise the dependence upon the function $f(\ )$, in fact only $f(2)$ appears, but no attempt has been made to determine the best values for the constants involved.

Proof. The main-spring of the proof is the two identities:

$$2\{f(n) - f(n+3)\} - \{f(n-6) - f(n+3)\}$$

$$= [f(n-5) - f(n-6)] + [f(n) - f(n-1)] + [f(n+4) - f(n+3)]$$

$$+ [f(n(n-1)) - f((n-5)(n+4))],$$

which is valid if n is not congruent to 2 (mod 3), so that $(n-5, n+4) = 1$; and

$$2\{f(n) - f(n+3)\} - \{f(n-6) - f(n+3)\}$$
$$= [f(n) - f(n-2)] + [f(n) - f(n-1)] + [f(n+3) - f(n)]$$
$$+ [f((n-1)(n-2)) - f((n+3)(n-6))],$$

which is valid if n is not divisible by 3, so that $(n+3, n-6) = 1$. In these identities each square bracket encloses an expression of the form $f(m+j) - f(m)$ where $1 \leq j \leq 20$.

Thus

(3) $$2\{f(n) - f(n+3)\} - \{f(n-6) - f(n+3)\} \leq 26 E(x^2)$$

holds for $7 \leq n \leq x$.

Suppose that ρ is the maximal value of $f(n) - f(n+3)$ over the range $7 \leq n \leq x$. If the integer n (≥ 17) gives a maximal value, choose the greatest positive integer k which satisfied $k \neq 0, -3$ (mod 11) and $11k \leq n - 6$. Then $n - 6 < 11(k+2)$, so that

$$11k \leq n - 6 < n + 3 < 11(k+3).$$

Hence

(4) $$f(n-6) - f(n+3) = f(k) - f(k+3) + [f(n-6) - f(11k)]$$
$$+ [f(11(k+3)) - f(n+3)].$$

Once again the square brackets contain expressions of the form $f(m+j) - f(m)$, this time with $0 \leq j \leq 23$.

If $k \geq 7$ then (3) and (4) together give

$$\rho \leq 49 E(x^2).$$

Direct computations show that

$$f(k) - f(k+3) \leq 33 \max(E(x^2), 0) - 4f(2),$$

uniformly for $1 \leq k \leq 6$. The most interesting of these values is perhaps $k = 1$. From the first of Ruzsa's identities with $n = 9$ we obtain

$$2\{f(9) - f(12)\} - \{f(3) - f(12)\}$$
$$= [f(4) - f(3)] + [f(9) - f(8)] + [f(13) - f(12)] + [f(72) - f(52)].$$

Here the left-hand side has the alternative representation

$$2\{f(9)-f(3)\}-f(4),$$

and since

$$f(3)-f(9)=-2f(2)+[f(6)-f(5)]+[f(10)-f(9)],$$

we obtain

$$f(1)-f(4)\leq 27E(x^2)-4f(2).$$

Note that

$$f(2)=f(2)-f(1)\leq E(x^2).$$

In particular, we see from (3) and (4) that

$$\rho\leq 53\max(E(x^2),0)-4f(2),$$

whether $k\geq 7$ or not.

It is not difficult to show that

$$f(n)-f(n+3)\leq 57\max(E(x^2),0)-4f(2)$$

uniformly for $7\leq n\leq 17$. Indeed, if $(n,2)=1$ then

$$f(n)-f(n+3)=f(n)-f(2n)+[f(2n)-f(n+3)].$$

Over the range $7\leq n\leq 17$ we have $2n-(n+3)\leq 14$, so that the right-hand side of this equation does not exceed

$$-f(2)+14E(x^2)\leq -4f(2)+17E(x^2).$$

If $(n,3)=1$ and $n\leq 16$, then

$$f(n)-f(n+3)=f(n)-f(3n)+[f(3n)-f(n+3)].$$

Here $3n-(n+3)\leq 29$, and the right-hand side is

$$\leq -f(3)+29E(x^2)=-f(4)+[f(4)-f(3)]+29E(x^2)$$

$$\leq -4f(2)+57E(x^2).$$

This leaves the case $n=12$, for which

$$f(12)-f(15)=f(3)-f(5)=f(3)-f(12)+[f(12)-f(5)]$$

$$\leq -4f(2)+34E(x^2).$$

Since

$$f(n)-f(n+1)=f(n)-f(n+3)+[f(n+3)-f(n+2)]+[f(n+2)-f(n)],$$

we have proved that

$$f(n)-f(n+1)\leq 59\max(E(x^2),0)-4f(2),$$

uniformly for $7\leq n\leq x$.

Further calculations show that

$$f(n)-f(n+1)\leq 15\max(E(5x^2),0)-2f(2)$$

for $1\leq n\leq 7$. The "worst" case seems to be

$$f(2)-f(3)=-f(2)+[f(10)-f(9)]+[f(18)-f(15)],$$

corresponding to $n=2$, when we have to go as far up as 18.

The proof of lemma (14.2) is complete.

In his paper [3], Wirsing requires that there be a positive non-decreasing function $\beta(x)$ so that

$$\beta(x^2)\leq 2^{6/5}\beta(x),$$

$$f(n+1)-f(n)\leq \beta(n),$$

$$f(2)\geq 0,$$

the first inequality to hold for all real $x\geq 1$, the second for positive integers n. His main conclusion, from which his results follow, is that for a suitable constant γ,

$$\left|\frac{f(m)}{\log m}-\frac{f(n)}{\log n}\right|\leq \gamma\left(\frac{\beta(m)}{\log m}+\frac{\beta(n)}{\log n}\right)$$

uniformly for $2\leq m\leq n\leq e^m$.

As a further example I shall obtain an inequality of this type which in particular is valid under the weaker requirement that

$$\beta(x^2) \le C\beta(x)$$

holds for *some* positive constant C. This will then allow the choice $\beta(x) = (\log x)^\tau$ for every fixed positive value of τ, rather than those not exceeding $\frac{6}{5}$, as permitted by Wirsing's hypothesis.

In view of Ruzsa's lemma we may assume that

$$|f(n+1) - f(n)| \le c_7 \beta(x)$$

for some positive constant c_7. This enables us to apply theorem (14.1).

Let us return to the general situation considered in theorem (14.1), and define

$$L(x) = \max_{n \le x^c} |f(an+b) - f(An+B)|.$$

Theorem (14.3). *There are positive constants c_8 and c so that*

$$\left| \frac{f(m)}{\log m} - \frac{f(n)}{\log n} \right| \le c_8 \left(\frac{L(m)}{\log m} + \frac{L(n)}{\log n} \right)$$

holds uniformly for all integers m and n which satisfy $2 \le m \le n \le e^m$ and are prime to $aA\Delta$.

Remarks. The constants c_8 and c depend only upon the integers a, b, A and B. If our inequality is only required over the range $c_9 \le m \le n \le e^m$, then for a suitably chosen c_8 (once again in terms of the a, b, A, B) it would be perhaps possible to obtain for c an absolute value.

The constant c need not have the same value as that which appears in the statement of theorem (14.1).

Proof. It will be enough to establish the theorem under the restriction that m be sufficiently large. The complete range may then be obtained by inflating the value of c.

Noting that

$$\max_{x < n \le x^c} \log\left(\frac{an+b}{An+B}\right) \ll x^{-1},$$

when $a = A$, we see from the third assertion of theorem (14.1) that

$$\psi(x) + \theta(x, F(x)) \ll L(x).$$

Inequalities in L^∞

By adjusting the value of c we obtain from the first assertion of that theorem the bound

(5) $$\frac{f(m)}{\log m} - \frac{f(n)}{\log n} \ll \frac{L(m)}{\log m}$$

uniformly for $m \le n \le m^3$.

More generally, suppose that

$$m^3 < n \le e^m.$$

Let $p_1 < p_2 < \cdots$ denote the increasing sequence of prime numbers which do not divide $aA\Delta$. Define

$$D_r = \prod_{i=1}^{r} p_i.$$

It follows from the Prime Number Theorem that

$$D_r \sim r \log r \sim p_r$$

as $r \to \infty$. In particular, if we choose r to be the largest integer for which

$$P_r = \prod_{\substack{i=1 \\ p_i > \sqrt{m}}}^{r} p_i$$

does not exceed n then, for all large m, $p_r \le 2m$ and $P_{r+1} < P_r^2$. Therefore

$$\frac{f(n)}{\log n} - \frac{f(P_r)}{\log P_r} \ll \frac{L(P_r)}{\log P_r} \ll \frac{L(n)}{\log n}$$

by means of the short range inequality (5) that we have already established.

Appealing to the additive property of the function $f(\ )$:

$$f(P_r) - \frac{f(m)}{\log m} \log P_r = \sum_{\substack{i=1 \\ p_i > \sqrt{m}}}^{r} \left(f(p_i) - \frac{f(m)}{\log m} \cdot \log p_i \right).$$

Typically

$$\sqrt{m} < p_i < (\sqrt{m})^3,$$

and to each summand we may again apply the inequality (5). This gives for the whole sum a bound

$$\ll \frac{L(m)}{\log m} \sum_{\substack{i=1 \\ p_i > \sqrt{m}}}^{r} \log p_i \ll \frac{L(m)}{\log m} \log P_r.$$

The proof of theorem (14.3) is complete.

Another application on these results will be given in the chapter on the Information Equation.

Chapter 15

Integers as Products

More Duality; Additive Functions as Characters

In this and the following four chapters I study the representation of integers by products of integers of a prescribed type. Besides being of interest in its own right, this study will motivate many interesting questions in the theory of additive arithmetic functions.

Let $a_1, a_2, \ldots$ be a sequence of positive integers. In order to prove that every positive integer n has a product representation

$$n = \prod_{j=1}^{k} a_j^{d_j}$$

for suitably chosen integers $k > 0$ and d_j, we consider the multiplicative group Q^* of positive rationals, its subgroup Γ generated by the a_j, and form the quotient group Q^*/Γ. Our general procedure will be to prove that this quotient group is:

(i) finitely generated;
(ii) of bounded order;
(iii) trivial.

A group is said to be of bounded order if there is an integer m so that the mth power of every element in its is the identity. Such a group need not be finite.

If for a commutative group both (i) and (ii) have been obtained then it will indeed be finite. We have preserved the above formulation since one can sometimes obtain (ii) without (i), and the method may still proceed. Moreover, our following arguments generalize almost at once to a wide class of modules, for which an analogue of (ii) can be readily formulated.

Let G and D be abelian groups (written additively). We can give a group structure $\hat{G}$ to the set of homomorphisms $f(\ )$ of G into D by defining

$$(f_1 + f_2)(g) = f_1(g) + f_2(g)$$

for every element g in G. We can regard this group $\hat{G}$ as a dual of G. In general it will not be the usual dual group. However, in order that it be useful it will suffice if the image group D has enough structure that $\hat{G}$ can distinguish between interesting subgroups of G, and if D is more amenable than G itself. For example it may be a field, or have a non-trivial topological structure.

If G is finite the possibilities for D are limited. There is essentially only the choice of the multiplicative group of roots of unity for D, and $\hat{G}$ will then be the usual dual group.

For infinite abelian groups, the case in which we are interested, other worthwhile choices for D can be made. For example with D the additive group of real numbers we are led to the study of additive arithmetic functions $f(\ )$ which satisfy $f(a_n) = 0$. The study of such functions for varying sequences $\{a_n\}$ represents an enormous program.

When the a_n are characterized only by their distributional properties, the methods of Probabilistic Number Theory may apply. For an account of this developed, but certainly not complete, theory, see my book [11], and the books of Kac [1] and Kubilius [1].

When the a_n have an algebraic structure very little is known, and it is towards this area that the methods and results of the present volume are directed.

In this chapter we consider some groups which can advantageously play the rôle of the image group D.

Divisible Groups and Modules

Let G be an abelian group, written additively, which becomes a module under the action of a principal ideal domain R. We shall write this action on the left side, thus rg is defined for r in R and g in G.

We say that G is a divisible (R-) module if for every element g in G and every non-zero member r of R we can find a further element h in G so that $g = rh$.

If $\mathbb{Z}$ denotes the ring of rational integers, then every abelian group can be viewed as a $\mathbb{Z}$-module, with an integer n acting upon a group element g by taking it to its nth power ng. The additive group of real numbers is clearly a divisible $\mathbb{Z}$-module.

Lemma (15.1). *Let H be a submodule of the R-module G. Then any homomorphism of H into a divisible R-module D can be extended to a homomorphism of G into D.*

Remark. In this and what follows the homomorphisms are module homomorphisms.

Proof. Consider the collection of all pairs (K, t) of submodules K,

Divisible Groups and Modules

containing H, which have a homomorphism t into D extending that defined on H. We partially order these pairs by

$$(K, t) > (K', t')$$

if K includes K' and t extends t'. It is readily checked that any chain has an upper bound, and by Zorn's lemma the collection contains a maximal pair, (L, τ) say.

Suppose that L is not G, and let g be an element of G not in L. Let M be the submodule generated by g and L.

If mg does not lie in L for any non-zero member m of R then M is the direct sum of L and the module generated by g. We may define a map $T: M \to D$ by

$$T(mg + \lambda) = \tau(\lambda)$$

for all λ in L.

Otherwise there will be a non-trivial ideal of elements m in R so that mg lies in L. Since R is principal this ideal will be generated by an element, π say.

We now appeal to the divisilibity of the module D, and let δ be an element of it for which $\tau(\pi g) = \pi \delta$. We then define

$$T(mg + \lambda) = m\delta + \tau(\lambda).$$

Note that if

$$m_1 g + \lambda_1 = m_2 g + \lambda_2$$

then $(m_1 - m_2)g$ belongs to L, so that in R $m_1 - m_2$ must be a multiple of π, say $k\pi$. Hence

$$T(m_1 g + \lambda_1) - T(m_2 g + \lambda_2) = (m_1 - m_2)\delta + \tau(\lambda_1) - \tau(\lambda_2)$$

$$= k\pi\delta + \tau(\lambda_1 - \lambda_2)$$

$$= k\pi\delta + \tau(\{m_2 - m_1\}g)$$

$$= k\pi\delta - k\tau(\pi g) = 0,$$

so that T is well defined.

In either case we obtain a genuine extension

$$(M, T) > (L, \tau),$$

contradicting the maximality of (L, τ).

Thus $L = G$ and the lemma is proved.

We say that a homomorphism is trivial on a set of elements of a module if it takes each of them to zero (the identity).

In what follows D will be a divisible module containing at least two elements.

Lemma (15.2). *Let G_1 and G_2 be submodules of an R-module G_1. If every homomorphism of G into D which is trivial on G_1 is also trivial on G_2, then to each element g in G_2 there is a non-zero member r of R so that rg lies in G_1.*

Proof. Suppose, to the contrary, that no product rg, with r in R, of the element g of G_2 lies in G_1. Let M be the module generated by g and G_1. Since M is the direct sum of G_1 and the module generated by g, we may define a homomorphism t of M into D by setting $t(g)$ to be any non-zero element in D, and defining

$$t(rg + \mu) = rt(g)$$

for each element μ of G_1.

By lemma (15.1) t may be extended to a homomorphism of G into D. Moreover, this new homomorphism is trivial on G_1 but not on G_2, contradicting the hypothesis of the lemma.

Lemma (15.2) is established.

Remark. In our applications it will not be assumed that the modules G_1 and G_2 have any non-trivial intersections.

An abelian group which is divisible as a $\mathbb{Z}$-module we shall call a *divisible abelian group* without mentioning its module structure. In this case we can reformulate lemma (15.2) as

Lemma (15.3). *Let G_1 and G_2 be subgroups of an abelian group G. Suppose that every homomorphism of G into a (non-trivial) divisible abelian group D which is trivial on G_1, is also trivial on G_2. Then every element in G_2 has a power in G_1.*

For groups which have torsion better can sometimes be done. Let p be a rational prime number and let G be a possibly infinite abelian group, each of whose non-trivial elements has order p. Let F_p be a finite field of p elements.

We can make F_p act on G by identifying F_p with the field of integer residue classes (mod p), $\mathbb{Z}/p\mathbb{Z}$, and using the rule

$$(n(\text{mod } p), g) \mapsto ng.$$

In view of the p-torsion this action is well defined. G now becomes a vector space over F_p.

The analogue of lemma (15.3) is

Lemma (15.4). *Let G_1 and G_2 be subgroups of an abelian group G with p-torsion. Suppose that every homomorphism of G into a non-trivial vector space over F_p which is trivial on G_1 is also trivial on G_2. Then G_2 is contained in G_1.*

Proof. Since F_p is a field, any vector space over F_p is F_p-divisible. According to lemma (15.2) with $R = F_p$, to each element g of G_2 there is a non-zero member r of F_p so that rg belongs to G_1. Once again using that F_p is a field, there is a member s of F_p so that $sr = 1$, and therefore $g = s(rg)$ itself belongs to G_1.

Remark. In our applications of lemma (15.4) groups G arise which need not have p-torsion; so we give them it by considering the factor groups G/pG, where pG denotes the subgroup of the pth-powers of elements in G.

Sets of Uniqueness

A sequence

$$A: a_1 < a_2 < \cdots$$

of positive integers with the property that every real-valued completely additive arithmetic function which vanished on them also vanished identically, was said by Kátai [1] to be a *set of uniqueness*. In particular, he proved [2] that if to the sequence

$$P: 3 < 4 < 6 < \cdots < p+1 < \cdots,$$

where the p are primes, we adjoin finitely many integers then we obtain a set of uniqueness. He conjectured that P itself was a set of uniqueness. This I established as true in Elliott [4].

It was proved by Wolke [3], and Dress and Volkmann [1], that in order for a sequence A to be such a set of uniqueness, it is necessary and sufficient that every positive integer n has a multiplicative representation

$$n^h = \prod a_{j_i}^{\varepsilon_i}$$

with $\varepsilon_i = \pm 1$. The h may vary with n. Their proofs employ vector spaces over the field of rational numbers.

It followed from my proof of Kátai's conjecture that (as Wolke, and Dress and Volkmann mentioned) there is a representation

(1) $$n^h = \prod (p_i + 1)^{\varepsilon_i}$$

with the p_i prime and $\varepsilon_i = \pm 1$.

Let

$$M(x) = \max_{1 \leq n \leq x} |f(n)|, \qquad E(x) = \max_{p \leq x} |f(p+1)|.$$

In a later paper (Elliott [7]) I proved that for completely additive functions $f(n)$ there are positive (absolute) constants so that

(2) $$M(x) \leq AE(x^B)$$

holds for $x \geq 2$.

Let Q^* be multiplicative group of positive rational numbers, written multiplicatively. Let G_1 be the subgroup generated by the integers of the form $p+1$ with $p \leq n^B$, and G_2 the subgroup generated by the positive integers not exceeding n.

According to the above inequality, any completely additive function which vanishes on G_1 and so has $E(n^B) = 0$, will vanish on G_2. An application of lemma (15.3) now shows that the primes in the representation (1) may be restricted to lie in the interval $2 \leq p_i \leq n^B$.

Assuming only that $f(\ )$ be additive, I established the weaker result

$$M(x) \leq AE(x^B) + AM((\log x)^C)$$

for some $C > 0$.

That (2) holds for all additive functions $f(\ )$ was proved by Wirsing [4]. He strengthened the representation (1) by restricting the primes p_i to lie in an interval $n < p_i \leq n^B$, and having both h and the total number of factors in the product bounded independently of n. His argument did not make use of additive functions. In particular, Wirsing's result showed that Q^*/P has bounded order.

For the sequence P this brings us to the end of stage (ii) of the general procedure discussed at the beginning of the present chapter.

In order to prove that every integer n has a representation

(3) $$n = \prod_i (p_i + 1)^{\varepsilon_i}, \qquad \varepsilon_i = \pm 1$$

it is sufficient, and also necessary, that for each prime q, a completely additive arithmetic function $f(\ )$, with values in the group $\mathbb{Z}/q\mathbb{Z}$, which satisfies $f(p+1) \equiv 0 \pmod{q}$ for all primes p must also satisfy $f(n) \equiv 0 \pmod{q}$ for every positive integer n.

Indeed, this condition is clearly necessary. Supposing it to be satisfied let p be a prime divisor of the h appearing in Wirsing's form of the representation (1). We apply lemma (15.4) with $G = G_2 = Q^*/pQ^*$, $G_1 = P/pQ^*$. For each positive integer n our temporary hypothesis thus guarantees the existence of a representation

$$n = z^p \prod_i (p_i + 1)^{\varepsilon_i}$$

with a rational z and primes p_i.

Then

$$n^{p^{-1}h} = z^h \prod_i (p_i + 1)^{\varepsilon_i},$$

and since z^h is representable as a product of the $(p_j + 1)$, we can obtain a representation of the form (1) with $p^{-1}h$ in place of h.

Stripping off the prime divisors of h in this manner we obtain (3).

Note that this argument is possible only because there is a uniform bound for our initial h, that is, the group Q^*/P has bounded order.

Further Results

Let q be a rational prime and let $p_1 < p_2 < \cdots$ be the rational primes. Define the integers

$$a_1 = p_1^{q^2}, \qquad a_{j+1} = (p_1 \cdots p_{j+1})^q (p_1 \cdots p_j)^{-1}, \qquad j = 1, 2, \ldots.$$

Let A be the subgroup of Q^* which is generated by these a_i.

If for any $i \geq 2$ and integer m, p_i^m belongs to A, then there will be a representation

$$p_i^m = \prod_{l=1}^{s} a_l^{d_l}$$

with integers d_l, and $s \geq 1$. Since each p_{j+1} occurs in a_{j+1} and in no a_w with $w \leq j$, we see that $s \leq i$ must hold. Then $m = qd_i$.

On the other hand (group-theoretically)

$$p_{j+1}^q \equiv (p_1 \cdots p_j)^{q-1} \pmod{A},$$

and an easy inductive proof shows that

$$p_i^{q^{i+1}} \equiv 1 \pmod{A}, \qquad i = 1, 2, \ldots.$$

We see that every element of the group Q^*/A has an order which is a power of q, and which is at least q.

Since every element of Q^*/A has a finite order all homomorphisms of it into the additive group of real numbers are trivial. Likewise it cannot have a non-trivial homomorphism into an F_p with $p \neq q$. Moreover,

$$p_j = \frac{p_1 \cdots p_j}{p_1 \cdots p_{j-1}} \equiv \frac{(p_1 \cdots p_{j+1})^q}{(p_1 \cdots p_j)^q} \equiv p_{j+1}^q \pmod{A}$$

for $j \geq 2$, and

$$p_1 = \frac{p_1 p_2}{p_2} \equiv \frac{(p_1 p_2 p_3)^q}{p_2} \pmod{A},$$

so that every element of Q^*/A is a qth-power. Thus it has no non-trivial homomorphisms into F_q.

Since no p_i belongs to A it is clear that the triviality of the homomorphisms into the additive group of the reals, and of those into the finite fields F_p is not enough to ensure the triviality of Q^*/A, and therefore the representation of integers as products of the a_j.

The reason for this is that these homomorphisms have ranges in groups which do not possess enough structure. By considering maps into more structured groups better may be done.

Let D_1 be a divisible R-module. Suppose further that for each prime element π of R there is a non-zero element g of D_1 so that $\pi g_1 = 0$. We shall call these modules *extra-divisible*.

Lemma (15.5). *Let G_1 and G_2 be submodules of an R-module G. Suppose that every homomorphism of G into D_1 which is trivial on G_1 is also trivial on G_2. Then G_2 is contained in G_1.*

Proof. For each g in G_2, lemma (15.2) guarantees that the ideal of elements r in R for which rg belongs to G_1 contains a non-zero element.

Suppose that for some g in G_2 this ideal is non-trivial, and is generated by α. Let π be a prime element of R which divides α and set $y = \pi^{-1}\alpha g$. Then y belongs to G_2 but not to G_1. Moreover, πy lies in G_1.

Let M be the module generated by G_1 and y. We define a map T of M into D_1 by choosing a non-zero element δ of D_1 which satisfies $\pi\delta = 0$ and setting

$$T(ry + \mu) = r\delta$$

for every r in R and μ in G_1. If

$$r_1 y + \mu_1 = r_2 y + \mu_2,$$

then $(r_1 - r_2)y$ belongs to G_1, so that π divides $r_1 - r_2$. Let $r_1 - r_2 = \rho\pi$. Then

$$T(r_1 y + \mu_1) - T(r_2 y + \mu_2) = (r_1 - r_2)\delta = \rho\pi\delta = 0,$$

and T is well defined.

By lemma (15.1) we may extend T to a homomorphism of G into D_1, which is then trivial on G_1 but not G_2. This contradicts the hypothesis of the lemma.

Lemma (15.5) is proved.

A candidate for D_1 is the multiplicative group of complex numbers which are roots of unity, or its isomorphic copy the additive group $Q/\mathbb{Z}$ of rationals (mod 1), viewed as $\mathbb{Z}$-modules.

A complex-valued arithmetic function $\varphi(n)$ is said to be *multiplicative* if it satisfies

$$\varphi(ab) = \varphi(a)\varphi(b)$$

for all pairs of positive coprime integers a, b and to be *completely multiplicative* if this relation holds for all positive integers a and b.

We can now state

Lemma (15.6). *Let $a_1, a_2, \ldots$, be a sequence of positive integers. In order that every positive integer may have a representation of the form*

$$n = \prod_{j=1}^{s} a_j^{d_j}$$

for some integers d_j, positive, negative or zero, it is necessary and sufficient that every completely multiplicative arithmetic function which is 1 on the a_j and whose values are roots of unity, be identically 1.

A form of this result would be implicit in Theorem 2 of Dress and Volkmann [1]. There they were interested in what properties a sequence a_j must have in order that one could reconstruct a complex-valued, completely multiplicative function $\varphi(\ )$ from its values $\varphi(a_j)$. In particular $\varphi(n)$ was allowed to be sometimes zero. However, the proof which they give is not complete. A detailed discussion of their argument may be found in my paper, Elliott [16].

A correct form of lemma (15.6) was first given by Meyer [1], save that his multiplicative functions assume arbitrary complex values. His proof of this result employs $\mathbb{Z}$-modules and rests implicitly upon the fact that as a $\mathbb{Z}$-module Q^* is free. Note that in his paper an additive function takes (the traditional) real values.

My present treatment differs in the following manner. Additive and multiplicative functions are regarded as being essentially group homomorphisms, and so differ only insofar as they assume values in non-isomorphic groups. The emphasis thus falls on the image groups (here the *divisible* and *extradivisible* groups) and the results may be used to study any infinite abelian group, whether it be free or not. All the proofs are module-theoretic.

Moreover, the method is localized. As I shall presently indicate, this brings an important advantage: it allows the construction of an algorithm for the representation of integers as products. For this last I do employ the freedom of Q^*.

Simultaneous Representation

Let $Q_2 = Q^* \oplus Q^*$ be the direct sum of two copies of the multiplicative group of positive rationals. Let a_n, $n = 1, 2, \ldots$, and b_m, $m = 1, 2, \ldots$, be two infinite sequences of positive integers, and let G be the subgroup generated by the pairs $a_n \oplus b_n$.

If $f(\ )$ is a homomorphism of Q_2 into, for example, the additive group of the real numbers, then we can decompose it as

$$f(\ ) = f_1(\ ) + f_2(\ ),$$

where

$$f_1(r \oplus s) = f(r \oplus 0), \qquad f_2(r \oplus s) = f(0 \oplus r)$$

for all r and s in Q^*. This naturally defines two maps of Q^* into $\mathbb{R}$.

The group Q_2/G is now studied by considering those additive functions $f_i(n)$, $i = 1, 2$, which satisfy

$$f_1(a_n) + f_2(b_n) = 0$$

for all n. If these must vanish identically, then to each pair of positive integers m_1, m_2 we can find further integers $k > 0$, $n_i > 0$, $\varepsilon_i = \pm 1$, $i = 1, \ldots, s$, so that (simultaneously)

$$m_1^k = \prod_{i=1}^{s} a_{n_i}^{\varepsilon_i},$$

$$m_2^k = \prod_{i=1}^{s} b_{n_i}^{\varepsilon_i}.$$

The preceding theory may thus be adapted to deal with the simultaneous representation of integers.

Algorithms

As an example suppose that a positive integer m has a product representation

$$m^v = \prod_{i=1}^{t} a_i^{d_i},$$

where the a_i are positive integers and the d_i are positive, negative or zero integers. Suppose further that the representing a_i are localized by $a_i \leq M$. For convenience of exposition let $m \leq M$ also hold. I now give an algorithm to determine values of v and the d_i.

Let $q_1, q_2, \ldots, q_k$ be the primes which divide m and the a_i, $1 \leq i \leq t$. Then there are the canonical representations.

$$m = \prod_{j=1}^{k} q_j^{\beta_j},$$

$$a_i = \prod_{j=1}^{k} q_j^{\alpha_{ij}}, \qquad i = 1, \ldots, t.$$

Corresponding to these representations there are vectors

$$\mathbf{y} = \begin{pmatrix} \beta_1 \\ \vdots \\ \beta_k \end{pmatrix}, \quad \mathbf{x}_i = \begin{pmatrix} \alpha_{i1} \\ \vdots \\ \alpha_{ik} \end{pmatrix}, \qquad i = 1, \ldots, t,$$

in Q^k, which satisfy

$$v\mathbf{y} = \sum_{j=1}^{t} d_j \mathbf{x}_j.$$

Choose a maximal set of the vectors $\mathbf{x}_j$ which are linearly independent in the vector space Q^k defined over Q. This may be effected by evaluating the minors of the matrix

$$(\alpha_{ij})^T, \quad 1 \leq i \leq t, \quad 1 \leq j \leq k.$$

By relabelling, if necessary, we can suppose that $\mathbf{x}_1, \ldots, \mathbf{x}_r$ is such a set.
There will now be a representation

$$w\mathbf{y} = \sum_{i=1}^{r} h_i \mathbf{x}_i$$

with integers w, h_i, $1 \leq i \leq r$. It will be enough to solve the r linear equations obtained by taking any r of the (t available) coordinate relations. Using the usual (Cramer) procedure the h_i and w may be chosen to be certain $r \times r$ minors of the matrix

$$\begin{pmatrix} \alpha_{11} & \cdots & \alpha_{r1} & \beta_1 \\ \vdots & & \vdots & \vdots \\ \alpha_{1r} & \cdots & \alpha_{rr} & \beta_r \end{pmatrix}$$

with possibly a change of sign.

Typically

$$\sum_{j=1}^{r} \alpha_{ij} \le \frac{1}{\log 2} \sum_{j=1}^{r} \alpha_{ij} \log q_j$$

$$\le \frac{1}{\log 2} \log a_i \le \frac{\log M}{\log 2},$$

so that

$$\sum_{j=1}^{r} \alpha_{ij}^2 \le \left(\frac{\log M}{\log 2} \right)^2.$$

By means of Hadamard's inequality for determinants we see that choices for w, h_i, $i = 1, \ldots, r$, and therefore for v and the d_i exist which in absolute value do not exceed

$$\left(\frac{\log M}{\log 2} \right)^k \le \left(\frac{\log M}{\log 2} \right)^{\pi(M)}.$$

Here $\pi(M)$ denotes the total number of primes not exceeding M.

One may similarly consider the vectors $\mathbf{y}$ and $\mathbf{x}_i$ when they are interpreted as being in F_p^k over the field F_p.

This algorithm need not give at once the minimal positive v so that m^v has a product representation of the asserted type.

The minimal value of v will be a divisor of the integer w so far obtained.

Viewing them as vectors in Q^k over the rational field Q we can represent the $\mathbf{x}_s$ with $s > r$ by

$$\mathbf{x}_s = \sum_{i=1}^{r} \rho_{si} \mathbf{x}_i$$

with certain rational numbers ρ_{si}.

Let δ be a divisor of w and let

$$\delta \mathbf{y} = \sum_{i=1}^{r} \lambda_i \mathbf{x}_i$$

with the appropriate rational λ_i.

In order to represent $\delta \mathbf{y}$ as a linear combination of integral multiples of the $\mathbf{x}_i$, $1 \le i \le t$, we seek integers u_i so that

$$\lambda_i = u_i + \sum_{s=r+1}^{t} u_s \rho_{si}, \qquad i = 1, \ldots, r.$$

Let d be the least common denominator of the fractions λ_i, ρ_{si}, $1 \le i \le r$, $r+1 \le x \le t$. Our problem is then equivalent to solving the simultaneous congruences

$$\sum_{s=r+1}^{t} u_s d \rho_{si} \equiv d \lambda_i \pmod{d}, \qquad i = 1, \ldots, r.$$

If this can be done at all it can be done with integer values of the u_s in the interval $0 \le u_s \le d-1$.

There are finitely many choices of the u_s, $s = r+1, \ldots, t$ which need to be considered to decide whether a given power m^δ of m has a product representation by the a_i. Since there are only finitely many divisors δ of w, the minimal v for which m^v is representable by the given a_i may be determined.

Let $a_1, a_2, \ldots$ be a sequence of rational numbers which generate the subgroup Γ of Q^*. Suppose that we can prove the triviality of Q^*/Γ in the three steps suggested at the beginning of this chapter, and that it can be done with localization. Then: Q^*/Γ is finitely generated now allows us to determine a positive integer s so that the sth-power of each positive integer m has a product representation by the a_i. This may not be the minimal such s.

Since there exist localized solutions to

$$m = z^p \prod_{i=1}^{r} a_i^{d_i}, \qquad p \text{ prime},$$

the algorithms given above enable us to find suitable integers r, d_i and z. This argument can be carried out with appropriate primes p until we obtain a representation of the form

$$m = t^s \prod_{i=1}^{k} a_i^{w_i}$$

for some integers $t > 0$, k and w_i.

We can now determine a representation for m.

Suppose now that a simultaneous representation

$$m_1^v = \prod_{i=1}^{r} a_i^{u_i}, \qquad m_2^v = \prod_{i=1}^{r} b_i^{u_i}$$

is possible. Let $p_1, \ldots, p_s$ denote the primes which divide m_1, m_2 the a_i and the b_j. Let the corresponding canonical representations be

$$m_1 = \prod_{j=1}^{s} p_j^{\delta_j}, \qquad m_2 = \prod_{j=1}^{s} p_j^{\lambda_j},$$

$$a_i = \prod_{j=1}^{s} p_j^{\alpha_{ij}}, \qquad b_i = \prod_{j=1}^{s} p_j^{\beta_{ij}},$$

for $i = 1, \ldots, r$. To obtain a simultaneous representation for m_1^v and m_2^v is the same as solving in integers v, u_i, $i = 1, \ldots, r$, the equation

$$v \begin{pmatrix} \delta_1 \\ \vdots \\ \delta_s \\ \lambda_1 \\ \vdots \\ \lambda_s \end{pmatrix} = \sum_{i=1}^{r} u_i \begin{pmatrix} \alpha_{i1} \\ \vdots \\ \alpha_{is} \\ \beta_{i1} \\ \vdots \\ \beta_{is} \end{pmatrix},$$

and the above algorithms may be employed.

In the next four chapters we pursue in detail the representation of a given integer by products of certain specified types.

Simultaneous representations are considered in Chapters 18 and 19. As is to be expected the complications increase so that our results are less complete. In particular, methods from both algebra and analysis are employed.

It is clear that the results of the present five chapters represent only a beginning, and that many interesting questions remain to be solved.

Chapter 16

The Second Intermezzo

Let Q^* be the multiplicative group of the positive rational numbers, and let $a>0$, $A>0$, b, B be integers which satisfy

$$\Delta = \det\begin{pmatrix} a & b \\ A & B \end{pmatrix} \neq 0.$$

For each k, let $\Gamma(k)$ denote the subgroup of Q^* which is generated by the positive fractions of the form

$$\frac{an+b}{An+B}, \quad n \geq k.$$

In the first intermezzo it was proved that the quotient group $Q^*/\Gamma(k)$ is finitely generated. Let T denote the subgroup of it formed by the elements of finite order—the torsion subgroup. Then from the general theory of abelian groups (see, for example, Lang [1], Theorem 8, p. 49), T is finite and there is a direct sum decomposition

$$Q^*/\Gamma(k) = T \oplus K,$$

where K is a free subgroup of $Q^*/\Gamma(k)$.

We now determine the membership of T and the rank of K.

A prime p will be called *exceptional* if there are integers $h_1 > h_2 \geq 0$, $t_1 > t_2 \geq 0$ so that

$$p^{h_1} \| a, \quad p^{h_2} \| b; \quad p^{t_1} \| A, \quad p^{t_2} \| B.$$

Given any sufficiently large power q^j of a prime q which is not exceptional, there will be at least one (and therefore infinitely many) integer(s) so that q^j appears in the canonical factorization of the fraction $(an+b)/(An+B)$.

Let ρ be the ratio $(a, b)/(A, B)$ of the highest common factors of a and b, A and B, respectively. We define successively ρ_0 to be that part of ρ

which is made up of powers of exceptional primes, and ρ_1 to be the positive integer for which $\rho_0 = \rho_1^m$ with m as large as possible. For example, if $a = 27$, $b = 36$, $A = 20$, $B = 12$, then $\rho = \tfrac{9}{4}$ and $\rho_1 = 1$. Another example is $a = 2^5 3^2 5^2 7$, $b = 2^4 3 \cdot 5 \cdot 7$, $A = 2 \cdot 3^2 5^4 7^3$ and $B = 3 \cdot 5^3 7^2$. The exceptional primes are 2, 3 and 5, so that $\rho = 2^4 5^{-2} 7$, $\rho_0 = 2^4 5^{-2}$ and $\rho_1 = 2^2 5^{-1}$.

Lemma (16.1). *In order that an integer n should have a representation*

(1)
$$n^v = \prod_{i=1}^{r} \left(\frac{an_i + b}{An_i + B} \right)^{d_i},$$

with integers d_i, $v > 0$ and $n_i > k$, it is necessary and sufficient that it be a product of an integer which is not divisible by an exceptional prime, and of a power of ρ_1.

Proof. Let $f(\ )$ be a real-valued completely additive function which satisfies

$$f(an + b) - f(An + B) = 0$$

for all sufficiently large positive integers n. From our solution of Kátai's problem, in Chapter 13, there will be a constant F so that

$$f(D) = F \log D$$

holds for all integers D which are prime to $aA\Delta$.

If, moreover, p is a prime which divides $aA\Delta$, and further satisfies

$$p^i \| (an + b), \qquad p^j \| (An + B)$$

for some n then, in the notation of theorem (10.2),

$$L(p^i, p^j, F) = 0.$$

For non-exceptional primes p we may by taking suitably large values of i or j, and appealing to the complete additivity of our function $f(\ )$, deduce that

$$f(p) = F \log p.$$

The assertion

$$L_t(F) + F \log(a/A) = C$$

of theorem 13, with the appropriate choice of $t = 1$ or 2, leads in the present circumstances to

$$f(\rho_0) = F \log \rho_0 - F \log(a/A).$$

Consider now any positive fraction of the form $(an+b)/(An+B)$. We can express it in the form

(2) $$\rho\left(\frac{a_1 n + b_1}{A_1 n + B_1}\right),$$

where $a_1 = a(a, b)^{-1}$ and so on. Only ρ may be divisible by exceptional primes, and such primes can be considered together as ρ_0. From what we have so far established

$$F \log\left(\frac{an+b}{An+B}\right) = F \log\left(\frac{a}{A}\right).$$

This can only hold for *all* large n if $F = 0$.

We have now shown that any real completely additive function which vanishes on $\Gamma(k)$ also vanishes on the non-exceptional primes, and on ρ_1. An application of lemma (15.2) of Chapter 15 now shows that the condition in lemma (16.1) is sufficient.

Conversely, if an integer n has a representation of the form (1) then let $n = n_1 n_2$ where n_2 contains all the exceptional prime factors of n. Since each fraction $(an+b)/(An+B)$ can be put into the form (2), we must have

(3) $$n_2^v = \rho_0^s = \rho_1^{ms}.$$

for some (possibly zero) integer s.

Let p_i, $i = 1, \ldots, w$ be the exceptional primes which divide ρ_1, and let

$$\rho_1 = \prod_{i=1}^{w} p_i^{\alpha_i}.$$

Clearly only these p_i can occur as factors of n_2. Let

$$n_2 = \prod_{i=1}^{w} p_i^{\beta_i}.$$

Since the integer m in the definition of ρ_1 is maximal, the exponents α_i, $i = 1, \ldots, w$, have highest common factor 1. There are thus integers z_i so that

$$\sum_{i=1}^{w} z_i \alpha_i = 1.$$

The rational integers are a unique factorization domain, and (3) allows us to assert that

$$v\left(\sum_{i=1}^{w} z_i \beta_i\right) = ms\left(\sum_{i=1}^{w} z_i \alpha_i\right) = ms.$$

In particular v divides ms, say $ms = zv$.
Therefore

$$n_2 = \rho_1^z,$$

and our proof of lemma (16.1) is complete.

We see from lemma (16.1) that the torsion group T is generated by the non-exceptional primes and by $\rho_1 (\bmod \Gamma(k))$.

Let there be r exceptional primes, p_l.
We shall now show that the free (component) group K has rank

$$\begin{cases} r & \text{if } \rho_1 \text{ has no exceptional prime divisors,} \\ r-1 & \text{if some exceptional prime divides } \rho_1. \end{cases}$$

In the second of these possibilities it is to be understoood that $r \geq 1$.
The following simple result will expedite our considerations.

Lemma (16.2). *Any $w+1$ elements in an abelian group with w or fewer generators, necessarily satisfy a non-trivial relation.*

Proof. Let g_i, $i = 1, \ldots, y$ with $y \leq w$, span the group, and let h_j, $j = 1, \ldots, w$, be further elements in the group.
Writing additively, there are integers θ_{ji} so that

$$h_j = \sum_{i=1}^{y} \theta_{ji} g_i, \quad j = 1, \ldots, w.$$

Since $w > y$ there are non-zero rational, and indeed integer, solutions to the system of equations

$$\sum_{j=1}^{w} z_j \theta_{ji} = 0, \quad i = 1, \ldots, y.$$

In terms of these integers z_j

$$\sum_{j=1}^{w} z_j h_j = \sum_{i=1}^{y} \left(\sum_{j=1}^{w} z_j \theta_{ji}\right) g_i = 0.$$

Lemma (16.2) is proved.

The Second Intermezzo 295

Returning to our group K let us assume that ρ_1 is divisible by at least one exceptional prime, the other case being simpler to decide. Any element g in K is the image of a positive rational number, r say. This rational may be written (uniquely) as the product of primes, non-exceptional and exceptional. If τ denotes the order of the group T, then g^τ has (in $Q^*/\Gamma(k)$) the form

$$\prod_{l=1}^{r} \bar{p}_l^{\delta_l},$$

where $\bar{p}$ denotes the image of p under the projection

$$Q^* \to Q^*/\Gamma(k).$$

Between the elements $\bar{p}_l$ there is the non-trivial relation $\bar{\rho} = 1$. Without loss of generality let p_1 divide ρ_1. Then there is a positive integer λ so that for every g in K,

$$g^{\tau\lambda} = \prod_{l=2}^{r} \bar{p}_l^{\nu_l}.$$

The subgroup of $Q^*/\Gamma(k)$ which is generated by the $\bar{p}_l$ with $2 \le l \le r$, contains the free subgroup $K^{\tau\lambda}$.

By lemma (16.2) $K^{\tau\lambda}$, and so K, has rank at most $r-1$.

In the other direction, suppose that K has a rank of less than $r-1$, so that $r \ge 2$ is assumed. Each $\bar{p}_l$ with $2 \le l \le r$ differs (multiplicatively) from a member of K by a torsion element. There is therefore a positive integer λ so that all of $\bar{p}_l^\lambda$, $l = 2, \ldots, r$, lie in K.

By lemma (16.2) our temporary hypothesis allows us to conclude the existence of a non-trivial relation

$$\prod_{l=2}^{r} \bar{p}_l^{\lambda z_l} = 1.$$

In other words there is a representation

$$\prod_{l=2}^{r} p_l^{\lambda z_l} = \prod \left(\frac{an_i + b}{An_i + B} \right)^{d_i}.$$

Once again (see (2)) there will be an integer s so that

$$\prod_{l=2}^{r} p_l^{\lambda z_l} = \rho_1^s.$$

Since not all the z_l are zero, s cannot be zero. But then p_1 divides the right-hand side of this equation and not the left.

With this contradiction we have proved that K has rank at least, and therefore exactly, $r-1$.

The example

$$a = A = 3, \quad b = 1, \quad B = 2,$$

was considered at the end of the first intermezzo, where it was proved that the group $Q^*/\Gamma(k)$ has at least one element of infinite order. In fact there is the single exceptional prime 3, and $\rho_1 = \rho = 1$. We see now that rank $K = 1$, and $Q^*/\Gamma(k)$ has precisely one infinite generator.

As a further example, let

$$a = 15, \quad b = 13; \quad A = 30, \quad B = 7.$$

In this case there are two exceptional primes, 3 and 5. Once again $\rho_1 = 1$, and the free summand K of $Q^*/\Gamma(k)$ has rank 2.

Note that in this second example the torsion group T is not trivial. Indeed if

$$n = \prod_i \left(\frac{15n_i + 13}{30n_i + 7}\right)^{d_i},$$

then there is an integer d so that

$$n \equiv (\tfrac{13}{7})^d \pmod 5.$$

Since the possibilities of $(\tfrac{13}{7})^d$ are only $\pm 1 \pmod 5$, no integer of the form $5h + 2$ will belong to $\Gamma(k)$, although some power of it must.

It is not difficult to show that torsion groups T of arbitrarily high order can occur.

Chapter 17

Product Representations by Values of Rational Functions

In this chapter I prove that every positive integer may be represented as a product of the values of certain completely reducible rational functions. This will be an example in the systematic application of the method of Chapter 15. An essential rôle will be played by a ring of operators.

Theorem (17.1). *Let*

$$R(x) = \prod_{i=1}^{h} (x + a_i)^{b_i}$$

be a rational function with integer roots $-a_i \leq 0$, *and non-zero exponents whose highest common factor* $(b_1, \ldots, b_h)$ *is* 1. *Let an integer* $k \geq 3$ *be given.*
    Then every positive integer n *has a representation of the form*

$$n = \prod_j R(n_j)^{\varepsilon_j},$$

where each $\varepsilon_j = \pm 1$, *and the* n_j *lie in an interval* $k \leq n_j \leq c_0 n$ *for some constant* c_0.

Remarks. The condition $(b_1, \ldots, b_h) = 1$ is necessary. If, for example, every b_i were even, then products of the $R(m)$ could only represent squares of integers.
    An algorithm can be given for the determination of the constant c_0.

A Ring of Operators

Let S be an R-module, containing at least two elements, defined over an integral domain R which has an identity. Consider the set of all doubly infinite sequences $(\ldots s_{-1}, s_0, s_1, s_2, \ldots)$ of elements of S. We introduce the shift operator E whose action takes a typical sequence $\{s_n\}$ to the new sequence $\{s_{n+1}\}$. If

$$F(x) = \sum_{j=1}^{r} c_j x^j$$

is a polynomial with coefficients in R, we extend this definition by defining

$$F(E)s_n = \sum_{j=1}^{r} c_j s_{n+j}.$$

In this way we define a ring of operators which is isomorphic to the ring of polynomials with coefficients in R. In what follows operator will mean a (polynomial) operator which belongs to this ring.

Let K be the quotient field of R.

Lemma (17.2). *Let $F(x)$ be a polynomial in $R[x]$ which factorizes into*

$$a \prod_{i=1}^{r} (x - \theta_i)$$

over some extension field of K. Then for each positive integer d

$$a^{rd} \prod_{i=1}^{r} (x - \theta_i^d)$$

also belongs to $R[x]$.

If, furthermore, R is integrally closed, then the polynomial

$$a^{rd} \prod_{i=1}^{r} (x^d - \theta_i^d)$$

is divisible by $F(x)$ in $R[x]$.

Remark. For the properties of integral closure, see Zariski and Samuel [1], Chapter V.

Proof. Consider the polynomial

$$\prod_{i=1}^{r} (x - y_i^d),$$

with the y_i distinct indeterminates over K. The coefficient b_j of x^j, $0 \le j \le r$, is a symmetric function if the y_i, of total degree $(r-j)d$. If σ_ν, $\nu = 0, \ldots, r$ denotes the elementary symmetric functions of the y_i, then b_j is a polynomial in these σ_ν, of degree at most rd. (See, for example, van der Waerden [1], Volume 1, Chapter 26.)

Specializing the y_i to θ_i, we see from our first hypothesis that every $a\sigma_\nu$ belongs to R. Hence $a^{rd}b_j$ belongs to R for every j, which justifies the first assertion of the lemma.

Consider next the polynomial

$$W(x) = a^{rd} \prod_{i=1}^{r} (x^d - \theta_i^d).$$

Clearly each factor $x^d - \theta_i^d$ is divisible by $x - \theta_i$ in some algebraic extension of K. By working in a large enough extension $F(x)$ will divide $W(x)$. Since K is a field $F(x)$ then divides $W(x)$ in $K[x]$.

For each root θ_i of $F(x) = 0$, $a\theta_i$ is integral over R. The coefficients of the polynomial

$$a^{d-1} \frac{x^d - \theta_i^d}{x - \theta_i}$$

are thus integral over R, and so are those of the polynomial $W(x)R(x)^{-1}$.

Since R is integrally closed in its quotient field, this last polynomial actually belongs to $R[x]$.

The lemma is proved.

In our next two lemmas and in their application, R will be a unique factorization integral domain with identity.

A function $f(n)$ is said to be *arithmetic* if it is defined on the positive natural integers.

We extend the sequence $f(1), f(2), \ldots$ of values of an arithmetic function to a doubly infinite sequence by setting $f(n) = 0$ if $n \leq 0$.

Note that if $f(\ )$ is an arithmetic function

$$Ef(2n) = f(2n+1).$$

If, however, we define a new arithmetic function $g(\ )$ by $g(n) = f(2n)$, then

$$Eg(n) = g(n+1) = f(2n+2).$$

In what follows $f(\ )$ will denote a completely additive arithmetic function.

Lemma (17.3). *In the above notation suppose that the additive arithmetic function $f(\ )$ satisfies*

$$\phi(E)f(n) = \text{constant}, \qquad k \leq n \leq H$$

for some operator $\phi(E)$.

Let

$$\phi(x) = a \prod_{i=1}^{s} (x - \omega_i)^{r_i},$$

with distinct ω_i, hold over some extension field of K. Let $t = r_1 + \cdots + r_s$ denote the degree of $\phi(x)$.

Let a positive integer d be given.

Then either there is a permutation σ of the ω_i with

(1) $$\sigma \omega_i = \omega_i^d, \qquad i = 1, \ldots, s,$$

or there is a further non-zero polynomial $\phi_1(x)$, defined over R and with degree less than that of $\phi(x)$, so that

(2) $$\phi_1(E)f(n) = \text{(another) constant}$$

holds over the interval $k \leq n \leq (H/d) - t$.

Proof. Consider the polynomial

$$G(x) = a^{td} \prod_{i=1}^{s} (x - \omega_i^d)^{r_i}$$

By lemma (17.2), $\phi(x)$ divides $G(x^d)$ in $R[x]$. Therefore

$$G(E^d)f(n) = \text{constant}$$

for $k \leq n \leq H - m_1$ where

$$m_1 = \deg\left(\frac{G(x^d)}{\phi(x)}\right) \leq t(d-1).$$

Let

$$G(x) = \sum_{j=0}^{t} c_j x^j.$$

Then

$$G(E^d)f(n) = \sum_{j=0}^{t} c_j f(n+dj),$$

so that for $k \le nd \le H - m_1$

$$\sum_{j=0}^{t} c_j f(n+j) = \sum_{j=0}^{t} c_j [f(d\{n+j\}) - f(d)]$$

$$= G(E^d)f(nd) - G(1)f(d) = \text{constant}.$$

In particular

$$G(E)f(n) = \text{constant}$$

over the range $k \le n \le (H/d) - t$.

If the roots of G are a permutation of the ω_i (in both cases neglecting the multiplicities r_i) we obtain the first of the two possibilities appearing in the statement of lemma (17.3). Otherwise $G(x)$ and $a^{td-1}\phi(x)$ have the same leading terms, but are distinct. With $\phi_1(x) = a^{td-1}\phi(x) - G(x)$ we then have the second of the possibilities.

Lemma (17.3) is proved.

Lemma (17.4). *Let*

$$\phi(E)f(n) = \text{constant}, \qquad k \le n \le H,$$

where $\phi(x)$ is a polynomial over R of degree t. Let d be an integer, $d \ge 2$. Then there are integers q, $0 \le q \le t$, and a non-zero element δ of R, such that

$$\delta(E-1)^q f(n) = 0 \quad \text{for} \quad k + t \le n \le Hd^{-t^2(t+1)} - 2t.$$

Moreover, if $H \ge 2^{11t^3} k^2$ and S is a field, then

$$f(n) = 0 \quad \text{for} \quad k + t \le n \le 2^{-6t^3} H.$$

Remark. The same value of δ may serve for all the H which satisfy the hypothesis of the lemma.

Proof. The hypotheses of lemma (17.3) are satisfied. Assume first that $\phi(0) \ne 0$. Suppose that a permutation σ with the property (1) of that lemma exists. Consider a cycle in the permutation, say

$$z_1 \to z_2 \to \cdots \to z_h \to z_1,$$

so that

$$z_j = \sigma z_{j-1}, \quad j = 1, \ldots, h.$$

Then by (1)

$$z_1 = \sigma z_h = z_h^d = (\sigma z_{h-1})^d \cdots = z_1^{d^h},$$

giving

$$z_1^{d^h - 1} = 1.$$

In this way every root ω_i of $\phi(x)$ is seen to be a root of unity, $\omega_i^{d_i} = 1$ say, and each d_i is a divisor of one of the numbers $d^w - 1$, $1 \leq w \leq s$.

Let

$$D = \prod_{w=1}^{s} (d^w - 1) < d^{s^2} \leq d^{t^2}.$$

Then $\omega_i^D = 1$ for every i. Clearly $\phi(x)$ divides the polynomial

$$a^{tD} \prod_{i=1}^{s} (x^D - \omega_i^D)^{r_i},$$

giving

$$a^{tD}(E^D - 1)^t f(n) = \text{constant}$$

over the interval $k \leq n \leq H - t(D-1)$. Arguing as in the proof of lemma (17.3) we replace n by Dn and reach

$$a^{tD}(E - 1)^t f(n) = \text{constant}$$

for $k \leq n \leq (H/D) - t$.

It is convenient at this point to consider the alternative (3) presented in lemma (17.3). This has the form of the hypothesis of the present lemma save that the degree of $\phi_1(x)$ is less than that of $\phi(x)$, and the range $[k, H]$ is reduced to $[k, (H/D) - t]$. Moreover, if $\phi(0) = 0$ then we can choose $\phi_1(x) = x^{-1}\phi(x)$ and replace the range by $[k+1, H-1]$.

Assume that $t \geq 1$. We may argue inductively to reach an integer q, $0 \leq q \leq t$, a non-zero element δ, such that

$$\delta(E - 1)^q f(n) = \text{constant}$$

holds for

$$k+t \leq n \leq \frac{H}{d^{t^2(t-q+1)}} - t\left(1 + \frac{1}{d^{t^2}} + \cdots + \frac{1}{d^{t^2(t-q)}}\right),$$

and certainly over the range

$$k+t \leq n \leq Hd^{-t^2(t+1)} - 2t.$$

If now S is a field, and $q \geq 1$, we set

$$s(n) = (E-1)^{q-1} f(n)$$

and over this same range have

$$s(n+1) - s(n) = c_0,$$

say, giving

$$s(n) = c_0 n + c_1$$

for certain constants c_0, c_1. Proceeding inductively in this manner we obtain a polynomial $g(y)$, of degree at most t, such that

$$f(n) = g(n) \quad \text{for} \quad k+t \leq n \leq Hd^{-t^2(t+1)} - 2t.$$

We next note that so long as

$$(k+t)^2 \leq n^2 \leq Hd^{-t^2(t+1)} - 2t$$

we have

$$g(n^2) - 2g(n) = f(n^2) - 2f(n) = 0.$$

If now $H \geq 2t(4t+k)^2 d^{t^2(t+1)}$, $t \geq 1$, and S has characteristic zero, the polynomial $g(x^2) - 2g(x)$, which is of degree at most $2t$, will have more than $2t$ distinct (integer) roots. It must therefore be identically zero, that is to say $g(x)$ must be a constant.

The same argument may still be made unless S is a field of finite characteristic $p \leq 2t$. For such fields we have

$$f(p) + f(n) = f(pn) = g(0) \quad \text{for} \quad k+t \leq pn \leq Hd^{-t^2(t+1)} - 2t.$$

Thus in every case

$$f(n) = \text{constant}, \quad k+t \le n \le (2t)^{-1} H d^{-t^2(t+1)} - 1.$$

Denoting this constant by c we have

$$2c = 2f(k+t) = f((k+t)^2) = c,$$

giving $c = 0$.

The proof of the lemma is now completed by setting $d = 2$ and treating the simple case $t = 0$ directly.

Proof of the theorem. Let Q^* denote the multiplicative group of positive rational numbers.

Let G_1 denote its subgroup generated by the fractions

$$R(l) \quad \text{with} \quad k \le l \le H,$$

and let G_2 be the subgroup generated by the integers in the interval

$$k + t \le n \le 2^{-6t^3} H.$$

Let W be the group G_2/G_1, viewed as the subgroup of Q^*/G_1 which is generated by the cosets $g \pmod{G_1}$ as g runs through the elements of G_2.

Suppose that f^* is a homomorphism of W into an R-divisible group Γ, where R is a principal ideal domain which acts upon W and Q^*/G_1. Then by lemma (15.1) there is an extension of f^* which maps the whole of Q^*/G_1 into Γ. Thus there is an additive function

$$f: Q^* \to \Gamma$$

which is trivial on the subgroup G_1, and which is consistent with f^* when suitably restricted.

For a homomorphism $f(\ )$ of Q^* into Γ to be trivial on G_1 we must have $f(R(l)) = 0$, that is

(3) $$\sum_{i=1}^{h} b_i f(l + a_i) = 0$$

for each integer l in $[k, H]$. Here we have assumed that the rational integers can be given a suitable interpretation in R.

(i) If now $H \geq 2^{11t^3}k^2$ and $\Gamma = S$ is the additive group $\mathbb{Q}/\mathbb{Z}$, regarded as a $\mathbb{Z}$-module, then we may apply lemma (17.4) with

$$\phi(x) = \sum_{i=1}^{h} b_i x^{a_i}, \qquad t = \max a_i,$$

to obtain integers $m > 0$ and q, $0 \leq q \leq t$, so that

$$m(E-1)^q f(n) = 0, \qquad k + t \leq n \leq H 2^{-t^2(t+1)} - 2t.$$

Note that S is not a field, so we are not permitted to appeal to the second assertion of that lemma.

However, if $q \geq 1$ the function

$$m(E-1)^{q-1} f(n)$$

is constant on the interval $[k+t, 2^{-6t^3} H]$ and, if

$$f(j) = 0, \qquad k+t \leq j \leq k+2t$$

will be zero there.

Arguing inductively we see that the assumption (3) now forces $mf(n)$ to be zero over the whole range $[k+t, 2^{-6t^3} H]$.

In our above notation: if f^* is trivial on the subgroup of W generated by the cosets j (mod G_1), $j = k+t, \ldots, k+2t$, then f^* is trivial on mW the group of mth-powers of the elements in W.

By lemma (15.5) the group mW is finitely generated, with the j (mod G_1) as generators. Moreover, the value of m does not depend upon the value of H.

(ii) We now apply lemma (15.2) directly, with D the additive group of the real numbers, regarded as a $\mathbb{Z}$-module. In this case the hypothesis (3) leads to the conclusion

$$f(n) = 0 \quad \text{on} \quad [k+t, 2^{-6t^3} H] \text{ and so } [0, 2^{-6t^3} H],$$

for the reals are a field and we may apply the full force of lemma (17.4).

Since the above integers j lie in the interval $[k+t, 2^{-6t^3} H]$ there are positive integers μ_j so that

$$j^{\mu_j} \equiv 1 \pmod{G_1}.$$

Let μ denote their product. Then each g in W satisfies

$$g^m \equiv \prod_{j=k+t}^{k+2t} j^{s_j} \pmod{G_1}$$

for some s_j, and so

$$g^{m\mu} \equiv 1 \pmod{G_1}.$$

Thus W has bounded order.

This brings us to the end of stage (ii) of the proof. We have shown that for each integer $n \geq k+t$ there is a representation

$$n^{m\mu} = \prod_i R(n_i)^{\varepsilon_i}$$

with $k \leq n_i \leq 2^{6t^3}(n + 2^{5t^3}k^2)$. Moreover, the value of the exponent $m\mu$ does not depend upon H. In view of our remarks concerning algorithms which were made at the end of Chapter 15, we can determine bounds for the μ_j, and so for μ.

(iii) To complete our proof we apply the argument of part (i) with $\Gamma = S = F_p$, a finite field of p elements, but with Q^* replaced by Q^*/pQ^*, G_j by G_j/pQ^*.

Once again S is a field, and for an f which takes values in F_p to vanish on G_1/pQ^* we must have

$$\phi(E)f(n) = 0 \quad \text{for} \quad k \leq n \leq H.$$

Here the polynomial $\phi(x)$ is interpreted by considering the coefficients as in the residue class field $\mathbb{Z}/p\mathbb{Z}$. Since as rational integers the b_i have highest common factor 1, $\phi(x)$ will not then vanish identically.

We conclude from lemma (17.4) that

$$f(n) = 0 \quad \text{on} \quad [0, 2^{-6t^3}H].$$

The application of lemma (15.4) now shows that $G_2/pQ^* \subseteq G_1/pQ^*$ for every prime p.

In particular each integer $n \geq k+t$ has a representation

$$n = z^p \prod_j R(r_j)^{\nu_j}$$

with $\nu_j = \pm 1$ and $k \leq r_j \leq 2^{6t^3}(n + 2^{5t^3}k^2)$. Clearly the primes which appear in a canonical factorization of the fraction z do not exceed

$$\max_{1 \leq i \leq h} (2^{6t^3}[n + 2^{5t^3}k^2] + a_i) < c_1 n.$$

If now $m\mu > 1$ and p divides $m\mu$, then

$$n^{p^{-1}m\mu} = z^{m\mu} \prod_j R(r_j)^{\nu_j} = \prod_i R(n_i)^{\varepsilon_i},$$

this time with

$$k \le n_i \le 2^{6t^3}(c_1 n + 2^{5t^3} k^2).$$

Note that if z has a prime factor s which is less than $k+t$, then we consider it as a ratio $(k+t)s/(k+t)$.

Arguing inductively we strip off the primes in $m\mu$ to reach

$$n = \prod_i R(n_i)^{\varepsilon_i} \quad \text{with} \quad k \le n_i \le c_1^{\nu+1} n,$$

where v denotes the total number of prime divisors of $m\mu$. With $c_0 = c_1^{\nu+1}$ the theorem is proved.

Since we have a bound for v, c_0 can be computed.

Practical Measures

Given a polynomial $R(x)$, one can obtain a value for the constant c_0 which appears in the statement of theorem (17.1) by determining exponents μ_j for which

(4) $$j^{\mu_j} \equiv 1 \pmod{G_1}.$$

As indicated in Chapter 15, this can be reduced to solving systems of linear equations. In this context the following remark is perhaps helpful.

Let $b = b_1$ and assume that $a_1 > a_2 > \cdots > a_h$. Then for each positive integer n

$$n^b = R(n) \prod_{i=2}^{h} (n - a_1 + a_i)^{-b_i}.$$

By raising everything to the bth-power we may continue, to eventually reach

$$n^{b^s} \equiv \prod_{p \le u} p^{v_p} \pmod{G_1},$$

where u is the larger of $k-1$ and $a_1 - a_h + 1$. Here it is assumed that the H implicit in the definition of G_1 is as large as n. The exponents s and v_p may all be determined.

In order to obtain a value for c_0 it would thus suffice to obtain exponents μ_j for which (4) holds when j is a prime number not exceeding $\max(k-1, a_1 - a_h + 1)$.

As an example let $k = 3$,

$$R(x) = (x+3)^5 (x+1)^{-2}.$$

Then we need only examine the primes not exceeding 3. By evaluating $R(n)$ for small n one obtains readily that

$$2^{67} = R(3)^2 R(5)^5, \qquad 3^{67} = R(3)^{13} R(5)^{-1}.$$

Chapter 18

Simultaneous Product Representations by Values of Rational Functions

In this chapter $R(x)$ will continue to denote a rational function

$$\prod_{i=1}^{h} (x+a_i)^{b_i}$$

with distinct integers $a_i \geq 0$, and integral exponents b_j which may be positive or negative, but whose highest common factor is 1. I prove

Theorem (18.1). *Let m_1, m_2 and t be positive integers. Then there is a (simultaneous) representation*

$$m_1 = \prod_{j=1}^{r} R(n_j)^{\varepsilon_j},$$

$$m_2 = \prod_{j=1}^{r} R(n_j+t)^{\varepsilon_j},$$

with positive integers n_j, and each $\varepsilon_j = \pm 1$.

The method of proof shows that there are infinitely many such representations. Moreover, it can be strengthened to give an upper bound for the n_j in terms of m_1, m_2 and t.

The algebraic argument of the previous chapter which was applied to establish the representation of a single integer by values of $R(x)$ does not extend in any obvious way to cover simultaneous representations. The present proof applies new ideas. In particular, studies are made of linear recurrences defined over modules, and of the asymptotic behaviour of elliptic power-sums.

Without loss of generality I shall assume that $a_h > a_{h-1} > \cdots > a_1$.

Let Q^* be the abelian group of positive rational fractions with multiplication as the rule of combination, and let Q_2 be the direct sum of two copies of Q^*.

Let Γ be the subgroup of Q_2 generated by the (direct) summands $R(n) \oplus R(n+t)$, $n = 1, 2, \ldots$.

I shall establish the theorem by proving, in three steps, that the quotient group $G = Q_2/\Gamma$ is trivial.

Step One: G is Finitely Generated

Linear Recurrences in Modules

Let H be a $\mathbb{Z}$-module, with the operation of $\mathbb{Z}$ on H written on the left. We shall study the solution-sequences $(\alpha_1, \alpha_2, \ldots)$ with the α_j in H, of the recurrence

$$(1) \quad \sum_{j=0}^{k} c_j \alpha_{n+j} = 0,$$

where the c_j are integers with highest common factor 1.

Without loss of generality $c = c_k > 0$ and $k \geq 2$ will be assumed.

As will be seen, the argument would extend readily to R-modules with R a subring of a (non-archimedean) valuation ring.

Lemma (18.2). *Let M be an integer so that $M\alpha_n = 0$ for $n = 1, \ldots, k$. Then*

$$Mc^n \alpha_n = 0$$

for all $n \geq 1$.

Proof. We argue by induction on n. In fact

$$Mc\alpha_{k+1} = -\sum_{j=0}^{k-1} c_j M\alpha_{j+1} = 0,$$

and so on, to give

$$Mc^{n-k}\alpha_n = 0 \quad \text{for} \quad n \geq k+1,$$

from which the desired result follows.

In view of this lemma we shall from now on assume that every element of the module H has finite order.

Let p be a (positive) rational prime.

For each positive integer n let $|n|_p = p^{-r}$, where p^r is the exact power of the prime p which appears in the canonical factorization of n in the rational integers. With this definition one begins the derivation of the well-known p-adic metric on the rational numbers. We shall do our best to construct a valuation on the $\mathbb{Z}$-module H.

If α is a non-zero element of H which has order m, and if p^s is the exact power of p which divides m, $s = 0$ being permissible, we define $v(\alpha) = p^s$. We set $v(0) = 1$.

The appropriate properties of this *pre-valuation* are embodied in

Lemma (18.3).

(i) $v(\alpha) \geq 1$ *always*,
(ii) $v(n\alpha) = v(\alpha)$ *if* $(n, p) = 1$,
(iii) $v(n\alpha) \leq \max(|n|_p v(\alpha), 1)$,
(iv) $v(\alpha + \beta) \leq \max(v(\alpha), v(\beta))$.

Proof. Assertions (i) and (ii) follow directly from the definition of the pre-valuation.

If α and β have orders u and v, respectively, then the least common multiple $[u, v]$ will annihilate $\alpha + \beta$:

$$[u, v](\alpha + \beta) = 0.$$

Thus

$$v(\alpha + \beta) \leq |[u, v]|_p^{-1} = \max(|u|_p^{-1}, |v|_p^{-1}) = \max(v(\alpha), v(\beta)),$$

giving (iv).

Let α be a non-zero element of order m. Let m and n be exactly divisible by p^s and p^r, respectively. If $r \geq s$ then

$$|n|_p v(\alpha) = p^{-r+s} \leq 1,$$

giving the inequality of (iii). Otherwise

$$(p^{-r} m) n \alpha = 0$$

and

$$v(n\alpha) \leq p^{s-r} = |n|_p v(\alpha),$$

from which the inequality (iii) is again obtained.

Returning to the recurrence (1) we note that not every coefficient c_j is divisible by our (arbitrary) prime p.

Lemma (18.4). *Let* $(\alpha_1, \alpha_2, \ldots)$ *be a solution to the equation* (1). *Then for every* $n \geq k+1$ *with* $v(\alpha_n) > 1$:

either: there is an integer j, $1 \leq j \leq k$, *so that*

(2) $$v(\alpha_n) \leq v(\alpha_{n-j}),$$

or: there is an infinite sequence

$$v(\alpha_n) \leq p^{-1} v(\alpha_{n+r_1}) \leq p^{-2} v(\alpha_{n+r_1+r_2}) \leq \cdots,$$

where each r_i satisfies $1 \leq r_i \leq k$.

Proof. Let $\mu = c_h$ be the coefficient c_j, with the maximum j for which $(p, c_j) = 1$. Then for (each) $n \geq k+1$

$$\mu \alpha_n = \sum_{j=1}^{k-h} d_j \alpha_{n+j} + \sum_{j=1}^{h} e_j \alpha_{n-j},$$

where the integers d_j are divisible by p. As usual, empty sums are deemed to be 0.

In view of lemma (18.3)

$$v(\alpha_n) = v(\mu \alpha_n) \leq \max\{\max_{1 \leq j \leq k-h} v(d_j \alpha_{n+j}), \max_{1 \leq j \leq h} v(e_j \alpha_{n-j})\}.$$

Suppose first that this upper bound is

$$v(e_j \alpha_{n-j})$$

for some j in the range $1 \leq j \leq h$. Then since $|e_j|_p \leq 1$,

$$v(\alpha_n) \leq \max(v(\alpha_{n-j}), 1).$$

By hypothesis $v(\alpha_n) > 1$, giving

$$v(\alpha_n) \leq v(\alpha_{n-j}),$$

the first possibility in the lemma.

Otherwise

$$v(\alpha_n) \leq v(d_j \alpha_{n+j}) \leq \max(|d_j|_p v(\alpha_{n+j}), 1)$$

for some j in the range $1 \leq j \leq k-h \leq k$. Once again $v(\alpha_n) > 1$, giving now

(3) $$v(\alpha_n) \leq p^{-1} v(\alpha_{n+r_1})$$

for some r_1 in the interval $1 \leq r_1 \leq k$.

We suppose r_1 to be the minimal integer for which this inequality is valid, and repeat the above argument with $n + r_1$ in place of n. Note that $v(\alpha_{n+r_1}) > 1$.

If in this manner we arrive at an inequality

$$v(\alpha_{n+r_1}) \leq v(\alpha_{n+r_1-j})$$

with $1 \leq j \leq h$, let $m = n + r_1 - j$.

For $m < n$ we get again an inequality of the form (2) in the statement of the lemma.

With $m = n$ we would have

$$v(\alpha_n) \leq p^{-1} v(\alpha_{n+r_1}) \leq p^{-1} v(\alpha_m) = p^{-1} v(\alpha_n),$$

which is impossible.

For $n < m < n + r_1$ we would get

$$v(\alpha_n) \leq p^{-1} v(\alpha_{n+(m-n)}),$$

contradicting the minimality of r_1.

Otherwise we shall obtain an analogue of the inequality (3):

$$v(\alpha_{n+r_1}) \leq p^{-1} v(\alpha_{n+r_1+r_2})$$

for some (minimal) r_2 in the interval $1 \leq r_2 \leq k$.

The proof now proceeds by induction.

Remarks. This lemma shows that in some sense the order of α_n either remains bounded, or grows exponentially. In particular, the result of lemma (18.2) is not unreasonable.

We come now to our application to the theorem. Let T be a positive real number, and let Δ be the subgroup of Q_2 which is generated by Γ (see earlier) and the finite collection

$$l \oplus 1, \quad 1 \oplus l, \quad 1 \leq l \leq T.$$

In order for the group Q_2/Γ to be finitely generated it will suffice if Q_2/Δ is trivial for a suitable chosen T. Any homomorphism of Q_2/Δ into $Q/\mathbb{Z}$ will have the form

$$y \oplus z \to f_1(y) + f_2(z),$$

where the $f_i(\ )$ are, in the usual notation of analytic number theory, completely additive arithmetic functions with values in $Q/\mathbb{Z}$. In view of lemma (15.5) we prove

Lemma (18.5). *With a suitably chosen (finite) T, any pair (f_1, f_2) of completely*

additive functions which take values in $Q/\mathbb{Z}$ and satisfies

(4) $$f_1(R(n))+f_2(R(n+t))=0$$

for all $n \geq 1$, together with

$$f_i(l)=0, \quad i=1,2, \quad 1 \leq l \leq T,$$

is necessarily trivial.

During the proof of this lemma we shall apply the following sieve result.

Lemma (18.6). *Let d be a positive integer. Then there is a constant g so that the number of integers m in the interval $n < m \leq n+y$ which have no prime factor q in the range $d < q \leq \sqrt{y}$ is at most*

$$\frac{gy}{\log y}$$

uniformly for all integers $n \geq 1$ and real $y \geq 2$.

Proof. We apply Selberg's sieve in the form of lemma (1.6). In the notation used there $k(n)=1$ for all n, the rôle of the a_n is played by the integers m in the (present) interval $n < m \leq n+y$, $z = y^{1/4}$, $r = y^{1/40}$ and Q is the product of all primes p in the interval $d < p \leq y^{1/40}$.

With these choices one may set $X = y$, $\eta(d) = d^{-1}$ to define error terms $R(N, d)$ which do not exceed 1 in absolute value.

It is now easy to check that for large enough values of y the remaining conditions of lemma (1.6) are satisfied, and we obtain for the number of integers m counted in the statement of lemma (18.6) the upper bound

(5) $$2y \prod_{d < p \leq y^{1/40}}\left(1-\frac{1}{p}\right) + 2 \sum_{\substack{d \mid Q \\ d \leq y^{1/2}}} 3^{\omega(d)}.$$

The sum which appears here does not exceed

$$2y^{1/2} \sum_{d \mid Q} \frac{3^{\omega(d)}}{d} \leq 2y^{1/2} \prod_{d < p \leq y^{1/40}}\left(1+\frac{3}{p}\right),$$

and after two applications of lemma (1.6) the bound (5) will not be more than $gy/\log y$ for some constant g depending only upon d.

We complete the proof of lemma (18.6) by adjusting g to take care of the finitely many values of y remaining.

Proof of lemma (18.5). In view of the additive nature of the f_i

$$f_i(R(n)) = \sum_{j=1}^{h} b_j f_i(n + a_j).$$

The hypothesis (4) of lemma (18.5) may thus be expressed in the form

$$\sum_{j=0}^{k} c_j \alpha_{n+j} = 0$$

for all $n \geq 1$, where $k = a_h$,

$$\alpha_n = f_1(n) + f_2(n + t),$$

and the integers c_j, not all zero, have highest common factor 1.

We aim to prove that $f_i(n) = 0$, $i = 1, 2$, and so $\alpha_n = 0$, for all n. By hypothesis this assertion is valid for $1 \leq n \leq T - t$.

Recall that $c = c_k$ is positive. If $T \geq k + t$ then $\alpha_n = 0$ for $1 \leq n \leq k$, and by lemma (18.2), $c^n \alpha_n = 0$ for all positive integers n.

If $c = 1$ then $\alpha_n = 0$ for all n, and this already leads to the complete result. Indeed, replacing n by nt we obtain

$$0 = \alpha_{nt} = f_1(nt) + f_2(t\{n + 1\})$$

$$= f_1(n) + f_2(n + 1),$$

since $T \geq t$ and $f_1(t) = 0 = f_2(t)$.

Writing β_n for $f_1(n) + f_2(n + 1)$ we see that for $s \geq 2$

$$f_2(s) = \beta_{s-1} - f_1(s - 1),$$

(6) $\qquad f_1(s) = f_1(s/2) \quad \text{if } s \text{ is even,}$

$$f_1(s) = \beta_s - f_2((s + 1)/2) \quad \text{if } s \text{ is odd,}$$

provided $T \geq 2$ so that $f_1(2) = 0 = f_2(2)$.

Together with $\beta_s = 0$ for $s \geq 1$ these relations clearly demonstrate (inductively) the triviality of the functions f_i.

Suppose now that $c > 1$. Choose a prime divisor p of c and define a pre-valuation $v(\ )$ on $Q/\mathbb{Z}$ in terms of p. We shall prove that if T is fixed at a large enough value, *independent of the definition of the f_i*, then $v(\alpha_n) = 1$ holds for all n.

We argue by contradiction, noting that $v(\alpha_n) = 1$ for $1 \leq n \leq T - t$. Assume that there is an integer $n \geq k + 1$ with $v(\alpha_n) > 1$. We apply lemma (18.4) with the least such n. This rules out the possibility (i) given by that lemma, and we must have an infinite chain of inequalities

$$(7) \qquad v(\alpha_n) \le p^{-1} v(\alpha_{n+r_1}) \le p^{-2} v(\alpha_{n+r_1+r_2}) \le \cdots,$$

with $1 \le r_i \le k$.

There must be an integer J, bounded only in terms of k and t, so that each of the integers

$$n + \sum_{i=1}^{J} r_i,$$

$$\left(n + \sum_{i=1}^{J} r_i\right) + t,$$

has a prime factor q in the range $2t \le q \le n/(2t)$. For otherwise the integers

$$n + \sum_{i=1}^{w} r_i + \begin{Bmatrix} 0 \\ t \end{Bmatrix}$$

for $w = 1, 2, \ldots, z$, will between them generate at least $z/2$ numbers m which have no such factors, and which lie in the interval $n < m \le n + kz + t$. According to lemma (18.6), either $n \le 2t(kz+t)^{1/2}$ or

$$z/2 \le \frac{g(kz+t)}{\log(kz+t)}.$$

We choose for z a value large enough that this last inequality fails, and then restrict T to exceed $2t(kz+t)^{1/2} + t$. Since $v(\alpha_n) > 1$ this will not allow the penultimate inequality.

Hence, writing δ for the sum $r_1 + \cdots + r_J$, we have

$$n + \delta = m_1 m_2,$$

$$n + \delta + t = m_3 m_4,$$

where $2t < m_i \le (n+\delta+t)/(2t)$ for $i = 1, \ldots, 4$. Therefore

$$\alpha_{n+\delta} = \sum_{i=1}^{2} f_1(m_i) + \sum_{j=3}^{4} f_2(m_j),$$

where for all large enough values of n

$$\max_{1 \le i \le 4} m_i \le \frac{n + Jk + t}{2t} < \frac{n-1}{t}.$$

According to our temporary hypothesis, $v(\alpha_u) = 1$ for $1 \le u \le n-1$, so that $v(\beta_s) = 1$ for $1 \le s \le (n-1)/t$. The relations (6) then allow us to assert that

$$v(f_i(s)) = 1$$

for $i = 1, 2$ and all s not exceeding $(n-1)/t$.

In particular we may conclude that $v(\alpha_{n+\delta}) = 1$.

Our chain of inequalities (7) now gives the impossible $v(\alpha_n) \le 1$.

We may carry out this argument using each of the prime divisors of c, and since the primes which divide the order of α_n also divide c, obtain that $\alpha_n = 0$ for every positive n.

Lemma (18.5) is now immediate, and with its proof we have completed step one.

Step Two: G is Finite

In this section I apply quite different ideas.

Lemma (18.7). *Let $h(y) = \alpha y^2 + \beta y$ be a quadratic polynomial with real coefficients, α irrational.*

Then the inequality

$$\sum_{m=1}^{M} \left| \sum_{n=1}^{N} c_n \exp(2\pi i h(n+m)) \right|^2 \le J \sum_{n=1}^{N} |c_n|^2,$$

with

$$J = \min\left(M + \sum_{j=1}^{N} \frac{1}{\|2j\alpha\|},\ N + \sum_{j=1}^{M} \frac{1}{\|2j\alpha\|} \right)$$

holds for all complex numbers c_n for all positive integers M and N.

Proof. Expanding the square and inverting the order of summation we see that the expression to be estimated is

$$\sum_{n_1=1}^{N} \sum_{n_2=1}^{N} c_{n_1} \bar{c}_{n_2} \sum_{m=1}^{M} \exp(2\pi i \{h(n_1+m) - h(n_2+m)\}).$$

If $n_1 = n_2$ the innersum is M. Otherwise (see the remark following lemma (1.7)) in absolute value it does not exceed

$$\left| \sum_{m=1}^{M} \exp(2\pi i (2(n_1 - n_2)\alpha m)) \right| \le (2\|2(n_1 - n_2)\alpha\|)^{-1}.$$

Using the fact that $|c_i c_j| \leq \frac{1}{2}(|c_i|^2 + |c_j|^2)$ we see that a permissible value for J is

$$M + \sum_{j=1}^{N} \frac{1}{\|2j\alpha\|}.$$

However we may dualize this inequality to obtain

$$\sum_{n=1}^{N} \left| \sum_{m=1}^{M} d_m \exp(2\pi i h(n+m)) \right|^2 \leq \left(M + \sum_{j=1}^{N} \frac{1}{\|2j\alpha\|} \right) \sum_{m=1}^{M} |d_m|^2$$

valid for all complex d_m. On the left-hand side of this inequality we have merely interchanged M and N and replaced c_n by d_n.

Lemma (18.8) (Weyl.)

$$\lim_{N \to \infty} N^{-1} \sum_{n=1}^{N} \exp(2\pi i h(n)) = 0.$$

Proof. For each positive integer m

$$S(N) = \left| \sum_{n=1}^{N} \exp(2\pi i h(n)) \right| \leq \left| \sum_{n=1}^{N} \exp(2\pi i h(n+m)) \right| + 2m.$$

Applying the Cauchy–Schwarz inequality and lemma (18.7) we see that

$$MS(N)^2 \leq 2N \left(N + \sum_{j=1}^{M} \frac{1}{\|2j\alpha\|} \right) + 4M^3.$$

Since α is irrational, for each fixed $M \geq 1$

$$\limsup_{N \to \infty} (N^{-1} S(N))^2 \leq 2M^{-1},$$

from which the desired result follows.

Remark. It is easy to modify this argument to obtain the same conclusion assuming only that one of α, β is irrational, a result also due to Weyl.

Elliptic Power-Sums

Lemma (18.9). *Let z_j, $j = 1, \ldots, k$, be complex numbers which satisfy $|z_j| = 1$. Let ρ_j, $j = 1, \ldots, k$ be further complex numbers, and assume that the function*

$$H(n) = \sum_{j=1}^{k} \rho_j z_j^{n^2}$$

is not zero for all positive integers n.
Then
$$\limsup_{n \to \infty} |H(n)| > 0.$$

Remark. The proof will show that if $H(n)$ vanishes for all positive n then either every $\rho_j = 0$, or some ratio z_i/z_j with $i \neq j$ is a root of unity.

Proof. We argue by induction on k. The case $k = 1$ is trivial.
Let $k \geq 2$. Without loss of generality $\rho_1 \neq 0$.
Suppose first that no z_j/z_1 is a root of unity. Then

$$H(n) = z_1^{n^2} \left(\rho_1 + \sum_{j=2}^{k} \rho_j (z_j z_1^{-1})^{n^2} \right),$$

where for $j \geq 2$,

$$z_j z_1^{-1} = \exp(2\pi i \theta_j)$$

for some irrational real number θ_j. Hence

$$\lim_{x \to \infty} x^{-1} \sum_{n \leq x} z_1^{-n^2} H(n) = \rho_1 + \sum_{j=2}^{k} \rho_j \lim_{x \to \infty} x^{-1} \sum_{n \leq x} \exp(2\pi i n^2 \theta_j)$$

$$= \rho_1,$$

each right-hand limit being zero by the result of Hermann Weyl. In this case we deduce that

$$\limsup_{n \to \infty} |H(n)| \geq |\rho_1| > 0.$$

Otherwise we can write

$$z_j = \lambda_j z_1, \quad j = 2, \ldots, m,$$

where z_j/z_1 is not a root of unity for $m < j \leq k$. We write $H(n)$ in the form

$$z_1^{n^2} \left(\sum_{j=1}^{m} \rho_j \lambda_j^{n^2} + \sum_{j=m+1}^{k} \rho_j (z_j z_1^{-1})^{n^2} \right) = z_1^{n^2} (H_1(n) + H_2(n))$$

say.

If $H_1(n) = 0$ for all n, then $H_2(n)$ is non-zero for at least one integer n, and we may apply our induction hypothesis to obtain the desired result.

If $H_1(n) \neq 0$ for some n, then the function

$$J(n) = \sum_{j=1}^{m} p_j \lambda_j^{n^2}$$

is periodic, of period q say, and there is a integer t so that $J(t) \neq 0$. Thus for all positive integers r

$$z_1^{-(t+rq)^2} H(t+rq) = J(t) + H_2(t+rq).$$

Once again $z_j/z_1 = \exp(2\pi i \theta_j)$ where θ_j is irrational for $j > m$, and by another appeal to the Weyl-sum estimate

$$\frac{1}{y} \sum_{r \leq y} \exp(2\pi i (r^2 q^2 + 2rtq)\theta_j) \to 0 \quad \text{as } y \to \infty.$$

Hence

$$\lim_{r \to \infty} \frac{1}{y} \sum_{r \leq y} H_2(t+rq) = 0,$$

and

$$\limsup_{n \to \infty} |H(n)| \geq \limsup_{r \to \infty} |H(t+rq)| \geq |J(t)| > 0.$$

This completes the proof of lemma (18.9).

The aim of this section is to prove

Lemma (18.10). *Any pair (f_1, f_2) of real-valued completely additive functions which satisfies*

$$f_1(R(n)) + f_2(R(n+t)) = 0$$

for all $n \geq 1$ must be trivial.

We can then appeal to lemma (15.2) to deduce that every element of the group G has torsion.

As in *part one*, with $\alpha_n = f_1(n) + f_2(n+t)$ we have

$$\sum_{j=0}^{k} c_j \alpha_{n+j} = 0.$$

Since the real numbers form a field this linear recurrence has a solution of the form

$$f_1(n) + f_2(n+t) = \alpha_n = \sum_{j=1}^{w} F_j(n)\delta_j^n, \quad n = 1, 2, \ldots,$$

where the δ_j lie in some algebraic extension of the rational field Q, and the $F_j(x)$ may be taken to be polynomials defined over this same extension field. Replacing n by tn and appealing to the additive nature of the f_i,

$$f_1(n) + f_2(n+1) = \sum_{j=1}^{w} F_j(tn)\delta_j^{tn} - \sum_{i=1}^{2} f_i(t).$$

This holds for all positive integers n, including even integers:

$$f_1(2n) + f_2(2n+1) = \sum_{j=1}^{w} F_j(2tn)\delta_j^{2tn} - \sum_{i=1}^{2} f_i(t).$$

By subtraction, writing f for f_2, we see that $f(2n+1) - f(n+1)$ and so $f(2n-1) - f(n)$ have representations of the same type:

$$f(2n-1) - f(n) = \sum_{j=1}^{v} P_j(n)\lambda_j^n.$$

Suppose now that the ratio $\lambda_1\lambda_2^{-1}$ is a root of unity, say $\lambda_1^l = \lambda_2^l$. If in this representation we replace n by ln then

$$f(2ln-1) - f(n) = \sum_{j=1}^{v} P_j(ln)\lambda_j^{ln} + f(l),$$

where the terms

$$P_1(ln)\lambda_1^{ln} + P_2(ln)\lambda_2^{ln}$$

may be coalesced into a single term of the same form.

Continuing in this manner we reach a representation

(8) $$f(D^2 n - 1) - f(n) = \sum_{j=1}^{r} S_j(n)\omega_j^n + \text{constant},$$

with D a positive integer, and where no ratio $\omega_i\omega_j^{-1}$ with $i \neq j$ is a root of unity. We shall prove that a representation of this type is only available to the trivial additive function f. In successive steps $f(n)$ is shown to be smaller and smaller.

Without loss of generality we may assume that

$$\omega = |\omega_1| = |\omega_2| = \cdots = |\omega_h| > |\omega_{h+1}| \geq \cdots \geq |\omega_r|.$$

Moreover, we may also assume that

$$d = \text{degree } S_1(x) \geq \text{degree } S_2(x) \geq \cdots \geq \text{degree } S_h(x).$$

Of course the polynomials $S_j(x)$ with $j > h$ (if there are any) may have degrees greater than d.

We begin our estimation of the size of $f(n)$ with

Lemma (18.11). *If $d \geq 1$ or $|\omega| > 1$ then there is a constant E so that*

$$|f(n)| \leq En^d \max(\omega, 1)^n$$

for all positive integers n.

Proof. It follows from the representation (8) that

$$|f(D^2 n - 1) - f(n)| \leq Ln^d \max(\omega, 1)^n$$

for some constant L and all $n \geq 1$. If $D = 1$ then we obtain the result of lemma (18.11) by summation over n. We shall therefore suppose that $D \geq 2$.

The proof now goes by induction on n. For simplicity of exposition let A denote D^2.

If a positive integer n is divisible by a prime divisor q of A, then there is an integer $n_1 = n/q \leq (1 - A^{-1})n$ so that

$$|f(n)| \leq |f(n_1)| + |f(q)|.$$

Otherwise n will have the form $Am + l$ where $1 \leq l \leq A$, $(l, A) = 1$. In this case let z be the unique integer in the interval $1 \leq z \leq A$ which satisfies $zl \equiv -1 \pmod{A}$, say with $zl = Au - 1$. Note that $A \geq 2$ and therefore $z \leq A - 1$ must hold. Then

$$f(n) = f(zn) - f(z) = f(A\{zm + u\} - 1) - f(z),$$

so that writing n_1 for $zm + u$

$$|f(n)| \leq |f(n_1)| + Ln_1^d \max(\omega, 1)^{n_1}.$$

Here the integer n_1 does not exceed $(1 - A^{-1})n + 1$.

In either case we have reduced ourself to estimating $f(\ )$ on a smaller integer, and an inductive argument is readily constructed.

Elliptic Power-Sums

Remark. Instead of this short direct argument we could have applied the result of Chapter 2, as exemplified in the first intermezzo.

Proof of lemma (18.10). If in the representation (8) we replace n by n^2, the term $D^2n^2 - 1$ factorizes into $(Dn-1)(Dn+1)$ and we obtain

$$\sum_{j=1}^{r} S_j(n^2)\omega_j^{n^2} = f(Dn+1) + f(Dn-1) - 2f(n) + N$$

for some constant N. In view of lemma (18.11) this right-hand side does not exceed a constant multiple of $n^d \max(\omega, 1)^{nD}$ in size.

Suppose that $\omega > 1$. Dividing both sides of the above equation by $n^{2d}\omega^{n^2}$ we obtain an asymptotic relation

$$\sum_{j=1}^{h} \rho_j z_j^{n^2} \to 0, \qquad n \to \infty,$$

since no matter what the values of d or D,

$$n^{-d} \max(\omega, 1)^{nD} \omega^{-n^2} \to 0$$

as n becomes unbounded. Here we have written z_j for $\omega_j \omega^{-1}$, and ρ_j is the coefficient of x^d in the polynomial $S_j(x)$.

In view of lemma (18.9), the elliptic power-sum

$$\sum_{j=1}^{h} \rho_j z_j^{n^2}$$

must be zero for all $n \geq 1$. But since not all the $\rho_j = 0$, and we have arranged that no ratio z_i/z_j with $i \neq j$ is a root of unity, this cannot be the case.

Thus $\omega \leq 1$, and every $|\omega_j| \leq 1$.

Suppose now, without loss of generality, that $\omega = 1$ but that $d \geq 1$. Then lemma (18.11) yields the bound

$$|f(n)| \leq En^d.$$

The argument given above will once again lead to a contradiction.

We can therefore write

$$f(D^2n - 1) - f(n) = \sum_{j=1}^{h} \rho_j \omega_j^n + Y + O(c^{-n}),$$

where Y and $c > 1$ are constants, and every $|\omega_j| = 1$.

In particular

(9) $$f(D^2n-1) - f(n) = O(1)$$

for all n.

Such functions were characterized in Chapter 14, in the example following theorem (14.1). There it was proved that on those integers n which are prime to D, $f(n)$ has the form $B \log n$ for a suitable constant B. The relation (9) then shows that the completely additive function $f(n) - B \log n$ is bounded for all positive n, and so identically zero.

Since

$$\log(D^2n - 1) - \log n = 2 \log D - (D^2n)^{-1} + O(n^{-2}),$$

we can define

$$\rho_0 = Y - 2 \log D, \qquad \omega_0 = 1,$$

and write

$$\sum_{j=0}^{h} \rho_j \omega_j^n = -\frac{B}{D^2 n} + O\left(\frac{1}{n^2}\right).$$

Suppose for the moment that B is non-zero. Replacing n by n^2 gives

$$V(n) = \sum_{j=0}^{h} \rho_j \omega_j^{n^2} = -\frac{B}{(Dn)^2} + O\left(\frac{1}{n^4}\right).$$

Here the expression on the right-hand side (and so also $V(n)$) does not vanish for all large n.

Another application of lemma (18.9) gives

$$\limsup_{n \to \infty} |V(n)| > 0,$$

which is not compatible with the bound $V(n) = O(n^{-2})$.

Hence $B = 0$, and we have proved that $f_2(n) = f(n) = 0$ for all positive integers n.

Returning to our first representation for α_n we now have the simpler

(10) $$f_1(n) = \sum_{j=1}^{w} F_j(n) \delta_j^n$$

valid for $n = 1, 2, \ldots$.

There are several ways to deduce that f_1 is trivial. For example, since $f_1(n)$ satisfies the linear recurrence

$$\sum_{j=1}^{h} b_j f_1(n+a_j) = 0$$

we may appeal to lemma (17.4) from the previous chapter.

Alternatively, we may treat the representation (10) as we did that of (8), after arranging that the ratios δ_i/δ_j, $i \neq j$ are not roots of unity. In this way $f_1(n)$ is seen to be bounded, and a (uniformly) bounded completely additive (real-valued) function is identically zero.

This completes the proof of lemma (18.10).

We have now proved that every element of the group $G = Q_2/\Gamma$ has finite order, and since we established in part one that G is finitely generated, it must in fact be finite.

This completes step two.

Remark. A variant of the argument of this section may be found in the final chapter on problems.

Step Three: G is Trivial

Once again the argument takes a different turn.

Lemma (18.12). *Any pair of completely additive functions (f_1, f_2) which take values in the finite field F_p and satisfies*

$$f_1(R(n)) + f_2(R(n)) = 0$$

for all positive integers is trivial.

Remark. This lemma remains valid if F_p, a finite field of p elements, is replaced by any finite field.

Proof. As in Step Two

$$\alpha_n = f_1(n) + f_2(n+t)$$

satisfies a linear recurrence

$$\sum_{j=0}^{k} c_j \alpha_{n+j} = 0.$$

Here the c_j are interpreted in F_p according to the map

$$c_j \mapsto c_j \pmod{p} \text{ in } \mathbb{Z}/p\mathbb{Z},$$

and since $(c_0, \ldots, c_k) = 1$, not all the c_j vanish $\pmod p$.

We obtain formally the same representation

$$f_1(n) + f_2(n+t) = \sum_{j=1}^{w} F_j(n) \delta_j^n$$

as in Step Two, and with f denoting f_2, reach

(11) $$f(2n-1) - f(n) = \sum_{j=1}^{v} P_j(n) \lambda_j^n,$$

where the λ_j and the coefficients in the polynomials P_j all belong to a finite algebraic extension of F_p, say F_q.

In particular, each λ_j^n is periodic in n, of period $q-1$. The $P_j(n)$ are periodic in n, of period p, so that the whole of the expression on the right-hand side of the above equation is periodic, with a period $p(q-1)$.

The function

$$f(2n-1) - f(n)$$

is therefore periodic, of period $d = p(q-1)$. This may not be its minimal period, but that will not matter in what follows.

Replacing n by $2n^2$ we see that the function

$$f(2n-1) + f(2n+1) - 2f(n) = f(4n^2 - 1) - f(2n^2) + f(2)$$

is also periodic, with the same period; and by linear combination the difference

(12) $$f(2n+1) - f(2n-1).$$

We shall denote this difference by $g(n)$.

Let

$$T = \sum_{n=1}^{d} g(n)$$

be a sum over a complete period (mod d). Then for any positive integer s

$$\sum_{n=1}^{pds} \{f(2n+1) - f(2n-1)\} = spT = 0.$$

But the sum telescopes to give

$$f(2pds+1) = 0$$

for all $s \geq 1$.

An additive function, with values in F_p, which satisfies

$$f(Dn+1) = 0$$

for some positive integer D and all $n \geq 1$, need not be identically zero on the integers prime to D. It will, however, be given by

$$\exp\left(\frac{2\pi i f(n)}{p}\right) = \chi(n)$$

for some (fixed) Dirichlet character χ (mod D). This will be established in lemma (19.3) in the next chapter. We shall not need this result here.

In fact the periodicity of (12) shows that $g(pds)$ has a period 1 in s; it is constant for all $s \geq 1$. With what we have already established, the replacement of n in (12) by pds shows that

$$f(2pds-1) = f(2pd-1) = \text{constant}$$

for all positive s.

Equation (11) with pds in place of n allows us to assert that if $\lambda_0 = 1$ and $P_0(x)$ is a suitable constant (polynomial), then there is a representation

$$f(s) = -\sum_{j=0}^{v} P_j(pds)\lambda_j^{pds}$$

valid for all $s \geq 1$. The expression on the right-hand side of this equation has period 1, so that $f(n)$ is a constant, μ say.

Since

$$-\mu = f(1^2) - 2f(1) = 0,$$

we have proved that the additive function $f_2 = f$ vanishes identically.

In particular

$$f_1(n) = \sum_{j=1}^{w} F_j(n)\delta_j^n$$

for all $n \geq 1$. It is easy to obtain from this representation that $f_1(n)$ is periodic and then a constant, and so zero.

In view of lemma (15.4) we see that whatever the choice of prime p, each element of the group G is a pth-power. This forces G to be trivial. For example, let G have order r, so that each element g of G satisfies $g^r = 1$. If p is a prime divisor of r then there is a further element γ of G so that $g = \gamma^p$. Hence

$$g^{rp^{-1}} = \gamma^r = 1.$$

Proceeding inductively we obtain $g = 1$, and the triviality of G.
Theorem (18.1) is proved.

Concluding Remarks

In order to localize the integer n_j in the representation of theorem (18.1) the foregoing arguments can be given a quantitative form. A sharper bound for elliptic power sums can be obtained by using a transcendence measure for the sum of two logarithms of algebraic integers.

Chapter 19

Simultaneous Product Representations with $a_i x + b_i$

In this chapter I establish simultaneous representations of the type mentioned in the preface.

Theorem (19.1). *Let $a > 0$, b, $A > 0$, B be non-zero integers which satisfy $(a, b) = 1 = (A, B)$ and $aB - Ab \neq 0$. Then there is a positive integer v with the following property:*

Given any pair of integers m_1 and m_2 for which $(m_1, a) = 1$ and $(m_2, A) = 1$ there is a simultaneous representation

$$m_1^v = \prod_{i=1}^{r} (an_i + b)^{\varepsilon_i},$$

$$m_2^v = \prod_{i=1}^{r} (An_i + B)^{\varepsilon_i},$$

with positive integer n_i and each $\varepsilon_i \pm 1$.

Remarks. The integers n_i may be required to exceed a given bound.

With the natural interpretation of the conditions $(m_1, a) = 1$ and $(m_2, A) = 1$ the m_i can be allowed to be positive rational fractions.

It is more complicated but not harder to establish a form of this theorem without the conditions $(a, b) = 1 = (A, B)$, and to allow bB to vanish. What is to be expected is indicated by the form of lemma (16.1) in the second intermezzo.

Proof. Let f_1 and f_2 be completely additive arithmetic functions that take values in a $\mathbb{Z}$-module, and satisfy

(1) $\qquad f_1(an + b) + f_2(An + B) = 0$

for all sufficiently large integers n. We shall reduce this to a similar equation involving a single function.

Replacing n by bBn and appealing to the additive nature of the f_i we obtain

$$f_1(aBn+1) + f_2(Abn+1) = c_1$$

for a constant c_1, which has in fact the value $-f_1(b) + f_2(B)$.

Replacing n by $(aB+1)n+1$ and again appealing to the additive nature of the f_i

$$f_1(aBn+1) + f_2(Ab[(aB+1)n+1]+1) = c_2.$$

Subtracting this equation from the previous one gives

$$f_2(Ab(aB+1)n + Ab + 1) - f_2(Abn + 1) = c_3,$$

an equation in one function.

Let the functions f_i assume values in the additive real numbers. This last equation is then of the type considered in the solution of Kátai's problem which I gave in Chapter 13. Note that

$$\det\begin{pmatrix} Ab(aB+1) & Ab+1 \\ Ab & 1 \end{pmatrix} = Ab(aB - Ab) \neq 0.$$

It follows from theorem 13 that

$$f_2(n) = 0 \quad \text{if} \quad (n, Ab) = 1.$$

The argument needed is almost exactly the same as that given in the proof of lemma (16.1), the fact that here the constant c_3 need not be zero being of no ultimate consequence.

If we interchange the rôles of a and A, b and B in the above argument, we instead obtain

$$f_1(n) = 0 \quad \text{if} \quad (n, aB) = 1.$$

Suppose now that p is a prime which divides b but not A. There are infinitely many integers α so that $p^\alpha \equiv 1 \pmod{A}$. For any positive integer s, the integer $(1+sA)Bp^\alpha$ will then have the form $At + B$, and the basic assumption (1) allows us to assert that

(2) $$\alpha f_2(p) = -f_1(at+b) - f_1(B(1+sA)).$$

We shall show that for suitably chosen bounded values of s, $f_1(at+b)$ is bounded for all such α.

Let q be a prime which divides B, and which divides the integer $at+b$, appearing in (2), to the exact power k.

If $q=p$ then q divides both $At+B$ and $at+b$, and so $Ab-aB$, to the power $\min(k,\alpha)$ at least. For all large enough α, k is forced to be bounded.

Suppose therefore that $q \neq p$. The congruence

$$at + b \equiv 0 \pmod{q^k}$$

can hold only if

$$a(1+sA)Bp^\alpha \equiv aB - Ab \pmod{q^k}$$

holds. Since $q \neq p$ there will be an integer r, and a bounded (for all α) integer h so that

(3) $$s \equiv r \pmod{q^{k-h}}.$$

Here r may depend upon α.

For given α and q one of the choices $s \equiv 1$ or $2 \pmod{q^{h+2}}$ is not a solution to the congruence (3), and for it $k \leq h+1$ must hold.

Given a value of α, we may apply the Chinese Remainder Theorem to obtain a bounded $s \pmod{B^l}$ for a suitably high fixed power of B, so that $at+b$ has only a bounded number of prime factors which divide aB.

Since in (2) $f_1(at+b)$ and so $\alpha f_2(p)$ are then bounded for an unbounded sequence of α-values, $f_2(p)=0$. Hence $f_2(n)$ vanishes for all integers n which are prime to A.

Likewise $f_1(n)$ vanishes for all n prime to a.

Let S_1 be the multiplicative group of rational numbers generated by the positive integers which are prime to a, and S_2 the similar group of rationals generated by the positive integers prime to A. We write $S_1 \oplus S_2$ for the direct sum of these groups.

Let Γ be the subgroup of $S_1 \oplus S_2$ which is generated by the elements of the form $(an+b) \oplus (An+B)$, where n runs through those positive integers for which both summands are positive.

Define the quotient group $G = (S_1 \oplus S_2)/\Gamma$.

The above argument shows that every homomorphism of G into the additive group of the real numbers is trivial. In view of lemma (15.3) this implies the existence of simultaneous representations of the type in the statement of theorem (19.1) save that it will not yet guarantee a uniform bound for the exponent v.

Such a bound will certainly exist if G is finitely generated, and so finite. Let (f_1, f_2) be a pair of homomorphisms of S_1 and S_2, respectively, into the additive group $\mathbb{Q}/\mathbb{Z}$ of the rationals (mod 1). This pair will correspond to a homomorphism of G into $\mathbb{Q}/\mathbb{Z}$ if

$$f_1(an+b) + f_2(An+B) = 0$$

whenever the positive integer n gives a positive value to both arguments. In particular, $\mathbb{Q}/\mathbb{Z}$ is a $\mathbb{Z}$-module and our early reduction argument applies, giving

$$f_2(Ab(aB+1)n + Ab+1) - f_2(Abn+1) = c_3$$

for all sufficiently large n. For ease of notation let

$$\beta = Ab(aB+1), \qquad \gamma = Ab+1, \qquad \delta = Ab.$$

If we assume that $f_i(n) = 0$ for all integers not exceeding a certain bound H, and for $i = 1, 2$, then we have

$$f_2(\beta n + \gamma) - f_2(\delta n + 1) = 0$$

for all sufficiently large values of n. Here H depends only upon the four parameters a, b, A and B.

According to the lemma of the first intermezzo, Chapter 4, every integer k has a representation

$$k = \frac{r}{s} \cdot \prod_j \left(\frac{\beta n_j + \gamma}{\delta n_j + 1}\right)^{\varepsilon_j},$$

where the n_j exceed (any given bound) L, each $\varepsilon_i = \pm 1$, and the integers r and s are made up of primes not exceeding a further number t, depending only upon a, b, A, B and L. Thus if H is large enough, and $(k, Ab) = 1$,

$$f_2(k) = \sum_j \varepsilon_j (f(\beta n_j + \gamma) - f(\delta n_j + 1)) = 0.$$

We may similarly prove that a large enough value of H forces f_1 to vanish on every positive integer which is prime to aB.

According to lemma (15.5) the group G is finitely generated, with $m \oplus 1$ and $1 \oplus n \pmod{\Gamma}$, $(m, a) = 1$, $(n, A) = 1$, $1 \leq m$, $n \leq H$, as generators.

G is indeed finite, and its order is a bound for v.

Theorem (19.1) is established.

In the example given in the preface $a = 3$, $b = 1$, $A = 5$, $B = 2$, so that $aB - Ab = 1$.

Theorem (19.1) does not give an explicit bound for v. To determine the best value a study of the value distribution of

$$f_1(an+b) + f_2(An+B)$$

when the f_i have ranges in finite fields is needed. At present I have only partial results. An idea of what is to be expected can be gained from the following

Theorem (19.2). *Let m_1 and m_2 be positive integers not divisible by 3. Then there is a simultaneous representation*

$$m_1 = \prod_{i=1}^{r} (3n_i - 17)^{\varepsilon_i},$$

$$m_2 = \prod_{i=1}^{r} (3n_i + 19)^{\varepsilon_i},$$

with integers $n_i \geq 7$, $\varepsilon_i = \pm 1$, if and only if

(4) $\qquad m_1 \equiv m_2 \equiv 1, 4 \text{ or } 7 \pmod{9}.$

is satisfied.

Remarks and examples. The integers of the form $3n-17$ and $3n+19$ represent 1, 4 or 7 (mod 9). These 3 classes (mod 9) form a group with respect to multiplication, in fact the group of reduced residue classes which are squares (mod 9). Since for each n_i, $3n_i - 17 \equiv 3n_i + 19 \pmod 9$, the congruence condition (4) is necessary.

Every integer t which is prime to 3 satisfies $t^6 \equiv 1 \pmod 9$. In the notation of theorem (19.1) we can take 6 as a value for v. However, in most cases better can be done.

For example, there is a representation

$$5^2 = \prod_{i=1}^{r} (3n_i - 17)^{\varepsilon_i},$$

$$7 = \prod_{i=1}^{r} (3n_i + 19)^{\varepsilon_i},$$

since the least power v for which $5^v \equiv 7 \pmod 9$ is $v = 2$.

A more complicated example is

$$8^2 = \prod_{j=1}^{k} (3t_j - 17)^{\varepsilon_j},$$

$$11^6 = \prod_{j=1}^{k} (3t_j + 19)^{\varepsilon_j}.$$

Here we note that $11^2 \equiv 4$, $11^4 \equiv 7$ and $11^6 \equiv 1 \pmod 9$. Since $8 \equiv -1 \pmod 9$ we shall have $8^v \equiv \pm 1 \pmod 9$ for any v. This forces us to take the sixth power of 11, and the second power of 8 is then the smallest that we can represent.

In theorem (19.2), as in theorem (19.1), it is possible to require that the n_i in the representation exceed a specified bound.

Calculations (mod 9) are aided by noting that 2 is a primitive root (mod 9). We need a preliminary result.

Lemma (19.3). *Let $g(\ )$ be a complex-valued multiplicative function with $g(an+b)$ a non-zero constant for (fixed) integers $a > 0$ and b, and all sufficiently large integers n.*

Then there is a Dirichlet character χ (mod a) so that

$$g(n) = \chi(n)$$

for all positive n which are prime to a.

Remark. This same conclusion in fact holds under the weaker assumption that those integers n for which $g(an+b)$ is a constant have asymptotic density 1. A proof, requiring more than that which follows, may be constructed by means of a theorem of Halász (see Elliott [11], Volume I, Chapter 6), and the sieve of Eratosthenes.

Proof of lemma (19.3). Let γ denote the constant value of $g(an+b)$.

Let t_1 and t_2 be positive integers, prime to a, which satisfy the condition $t_1 \equiv t_2 \pmod a$. By the Chinese Remainder Theorem there is a (large) positive integer c so that

$$ct_1 \equiv b \pmod a, \qquad c \equiv 1 \pmod{t_1 t_2}.$$

Then

$$g(c)g(t_1) = g(ct_1) = g(am+b) \quad \text{for some } m$$

$$= \gamma = g(c)g(t_2).$$

Since γ is non-zero, $g(c) \neq 0$ and $g(t_1) = g(t_2)$.

This shows that $g(\ )$ is periodic, of period a, on the integers prime to a. Now choose a positive integer s to satisfy

$$s \equiv t_1 \pmod{a}, \quad s \equiv 1 \pmod{t_2}.$$

Then

$$g(t_1 t_2) = g(st_2) = g(s)g(t_2) = g(t_1)g(t_2).$$

This proves that g is completely multiplicative on the integers prime to a.

The map $n \pmod{a} \mapsto g(n)$ thus defines a homomorphism of the multiplicative reduced residue class group (mod a) into the multiplicative complex numbers—it is a Dirichlet character.

Lemma (19.3) is proved.

Proof of theorem (19.2). Consider a pair (f_1, f_2) of completely additive functions which take values in a finite field F_p, of p elements, and which satisfy

(5) $$f_1(3n - 35) + f_2(3n + 1) = 0$$

for all $n \geq 13$. Note that

$$3(n - 6) - 17 = 3n - 35,$$

$$3(n - 6) + 19 = 3n + 1.$$

Replacing n in (5) by $3n^2 + 2n$ we employ the identities

$$3(3n^2 + 2n) + 1 = (3n + 1)^2,$$

$$3(3n^2 + 2n) - 35 = (3n - 5)(3n + 7),$$

and appeal to the additive nature of the f_i to obtain

$$f_1(3n - 5) + f_1(3n + 7) + 2f_2(3n + 1) = 0.$$

Eliminating $f_2(3n + 1)$ between this equation and that at (5) gives

$$f_1(3n - 5) + f_1(3n + 7) - 2f_1(3n - 25) = 0,$$

and equation which involves the single function f_1.

If we define

$$h(n) = f_1(3n - 5),$$

then we can write this equation in the form

$$h(n) + h(n+4) - 2h(n-10) = 0, \qquad n \geq 13,$$

or in terms of the shift operator of Chapter 17

$$(E^{10} + E^{14} - 2)h(n) = 0, \qquad n \geq 3.$$

Since $h(n)$ takes values in a field we may formally solve this linear recurrence to obtain a representation

$$h(n) = \sum_{j=1}^{w} P_j(n)\theta_j^n, \qquad n \geq 3,$$

where the θ_j and the coefficients of the polynomials $P_j(x)$ all lie in a finite algebraic extension of F_p, say F_q.

We employ a now-familiar argument. As a function of n, each θ_j^n has a period $q-1$, whilst each $P_j(n)$ has a period p. Thus $h(n)$ has a period $p(q-1)$. Writing D for $p(q-1)$ we see that

(6) $$f_1(3Ds - 5) = \text{constant}$$

for all $s \geq 2$.

Define the completely multiplicative functions

$$g_j(n) = \exp(2\pi i p^{-1} f_j(n)), \quad j = 1, 2.$$

From (6) and lemma (19.3) we see that there is a Dirichlet character $\chi_1 \pmod{3D}$ so that

$$g_1(n) = \chi_1(n) \quad \text{if} \quad (n, 3D) = 1.$$

In particular

$$g_1(3Ds + 1) = \chi_1(1) = 1.$$

From the basic relation (5) with n replaced by $Ds + 12$ we now obtain

$$g_2(3Ds + 37) = 1$$

for all $s \geq 2$. Another application of lemma (19.3) assures the existence of a (further) Dirichlet character $\chi_2 \pmod{3D}$ so that

$$g_2(n) = \chi_2(n) \quad \text{if} \quad (n, 3D) = 1.$$

Note that by (5)

(7) $$g_1(3n-35)g_2(3n+1) = 1, \qquad n \geq 13,$$

so that each $g_j(r)$ is non-zero whenever r is not divisible by 3.

We must now consider the behavior of these g_j on the integers which are not prime to D.

Let D_1 be the integer D stripped of its powers of 3, and let $D = 3^v D_1$. Suppose that n_1 and n_2 are integers prime to 3 which satisfy $n_1 \equiv n_2 \pmod{3^{v+1}}$. We can find an integer μ to satisfy the congruence

$$4D_1 \mu n_1 \equiv 1 \pmod{3}.$$

Then $4D_1 \mu n_1$ has the form $3k+1$ for some integer k and we apply (7):

$$g_2(n_1)g_2(4\mu D_1) = g_2(4\mu D_1 n_1) = \bar{g}_1(4\mu D_1 n_1 - 36)$$
$$= \bar{g}_1(4)\bar{g}_1(\mu D_1 n_1 - 9).$$

This chain of equalities remains true if we everywhere replace n_1 by n_2. Note that $(\mu D_1 n_1 - 9, 3D) = 1$, for $j = 1, 2$, and that

$$\mu D_1 n_1 - 9 \equiv \mu D_2 n_2 - 9 \pmod{3D}.$$

Since, on the integers prime to $3D$, $g_1(\ )$ coincides with the values of a character (mod $3D$), we deduce that $g_2(n_1) = g_2(n_2)$.

Thus there is a character (mod 3^{v+1}) with whose value $g_2(\ )$ coincides on integers prime to 3.

Similarly $g_1(\ )$ coincides with the value of a (possibly different) character (mod 3^{v+1}) on the integers prime to 3.

For simplicity we shall write $g_j(n) = \chi_j(n)$ for $(n, 3) = 1$; that is as if D were 3^v.

The basic relation (5) has come down to

(8) $$\chi_1(3n-35)\chi_2(3n+1) = 1$$

for all integers $n \geq 13$.

Our next step is to show that these characters have degrees which differ at most by a factor 2. Let χ_j have degree d_j, $j = 1, 2$. Then

(9) $$\chi_2^{d_1}(3n+1) = 1$$

for all $n \geq 13$, and so for all integers n.

Suppose for the moment that $\chi_2^{d_1}$ is non-principal. By raising it to an appropriate power we obtain a non-principal character χ, of prime order l, so that $\chi(3n+1) = 1$ for all $n \geq 1$.

Thus every residue class of the form $3n+1$ must be an lth-power (mod $3^{\nu+1}$). Since there are 3^ν classes of the former type and $l^{-1}\varphi(3^{\nu+1}) = l^{-1} 2 \cdot 3^\nu$ of the latter, $l = 2$.

Looking back at (9) it is clear that the order of $\chi_2^{d_1}$ must be a power of 2, 2^r say, where $r = 0$ is allowed.

The form of $\chi_2(n)$ is $\exp(2\pi i a d_2^{-1} \operatorname{ind} n)$, where the index is taken with respect to a primitive root (mod $3^{\nu+1}$), and $(a, d_2) = 1$. Thus

$$\frac{ad_1}{d_2} \equiv \frac{1}{2^r} \pmod{1},$$

which we may also write as

(10) $\qquad 2^r a d_1 \equiv d_2 \pmod{2^r d_2}$

those relations being in $\mathbb{Q}/\mathbb{Z}$ and $\mathbb{Z}/2^r d_2 \mathbb{Z}$, respectively. There will be a positive integer b so that

(11) $\qquad ab \equiv 1 \pmod{d_2},$

and we altogether obtain

$$2^r d_1 \equiv 2^r a b d_1 \equiv b d_2 \pmod{2^r d_2}.$$

If $r \geq 1$ then from (10) d_2 is even, and therefore by (11), b is odd. In this case there will be an integer M_1 so that

$$2^r d_1 = d_2(1 + 2M_1).$$

Otherwise $r = 0$ and there will be an integer L_1 so that

$$d_1 = L_1 d_2.$$

A similar argument can be made with the rôles of the characters χ_1 and χ_2 interchanged. This leads either to an $s \geq 1$ for which (in an obvious notation)

$$2^s d_2 = d_1(1 + 2M_2),$$

or to an integer L_2 so that

$$d_2 = L_2 d_1.$$

Examining the consequences of combining these pairs of possible conditions shows that the only possibilities are $d_1 = 2^s d_2$ for some $s \geq 0$, or $d_2 = 2^r d_1$ for some $r \geq 0$. For example, we cannot have relations with both r and s positive, for this would give

$$2^{r+s} d_1 = (1 + 2M_1)(1 + 2M_2) d_1,$$

and cancelling d_1 we obtain the inequality of two integers, one odd and one even.

For the sake of argument we shall treat the case when $d_2 = 2^r d_1$, the other case giving a similar outcome.

Let the character χ_1 be given by

$$\chi_1(n) = \exp(2\pi i c d_1^{-1} \text{ ind } n),$$

where $(c, d_1) = 1$, $c > 0$.

Then for integers m which are prime to 3

$$\chi_1(m) = \chi_2(m)^{2^r bc}.$$

Writing k for $2^r bc$ we can combine this relation with that of (8) to obtain

$$\chi_1(3n - 35) = (\chi_2(3n - 35))^k = (\bar{\chi}_1(3n + 1))^k.$$

Let σ denote the operation of raising χ_1 to its kth power and then forming the complex conjugate. In this notation

$$\chi_1(3m + 1) = \sigma \chi_1(3[m + 12] + 1)$$

for all m, positive, negative or zero. This is a relation which can be applied repeatedly, to give

$$\chi_1(3m + 1) = \sigma^t \chi_1(3[m + 12t] + 1)$$

for all m, and all $t \geq 0$.

Setting $m = -12t$ we obtain $\chi_1(-36t + 1) = 1$ and therefore

$$\chi_1(36t - 1) = \bar{\chi}_1(-1),$$

for all $t \geq 1$.

In particular $g_1(36t - 1)$ is a constant for all large t, and so by lemma (19.3), $g_1(\ )$ coincides with a Dirichlet character (mod 36) on the integers not divisible by 2 or 3.

This together with the relation (7) shows that when $(n, 6) = 1$ the values $g_2(n)$ coincide with the values of a (possibly different) character (mod 36).

There is one more reduction of this type, and the modulus goes down to 9. Let $n_1 \equiv n_2 \pmod 9$, $(n_1, 3) = 1$. Let w be a positive integer so that $wn_1 \equiv 1 \pmod 3$. Then $16wn_1$ has the form $3k+1$, and we may apply (7):

$$g_2(16w)g_2(n_1) = g_2(16wn_1) = \bar{g}_1(16wn_1 - 36)$$
$$= \bar{g}_1(4)\bar{g}_1(4wn_1 - 9).$$

Since $(4wn_j - 9, 36) = 1$ and

$$4wn_1 - 9 \equiv 4wn_2 - 9 \pmod{36},$$

the last term in the above chain of equalities is unaffected when n_2 is replaced by n_1. Hence $g_2(n_1) = g_2(n_2)$; which is to say that $g_2(n)$ has the period 9 on the integers prime to 3.

It is now clear that the $g_j(\ )$ coincide, on the integers prime to 3, with the values of (possibly different) Dirichlet characters $\chi_j \pmod 9$. Moreover these characters satisfy the relations (8), which becomes

$$\chi_1\chi_2(3n+1) = 1$$

for all n.

If the character $\chi_1\chi_2$ has order h then every residue class of the form $3n+1$ is an hth power (mod 9). There are 3 residue classes of the former type, and $h^{-1}\varphi(9) = 6/h$ of the latter (assuming, as we may, that h is a divisor of $\varphi(9)$). This is possible only if $h = 1$ or 2.

Thus either χ_2 is the complex conjugate character to χ_1, or it is the complex conjugate character multiplied by the quadratic character.

Since the original additive functions f_j took values in F_p for a prime p, some further control is exercised over the orders of these characters χ_j. This will not be needed here. What matters is that if m_1 and m_2 are a pair of integers which satisfy the condition (4) in the statement of theorem (19.2), then

$$\exp(2\pi i p^{-1}\{f_1(m_1) + f_2(m_2)\}) = g_1(m_1)g_2(m_2) = \chi_1(m_1)\chi_2(m_2) = 1,$$

so that

(12) $$f_1(m_1) + f_2(m_2) = 0$$

in F_p.

To complete the proof of theorem (19.1), let T be the multiplicative group of positive rationals which is generated by the positive integers which belong to one of the classes 1, 4 or 7 (mod 9). It will be more convenient to work in $T \oplus T$ rather than the direct sum of two copies of the positive rationals prime to 3, as would be the case in the proof of theorem (19.1).

Let Γ be the subgroup of $T \oplus T$ generated by the elements of the form

$$(3n-35) \oplus (3n+1), \qquad n = 13, 14, \ldots,$$

and form the quotient group $G = (T \oplus T)/\Gamma$.

In view of theorem (19.1), G is finite.

Suppose that this group has order $k > 1$, and that p is a prime divisor of k. Let pG be the group of pth-powers of the elements in G.

For odd primes p any homomorphism τ of G/pG into the finite field F_p may be extended to a homomorphism of Q^* the whole of the positive rationals, and so give rise to a pair (f_1, f_2) of completely additive arithmetic functions which take values in F_p and satisfy

$$f_1(3n-35) + f_2(3n+1) = 0.$$

Indeed, these functions will initially be defined on the integers generated by T. This includes every prime l which is $\equiv 1, 4$ or $7 \pmod 9$, but only the squares of the primes which are $\equiv 2, 5$ or $8 \pmod 9$. For such a prime l we define $f_j(l)$ by $2f_j(l) = f_j(l^2)$ which lies in F_p. Since $2 \neq 0$ in F_p we gain in this way an additive function on the whole of the positive integers.

In particular, our above analysis applies, and the homomorphism τ is seen to be trivial. Thus for any element δ of G there will be a further element γ so that $\delta = \gamma^p$. As a consequence

$$\delta^{kp^{-1}} = \gamma^k = 1.$$

In this way we can strip off all the odd prime divisors in the exponent k. In orther words every element of G will have an order which is a power of 2. By a well-known theorem of Sylow the order of G must also be a power of 2, but we shall not need this.

When $p = 2$ an extension of the functions f_j from T to the whole of the positive rationals may not be possible. Instead we prove directly that any homomorphism τ of $G/2G$ into F_2 is trivial. With the arguments which we have to hand this is not difficult.

Let f_1 and f_2 be additive functions, induced by τ, which are defined on the integers in T, map it into F_2, and satisfy

$$f_1(3n-35) + f_2(3n+1) = 0$$

for all $n \geq 13$.

We may still apply the argument which obtains from this an equation involving the single function f_1 (see (5) and what follows). In our present circumstances this gives

$$f_1(3n-5) + f_1(3n+7) = 0, \qquad n \geq 2,$$

or, what is the same,

$$f_1(3n+4) = f_1(3[n-4]+4)$$

for all $n \geq 3$.

In order to justify the claim, made in the remarks following the statement of theorem (19.2), that one may restrict the integers n_i from below, let us assume that this latest relation holds only for $n \geq d$. Iterating it t times gives

$$f_1(3n+4) = f_1(3[n-4t]+4)$$

so long as $n \geq 4t+d-4$. In particular, with $n = 4(t+d)$ we obtain

$$f_1(3t+3d+1) = f_1(3d+1)$$

and then, with $t = s(3d+1)$,

$$f_1(3s+1) = 0$$

for all $s \geq 1$.

Thus f_1 vanishes on every integer m_1 which belongs to one of the residue classes 1, 4 or 7 (mod 9).

This together with the basic relation (5) shows that

$$f_2(3n+1) = 0$$

first for all sufficiently large n, and then, by raising the argument to a high enough odd power, for all positive n.

The homomorphism τ is clearly trivial on $G/2G$. We may now argue as when considering the odd prime factors of the order k of G, to prove that in fact $k = 1$.

The group G is trivial.

Theorem (19.2) is established.

Chapter 20

Information and Arithmetic

In this chapter I consider what form a mathematical information function can take. I shall confine myself to the collection of all finite probability spaces. A typical space is modelled by a set of n points, together with n non-negative real numbers whose sum is 1. These numbers play the rôle of the probabilities with which the various (point) events occur. If the appropriate assumptions are made concerning atomicity there is no problem in extending the following results to other probability spaces.

An information function, defined on the collection of all possible pairs $(A, P(\ ))$ of events A taken from the various spaces, with attached probability measure P, is a (real-valued) function

$$\mathscr{F}(P(A))$$

whose value depends only upon $P(A)$, the probability with which the event occurs. This is a somewhat inhibiting requirement, but leads to interesting mathematical problems.

Given mutually exclusive events A, B in a space with measure $P(\ )$ we define

$$I(A, B) = \mathscr{F}(P(A)) + P(\bar{A})\mathscr{F}(P(B|\bar{A}))$$

if $P(\bar{A}) > 0$, and

$$I(A, B) = \mathscr{F}(P(A))$$

if $P(\bar{A}) = 0$. Here $\bar{A}$ denotes the event complementary to A in this same space, and

$$P(B|\bar{A}) = \frac{P(B)}{P(\bar{A})}$$

is the usual probability measure $P(\ )$ conditioned by the event $\bar{A}$. Since B is contained in $\bar{A}$ it is natural to treat the event B with this conditional measure. Not so justified is the choice of the scaling factor $P(\bar{A})$, and indeed other worthwhile choices are possible.

Two further requirements are now made of information functions: That $I(A, B)$ be symmetric in A and B; and that

(1) $$\mathcal{F}(P(\phi)) = \mathcal{F}(P(\Omega)).$$

This last states that in any space the null event and the total events, which occur almost never and almost always, respectively, are to carry the same information. This information will presently be shown to be zero.

A space with a probability measure which takes only rational values will be called a *rational probability space*. To begin with I shall consider only such spaces.

Let x, y be non-negative rational numbers whose sum does not exceed 1. We may clearly construct a finite probability space containing exclusive events A, B which occur with probability x and y respectively. In fact it only needs a space of three points. Then

$$I(A, B) = \mathcal{F}(x) + (1-x)\mathcal{F}\left(\frac{y}{1-x}\right)$$

and the symmetry condition gives

(2) $$\mathcal{F}(x) + (1-x)\mathcal{F}\left(\frac{y}{1-x}\right) = \mathcal{F}(y) + (1-y)\mathcal{F}\left(\frac{x}{1-y}\right).$$

This equation is certainly meaningful if neither of x or y is 1. In particular $x = 0$ gives $y\mathcal{F}(0) = 0$ for $0 \le y < 1$, and so $\mathcal{F}(0) = 0$. The third condition (1) to be met by information functions now yields

(3) $$\mathcal{F}(0) = 0 = \mathcal{F}(1)$$

and the equation (2) remains valid even if x or $y = 1$ so long as we interpret $0\mathcal{F}(0/0)$ to be 0.

If we do not assume condition (1) to be satisfied then the new information function

$$\tilde{\mathcal{F}}(x) = \mathcal{F}(x) - x\mathcal{F}(1)$$

clearly satisfies (1) and the information equation (2).

Setting $y = 1 - x$ in the equation (2) we obtain

(4) $$\mathcal{F}(x) = \mathcal{F}(1-x).$$

Information and Arithmetic 345

We are reduced to the solution of the equation (2), valid for all rationals with $x + y \leq 1$, subject to the boundary condition (3).

Theorem (20.1). *Let*

$$\liminf_{n \to \infty} \frac{\mathscr{F}(1/n)}{\log n} \geq 0$$

hold for rational probability spaces.
Then there is a constant A so that

$$\mathscr{F}(r) = A(r \log r + (1-r) \log(1-r))$$

for all rational numbers $0 \leq r \leq 1$.

In particular, the conclusion of this theorem holds under the natural assumption that no event decreases the information, so that $\mathscr{F}(1/n) \geq 0$.

It is traditional to require $\mathscr{F}(1/2) = 1$, which would fix the value of A at $-1/\log 2$. We shall not make this requirement, but define Shannon's (entropy) function

$$S(z) = -\frac{1}{\log 2}(z \log z + (1-z) \log(1-z))$$

for real z, $0 \leq z \leq 1$.

Our next result gives a quantitative local characterization of the information function.

Define

$$\mu(x) = \sup_{x \leq w \leq x^c} \frac{1}{W} \sum_{n \leq w} \left| \mathscr{F}\left(\frac{1}{n}\right) \right|^2.$$

Theorem (20.2). *There is a positive (absolute) constant c, and functions $G(x)$, $E(x)$ so that*

$$\frac{1}{x^2} \sum_{n \leq x} \sum_{m=1}^{n} \left| \mathscr{F}\left(\frac{m}{n}\right) - G(x) S\left(\frac{m}{n}\right) - E(x) \right|^2 \ll \mu(x)$$

for all $x \geq 2$.
With this same constant c there are arithmetic functions $G(n)$, $M(n)$ so that

$$\frac{1}{n} \sum_{m=1}^{n} \left| \mathscr{F}\left(\frac{m}{n}\right) - G(n) S\left(\frac{m}{n}\right) - M(n) \right|^2 \ll \mu(n)$$

for all $n \geq 2$.

If

$$\frac{1}{x} \sum_{n \leq x} \left| \mathcal{F}\left(\frac{1}{n}\right) \right|^2 \to 0, \qquad x \to \infty,$$

then there is a constant G so that

$$\mathcal{F}(r) = GS(r)$$

for all rationals $0 \leq r \leq 1$.

Remarks. In the bounds the implied constants are absolute.

Whilst these results are typical I shall later in this chapter consider information functions as algebraic objects.

Transition to Arithmetic

In this section, $\mathbf{a} = (a_1, \ldots, a_n)$, $\mathbf{b} = (b_1, \ldots, b_m)$ will denote vectors with at least two coordinates, each of which is a non-negative real number.

For each such vector $\mathbf{a}$ define

$$J(\mathbf{a}) = J(a_1, \ldots, a_n) = \sum_{k=2}^{n} (a_1 + \cdots + a_k) \mathcal{F}\left(\frac{a_k}{a_1 + \cdots + a_k}\right),$$

interpreting $0\mathcal{F}(x/0)$ to be 0. If $\mathbf{a}, \mathbf{b} = (b_1, \ldots, b_m)$ are two vectors, $J(\mathbf{a}, \mathbf{b})$ denotes

$$J(a_1, \ldots, a_n, b_1, \ldots, b_m),$$

whilst $J(a, \mathbf{b})$ denotes

$$J(a, b_1, \ldots, b_m),$$

and so on.

The following property of the function J will yield all its remaining properties that we need.

Lemma (20.3). *For $2 \leq m \leq n-1$*

$$J(a_1, \ldots, a_n) = J(a_1 + \cdots + a_m, a_{m+1}, \ldots, a_n) + J(a_1, \ldots, a_m).$$

Proof. If $m = 2$ then the desired result follows immediately from the definition of J. The cases $m \geq 3$ are then readily obtained by induction.

Lemma (20.4). $J(\mathbf{a})$ *is invariant under a permutation of the coordinates of* $\mathbf{a}$.

Proof. If $\mathbf{a}$ has two coordinates the desired result follows from the relation (4). If it has three coordinates then it follows from the basic equation (3) with

$$x = a_3(a_1 + a_2 + a_3)^{-1}, \qquad y = a_2(a_1 + a_2 + a_3)^{-1}.$$

Note that if $a_1 + a_2 + a_3 = 0$ then every $a_i = 0$, so that $J(\mathbf{a}) = 0$.

We now give a proof by induction on n, the number of coordinates of $\mathbf{a}$. It will be enough to prove invariance under the interchange of any two of the coordinates of $\mathbf{a}$. By lemma (20.3) and the inductive hypothesis we lose no generality if we take this interchange to be between a_2 and a_3.

Then from lemma (20.3)

$$J(\mathbf{a}) = J(a_1 + a_2 + a_3, a_4, \ldots) + J(a_1, a_2, a_3),$$

and each term on the right-hand side of this equation is invariant.

Before stating the next result we note that for any $\lambda \geq 0$

$$J(\lambda \mathbf{a}) = \lambda J(\mathbf{a}).$$

It will also be convenient to introduce the function

$$L(\mathbf{a}) = a_1 + a_2 + \cdots + a_n.$$

Lemma (20.5). *The relation*

$$J(p_1 \mathbf{q}_1, p_2 \mathbf{q}_2, \ldots, p_n \mathbf{q}_n) = J(p_1 L(\mathbf{q}_1), \ldots, p_n L(\mathbf{q}_n)) + \sum_{j=1}^{n} p_j J(\mathbf{q}_j)$$

holds for all vectors $\mathbf{p} = (p_1, \ldots, p_n)$, $\mathbf{q}_j$, $j = 1, \ldots, n$.

Proof. By lemma (20.3) the expression on the left-hand side of the above equation can also be written in the form

$$J(p_1 L(\mathbf{q}_1), p_2 \mathbf{q}_2, \ldots, p_n \mathbf{q}_n) + J(p_1 \mathbf{q}_1).$$

It is now easy to apply the invariance of J under permutation of the coordinates of its argument, and give a proof by induction.

Let $\mathbf{a} \otimes \mathbf{b}$ denote the tensor product of the vectors $\mathbf{a} = (a_1, \ldots, a_n)$, $\mathbf{b} = (b_1, \ldots, b_m)$, that is to say the vector with coordinates $a_i b_j$, $i = 1, \ldots, n$, $j = 1, \ldots, m$.

Lemma (20.6).

$$J(\mathbf{a} \otimes \mathbf{b}) = J(\mathbf{a})L(\mathbf{b}) + J(\mathbf{b})L(\mathbf{a}).$$

Proof. This result follows from lemma (20.5) when

$$\mathbf{q}_1 = \mathbf{q}_2 = \cdots = \mathbf{q}_n = \mathbf{b}, \qquad \mathbf{p} = \mathbf{a}.$$

In particular, if $L(\mathbf{a}) = 1 = L(\mathbf{b})$ we obtain

$$J(\mathbf{a} \otimes \mathbf{b}) = J(\mathbf{a}) + J(\mathbf{b}).$$

Define the real-valued arithmetic function

$$f(n) = J\underbrace{\left(\frac{1}{n}, \ldots, \frac{1}{n}\right)}_{n \text{ times}}, \qquad n \geq 2$$

$$f(1) = 0.$$

Lemma (20.7). *The function $f(n)$ is completely additive. Moreover*

$$\mathcal{F}\left(\frac{m}{n}\right) = f(n) - \frac{m}{n}f(m) - \left(1 - \frac{m}{n}\right)f(n-m)$$

for all integers satisfying $1 \leq m \leq n$.

Proof. The first assertion follows from lemma (20.6) with

$$\mathbf{a} = \underbrace{\left(\frac{1}{n}, \ldots, \frac{1}{n}\right)}_{n \text{ times}}, \qquad \mathbf{b} = \underbrace{\left(\frac{1}{m}, \ldots, \frac{1}{m}\right)}_{m \text{ times}}.$$

To obtain the second assertion when $2 \leq m \leq n-2$ we apply lemma (20.5) with $p_1 = m/n$, $p_2 = 1 - (m/n)$,

$$\mathbf{q}_1 = \underbrace{\left(\frac{1}{m}, \ldots, \frac{1}{m}\right)}_{m \text{ times}}, \qquad \mathbf{q}_2 = \underbrace{\left(\frac{1}{n-m}, \ldots, \frac{1}{n-m}\right)}_{n-m \text{ times}}.$$

The case $m = n$ is trivially valid, whilst for $m = n-1$ we can employ $\mathcal{F}(1 - 1/n) = \mathcal{F}(1/n)$, and so reduce to the case $m = 1$.

In particular, with $m=1$ we obtain

(5) $$\mathcal{F}\left(\frac{1}{n}\right) = f(n) - \left(1 - \frac{1}{n}\right)f(n-1), \qquad n \geq 2.$$

Proof of theorem (20.1). Define the completely additive function $h(n) = -f(n)$. Then by hypothesis

(6) $$h(n) - \left(1 - \frac{1}{n}\right)h(n-1) \leq o(\log n), \qquad n \to \infty.$$

In particular

$$h(n) \leq \left(1 - \frac{1}{n}\right)h(n-1) + c_1 \log n$$

for some constant c_1 and all odd $n \geq 3$. This together with

(7) $$h(n) = h(n/2) + h(2)$$

which is valid for even n, leads by a simple inductive proof to a bound

$$h(n) \leq c_2 (\log n)^2.$$

Likewise our original inequality (6) gives

$$h(n) \geq \frac{n+1}{n} h(n+1) - 2c_1 \log n$$

and so

$$h(n) \geq \left(1 + \frac{1}{n}\right) h\left(\frac{n+1}{2}\right) + \left(1 + \frac{1}{n}\right) h(2) - 2c_1 \log n$$

for all odd $n \geq 3$. Together with (7) we obtain a bound

$$h(n) \geq -c_3 (\log n)^2.$$

Note that if a sequence of positive integers satisfies

$$n_{j+1} \leq (n_j + 1)/2, \qquad j = 1, 2, \ldots, k,$$

and $n_{k+1} \geq 2$, then

$$n_j \geq 2(\tfrac{3}{2})^{k-j}, \qquad j = 1, \ldots, k,$$

so that

$$\prod_j \left(1 + \frac{1}{n_j}\right) \le \exp\left(\sum_j \frac{1}{n_j}\right) \le \exp\left(\tfrac{1}{2} \sum_{s=0}^{\infty} (\tfrac{2}{3})^s\right) = c_4.$$

Hence $h(n) = O((\log n)^2)$, and after another appeal to (5)

$$h(n+1) - h(n) \le o(\log n), \qquad n \to \infty.$$

We can now apply lemma (14.2) to obtain

$$h(n+1) - h(n) = o(\log n), \qquad n \to \infty.$$

In view of our results from Chapter 14 there is a constant A so that

$$f(n) = A \log n, \qquad n \ge 1.$$

The representation of $\mathscr{F}(m/n)$ follows from lemma (20.7), and the proof of theorem (20.1) is complete.

In order to prove theorem (20.2) we relate the mean square of $f(n) - f(n-1)$ to that of $\mathscr{F}(1/n)$.

Lemma (20.8).

$$\sum_{2 \le n \le x} |f(n) - f(n-1)|^2 \le c_1 \sum_{2 \le n \le x} \left|\mathscr{F}\left(\frac{1}{n}\right)\right|^2.$$

Proof. If n is even we have

$$f(n) = f(n/2) + \mathscr{F}(\tfrac{1}{2}).$$

If n is odd we employ (5):

$$f(n) = \left(1 - \frac{1}{n}\right) f(n-1) + \mathscr{F}\left(\frac{1}{n}\right).$$

Arguing inductively we can thus obtain

$$|f(n)| \le \sum_{m_j \le n} \left|\mathscr{F}\left(\frac{1}{m_j}\right)\right| + O(|\mathscr{F}(\tfrac{1}{2})| \log n), \qquad n \ge 2,$$

Transition to Arithmetic

where the sequence of integers m_j satisfies $m_1 = n$ or $n/2$ and

$$m_{j+1} = \begin{cases} m_j/2 \\ \text{or } (m_j - 1)/2. \end{cases}$$

In particular, $m_{j+1} \leq m_j/2$, so that this sequence contains $O(\log n)$ members.

Squaring, employing the Cauchy–Schwarz inequality, dividing by n^2 and summing over $2 \leq n \leq x$ gives

$$(8) \quad \sum_{n \leq x} \left|\frac{f(n)}{n}\right|^2 \ll \sum_{n \leq x} \left(\frac{\log n}{n^2} \sum_{m_j \leq n} \left|\mathcal{F}\left(\frac{1}{m_j}\right)\right|^2 + \left(\frac{\log n}{n}\right)^2 \left|\mathcal{F}\left(\frac{1}{2}\right)\right|^2\right).$$

The series

$$\sum_{n=2}^{\infty} \left(\frac{\log n}{n}\right)^2$$

converges. Inverting the order of summation in the double sum at (8) gives an amount of not more than

$$\sum_{m \leq x} \left|\mathcal{F}\left(\frac{1}{m}\right)\right|^2 \sum_{n \leq x}' \frac{\log n}{n^2},$$

where for each m the variable n in the inner sum is to run through those integers which can be reached from m by operations of the type $t \to 2t$ and $t \to 2t+1$. Thus

$$n = \cdots 2(2(2m + \varepsilon_1) + \varepsilon_2) + \cdots,$$

where each $\varepsilon_i = 0$ or 1. If r such choices ε_i are made then $n \geq 2^r m$. Moreover, the binary expansion of n will be $\alpha_h \cdots \alpha_1 \varepsilon_1 \varepsilon_2 \cdots \varepsilon_r$ where $\alpha_h \cdots \alpha_1$ is the binary expansion of m_1, so that n can be reached in only one way. Thus

$$\sum_{n \leq x}' \frac{\log n}{n^2} < \sum' \frac{2}{n^{3/2}} \leq \sum_{r=0}^{\infty} \frac{2^{r+1}}{(2^r m)^{3/2}} < \frac{8}{m},$$

and the above double sum does not exceed

$$\sum_{m \leq x} \frac{8}{m} \left|\mathcal{F}\left(\frac{1}{m}\right)\right|^2.$$

Altogether

$$\sum_{2 \leq n \leq x} \left|\frac{f(n)}{n}\right|^2 \ll \sum_{m \leq x} \left|\mathcal{F}\left(\frac{1}{n}\right)\right|^2.$$

From this weak result and the relation (5) applied many times we readily obtain the inequality of the lemma.

Remark. If we restrict the integers n in the summations at (8) by $y < n \leq x$, then the above argument easily gives

$$\sum_{y < n \leq x} \left|\frac{f(n)}{n}\right|^2 \ll \frac{1}{\sqrt{y}} \sum_{n \leq x} \left|\mathscr{F}\left(\frac{1}{m}\right)\right|^2.$$

Fixing y at a suitably large (absolute) value we see that

$$\sum_{n \leq x} |f(n) - f(n-1)|^2 \geq \tfrac{1}{3} \sum_{m \leq x} \left|\mathscr{F}\left(\frac{1}{m}\right)\right|^2 - c_2 \sum_{m \leq y} \left|\mathscr{F}\left(\frac{1}{m}\right)\right|^2,$$

showing that the inequality of lemma (20.8) cannot be much improved.

Proof of theorem (20.2). From theorem (10.1) and lemma (20.8) there is a positive (absolute) constant c and a function $F(x)$ so that

(9) $$\sum_{q \leq x} \frac{1}{q} (f(q) - F(x) \log q)^2 \ll \mu(x).$$

Here q denotes a prime-power. In the notation of theorem (10.1) $a = A = 1$, $b = 1$, $B = 0$ and $\Delta = 1$, so that the condition $(q, aA\Delta) = 1$ is always satisfied.

Let $G(x) = -F(x) \log 2$. Then in terms of the Shannon function $S(z)$ we may write

$$\mathscr{F}\left(\frac{m}{n}\right) - G(x) S\left(\frac{m}{n}\right) - f(n) + F(x) \log n$$

$$= -\frac{m}{n}(f(m) - F(x) \log m) - \left(1 - \frac{m}{n}\right)(f(n-m) - F(x) \log(n-m)).$$

Define the mean

$$E = E(x) = \sum_{q \leq x} \frac{1}{q}(f(q) - F(x) \log q).$$

Then by the Turán–Kubilius inequality (lemma (1.3)) and (9)

$$\sum_{m \leq x} |f(m) - F(x) \log m - E|^2 \ll x\mu(x).$$

Similarly

$$\sum_{m=1}^{n-1} |f(n-m) - F(x)\log(n-m) - E|^2 \ll x\mu(x).$$

Hence

$$\sum_{m=1}^{n} \left| \mathcal{F}\left(\frac{m}{n}\right) - G(x)S\left(\frac{m}{n}\right) - F(n) + F(x)\log n + E \right|^2 \ll x\mu(x).$$

With $x = n$, $M(n) = f(n) - F(n)\log n - E(n)$ this gives the second result of theorem (20.2).

To obtain the first inequality of theorem (20.2) we apply the Cauchy–Schwarz inequality, sum over the integers not exceeding x, and appeal again to the Turán–Kubilius inequality.

If

$$\frac{1}{x} \sum_{n \leq x} \left| \mathcal{F}\left(\frac{1}{n}\right) \right|^2 \to 0, \qquad x \to \infty,$$

then by lemma (20.8)

$$\frac{1}{x} \sum_{n \leq x} |f(n+1) - f(n)|^2 \to 0, \qquad x \to \infty.$$

In particular, $\mu(x) \to 0$ as $x \to \infty$. It follows readily from (9) that

$$F = \lim_{x \to \infty} F(x)$$

exists (finitely) and that $f(q) = F \log q$ for every prime-power q. This gives the third result of theorem (20.2), and so completes its proof.

Information as an Algebraic Object

To give an algebraic formulation to the information function we regard it as a map from the rationals r in the interval $0 \leq r \leq 1$ into an abelian group G. In order to make sense of the basic information equation (2) we shall assume that the positive rationals act upon the elements of G in the usual way: if λ, μ are rationals, g_1, g_2 lie in G, then $(\lambda + \mu)g_1 = (\lambda g_1) + (\mu g_1)$, $\lambda(\mu g_1) = (\lambda \mu)g_1$, $\lambda(g_1 + g_2) = (\lambda g_1) + (\lambda g_2)$.

The argument given above formally proceeds until we reach the additive function $f(n)$. Regarding $f(\ )$ as extended to the group of positive rationals Q^* we must now decide what form a homomorphism $f: Q^* \to G$ may take when there is some information concerning the "difference"

(10) $$f(n) - \left(1 - \frac{1}{n}\right)f(n-1) = \mathscr{F}\left(\frac{1}{n}\right).$$

We do not pursue this question in general, but confine ourselves to an example. Let G be the additive group of the p-adic numbers, for some fixed rational prime p. Suppose that (in analogy with the case of real-valued information functions) $\mathscr{F}(r)$ is continuous at the origin. That is to say

$$\mathscr{F}\left(\frac{1}{n}\right) \to 0, \qquad n \to \infty,$$

in the p-adic topology, as $n \to \infty$. Then

$$nf(n) - (n-1)f(n-1) \to 0$$

as $n \to \infty$, and the series

$$\sum_{n=2}^{\infty} (nf(n) - (n-1)f(n-1))$$

converges. Thus

$$A = \lim_{n \to \infty} nf(n)$$

exists.

Replacing n by n^k, we see that if $f(n) \neq 0$ then

$$\lim_{k \to \infty} kn^k = A/f(n)$$

exists. By choosing k to run through increasing powers of p we can evaluate this limit to be 0. On the other hand, by choosing k to run through integers not divisible by p we see that it must be 1 if $(n, p) = 1$.

This contradiction proves that $f(n) = 0$ when $(n, p) = 1$.

Returning to (10) we see that

$$sf(p) = f(p^s) = \mathscr{F}(p^{-s}) \to 0, \qquad s \to \infty.$$

In particular

$$f(p) = (s+1)f(p) - sf(p) \to 0, \qquad s \to \infty.$$

We have now shown that $f(n) = 0$ for all positive integers n. Bearing in mind our remarks concerning the need for the condition (1), we obtain for a p-adic valued information function only the candidate

$$\mathscr{F}(r) = Br, \qquad 0 \le r \le 1,$$

where B denotes a constant and r a rational number.

An extensive treatment of possible measures of information, including the derivation of the information equation (2), and earlier forms of theorem (20.1), may be found in Aczél and Daróczy [1].

Chapter 21

Central Limit Theorem for Differences

In this chapter I combine the results of Chapter 10 with methods from the probabilistic theory of numbers to study the value distribution of the differences of real additive functions.

Let

$$\nu_x(n;\ldots)$$

denote the frequency $[x]^{-1}N$, where N is the number of integers n not exceeding x for which property ... holds.

Let $\beta(x)$ be a positive measurable real-valued function, defined for $x \geq 2$, which satisfies $\beta(x) \to \infty$, and for each positive y, $\beta(x^y)/\beta(x) \to 1$, as $x \to \infty$.

Theorem (21.1). *In order that the frequencies*

(1) $$\nu_x(n; f(n+1) - f(n) \leq z\beta(x))$$

should possess a limiting distribution as $x \to \infty$, and their means and variances be uniformly bounded, it is both necessary and sufficient that for some constant A there be a non-decreasing function $K(u)$ of bounded total variation so that on bounded intervals

(2) $$\sum_{\substack{q \leq x \\ h(q) \leq u\beta(x)}} \frac{1}{q} \left(\frac{h(q)}{\beta(x)}\right)^2 + \sum_{\substack{q \leq x \\ -h(q) \leq u\beta(x)}} \frac{1}{q} \left(\frac{h(q)}{\beta(x)}\right)^2 \Rightarrow K(u), \qquad x \to \infty,$$

where $h(q) = f(q) - A \log q$, q denotes a prime-power, and the convergence is the usual one of probability (measure theory), and that the sums in (2) be bounded uniformly in u and x.

The characteristic function of the limit law will then be

$$\exp\left(\int_{-\infty}^{\infty} (e^{itu} - 1 - itu) u^{-2} \, dK(u)\right).$$

Remarks. An important feature of this result is that the conditions for convergence are both necessary and sufficient. They are the first of this type where no *a priori* assumption is made concerning the size of the additive function $f(\ )$ on the prime-powers.

With appropriate changes the method of the present chapter will apply to the study of differences

$$f_1(an+b) - f_2(An+B)$$

of additive functions f_1 and f_2, with integers $a > 0$, $A > 0$, b and B satisfying $aB - Ab \neq 0$.

As will be seen, we are presently forced to consider concomitant bounds upon the mean and variance of the frequencies (1). It is as if we were operating in a theory of infinitely divisible distributions where only the Kolmogorov representation, rather than the full Lévy–Khinchine representation, was available (see, for example, Gnedenko and Kolmogorov [1], Chapter 3). However, the difficulties which remain in order to fully understand the asymptotic behaviour of the frequencies (1) are number-theoretic rather than probabilistic.

It turns out that the possible limit laws for the frequencies (1) belong to the class L of Khinchine. These laws were characterized by Lévy (see, for example, Gnedenko and Kolmogorov [1], Chapter 6, § 30). It was proven by Zolotarev that every proper law of class L is absolutely continuous, an account of his proof may be found in Elliott [11], Volume II, Chapter 17. In particular, the "*weak convergence*" which occurs at (1) in the statement of the theorem may for proper limit laws be replaced by the more usual *convergence*.

As an example in the application of the theorem:

$$\nu_x(n; f(n+1) - f(n) \leq z\beta(x)) \Rightarrow \frac{1}{\sqrt{2\pi}} \int_{-\infty}^{z} e^{-u^2/2}\, du$$

as $x \to \infty$, the limit law being the normal law with mean zero and variance 1, together with

$$(x\beta(x))^{-1} \sum_{n \leq x} (f(n+1) - f(n)) \to 0$$

and

$$(x\beta(x)^2)^{-1} \sum_{n \leq x} (f(n+1) - f(n))^2 \to 1$$

if and only if there is a constant A so that

$$\sum_{q \leq x} \frac{1}{q} \left(\frac{f(q) - A \log q}{\beta(x)} \right)^2 \to \tfrac{1}{2}$$

and for each fixed $\varepsilon > 0$

$$\sum_{\substack{q \leq x \\ |f(q) - A \log q| > \varepsilon \beta(x)}} \frac{1}{q} \left(\frac{f(q) - A \log q}{\beta(x)} \right)^2 \to 0$$

as $x \to \infty$.

With $A = 0$, $\beta(x) = (2 \log \log x)^{1/2}$ these conditions are met by the well-known function $\omega(n)$, which counts the number of distinct prime divisors of n. Thus

$$\nu_x(n; \omega(n+1) - \omega(n) \leq z(2 \log \log x)^{1/2}) \Rightarrow \frac{1}{\sqrt{2\pi}} \int_{-\infty}^{z} e^{-u^2/2} \, du,$$

a result which was obtained by LeVeque [1].

We shall obtain somewhat more than is stated in the theorem.

This ends our remarks.

Lévy Metric

For distribution functions F, G on the real line we define $\rho(F, G)$ to be the greatest lower bound of those numbers h with which

$$G(z - h) - h \leq F(z) \leq G(z + h) + h$$

holds for all z. This defines a metric on the space of distribution functions on the real line. It was introduced by Lévy.

The first of the three main ingredients of the proof of theorem (21.1) is a result from probabilistic number theory.

Lemma (21.2). *Let integers $a > 0$, b, $A > 0$, B satisfy $\Delta = aB - Ab \neq 0$. Let p_0 be a prime which exceeds $\max(a, A, |b|, |B|, |\Delta|)$. Let $f_1(n)$ be a strongly-additive arithmetic function which satisfies $f_1(p) = 0$ when $p \leq p_0$. Define independent random variables Y_p, one for each prime in the range $p_0 < p \leq x$, to satisfy*

$$Y_p = \begin{cases} f_1(p) & \text{with probability } 1/p, \\ -f_1(p) & \text{with probability } 1/p, \\ 0 & \text{with probability } 1 - 2/p. \end{cases}$$

Let

$$G_x(z) = \nu_x(n; f_1(an+b) - f_1(An+B) \le z\beta(x)),$$

$$Q_x(z) = P\left(\sum_{p_0 < p \le x} Y_p \le z\beta(x)\right).$$

Then

$$\left.\begin{array}{l} G_x(z) - Q_x(z+2u) \\ -G_x(z) + Q_x(z-2u) \end{array}\right\}$$

$$\le c\left(\sum_{\substack{x^\varepsilon < p \le x \\ |f_1(p)| > \varepsilon u \beta(x)}} \frac{1}{p} + \exp\left(-\frac{1}{8\varepsilon}\log\frac{1}{\varepsilon}\right) + x^{-1/15}\right),$$

and in particular

$$\rho(G_x, Q_x) \le c\left(\sum_{\substack{x^\varepsilon < p \le x \\ |f_1(p)| > u\beta(x)}} \frac{1}{p} + \frac{u}{\varepsilon} + \exp\left(-\frac{1}{8\varepsilon}\log\frac{1}{\varepsilon}\right) + x^{-1/15}\right)$$

uniformly for all $u > 0$, $0 < \varepsilon \le 1$, $x \ge 2$ and functions $f_1(n)$. The constant c depends at most upon the integers a, b, A and B.

Proof. A proof of this lemma may be obtained by directly modifying the proof of theorem (12.15) given on pages 48–50 of my book: Elliott [11], Volume II.

Remark. In this lemma $\beta(x)$ is assumed only to be positive for $x \ge 2$.

We shall apply the following result from the theory of probability proper.

Lemma (21.3). *In order that for some suitably chosen constants A_n the distribution laws of the sums*

$$\xi_{n1} + \cdots + \xi_{ns_n} - A_n$$

of independent, infinitesimal random variables converge to a limit, it is necessary and sufficient that there exist non-decreasing functions

$$M(u) \text{ with } M(-\infty) = 0 \quad \text{and} \quad N(u) \text{ with } N(+\infty) = 0$$

defined on the intervals $(-\infty, 0)$ and $(0, +\infty)$, respectively, and a constant $\sigma > 0$, such that

(i) At every continuity point of $M(u)$ and $N(u)$,

$$\lim_{n \to \infty} \sum_{k=1}^{s_n} F_{nk}(u) = M(u), \qquad u < 0,$$

$$\lim_{n \to \infty} \sum_{k=1}^{s_n} \{F_{nk}(u) - 1\} = N(u), \qquad u > 0,$$

(ii)

$$\lim_{\varepsilon \to 0} \liminf_{n \to \infty} \sum_{k=1}^{s_n} \left\{ \int_{|z| < \varepsilon} z^2 \, dF_{nk}(z) - \left(\int_{|z| < \varepsilon} z \, dF_{nk}(z) \right)^2 \right\}$$

$$= \lim_{\varepsilon \to 0} \limsup_{n \to \infty} \sum_{k=1}^{s_n} \left\{ \int_{|z| < \varepsilon} z^2 \, dF_{nk}(z) - \left(\int_{|z| < \varepsilon} z \, dF_{nk}(z) \right)^2 \right\} = \sigma^2,$$

where

$$F_{nk}(z) = P(\xi_{nk} \leq z).$$

Proof. A proof of this result of Gnedenko may be found in Gnedenko and Kolmogorov [1], Chapter 4, theorem 1, pp. 116–120.

Remark. The random variable ξ_{nk}, $k = 1, \ldots, s_n$ are said to be infinitesimal if for each fixed $\varepsilon > 0$

$$\lim_{n \to \infty} \max_{1 \leq k \leq s_n} P(|\xi_{nk}| > \varepsilon) = 0.$$

The second ingredient is an argument taken from my paper (Elliott [8]) (see also Elliott [11], Volume II, pp. 131–134).

Let $h(n)$ be an additive arithmetic function, and define

$$s(u) = \limsup_{x \to \infty} \sum_{\substack{p \leq x \\ |h(p)| > u\beta(x)}} \frac{1}{p}.$$

Lemma (21.4). *Let $\beta(x) \to \infty$ as $x \to \infty$, be measurable and satisfy*

$$\lim_{x \to \infty} \beta(x^y)/\beta(x) = 1$$

Central Limit Theorem for Differences

for each positive value of y. Define independent random variables Z_p, one for each prime not exceeding x, to satisfy

$$Z_p = \begin{cases} h(p) & \text{with probability } 1/p, \\ -h(p) & \text{with probability } 1/p, \\ 0 & \text{with probability } 1-2/p. \end{cases}$$

Assume that $s(u)$ exists for some positive u_0, and that

$$\lim_{u \to \infty} s(u) = 0.$$

Then

(3) $\quad \nu_x(n; h(an+b) - h(An+B) - \alpha(x) \le z\beta(x)) \Rightarrow F(z)$

as $x \to \infty$ if and only if

(4) $\quad P\left(\sum_{p \le x} Z_p - \alpha(x) \le z\beta(x)\right) \Rightarrow F(z)$

as $x \to \infty$.

Remark. In this formulation $\alpha(x)$ may be any real-valued function defined for $x \ge 2$, and $F(z)$ any distribution function on the line.

Let $F_x(z)$ and $P_x(z)$ denote the frequency and distribution functions which appear at (3) and (4), respectively. Before giving the proof of lemma (21.4) we note that if $h(p) = f_1(p)$ for all p then

(5) $\quad \rho(F_x, G_x) \to 0, \quad \rho(P_x, Q_x) \to 0$

as $x \to \infty$.

Suppose, for example, that $\delta > 0$ is given. Let r be a further real number, to be fixed presently. Then for all large x

$$|f_1(an+b) - h(an+b)| > \delta\beta(x)$$

can only hold if $an+b$ is divisible by the square of a prime $p > r$, or by the rth power of a prime $p \le r$, since

$$\lim_{x \to \infty} \beta(x)^{-1} \sum_{1 \le k < r} \sum_{p \le r} |h(p^k)| = 0.$$

The frequency of such special integers $an+b$ does not exceed a constant multiple of

$$\sum_{p \leq r} \frac{1}{p^r} + \sum_{p > r} \frac{1}{p^2},$$

an amount which can be made as small as desired by a suitable choice for r.

A similar argument can be made with the rôles of a, b played by A, B.

It follows that for a suitable constant c_1, depending only upon a, b, A and B,

$$\limsup_{x \to \infty} \rho(F_x, G_x) \leq c_1 \delta.$$

Since δ may be arbitrarily small,

$$\lim_{x \to \infty} \rho(F_x, G_x) = 0.$$

The second assertion of (5) is obtained in a similar manner.

Proof of lemma (21.4). Assume first that for some $\alpha(x)$ the frequencies $F_x(z)$ at (3) converge (weakly) to $F(z)$ as $x \to \infty$.

We begin by showing that $s(u)$ is well defined for every positive u. Let

$$\sigma(x, u) = \sum_{\substack{p \leq x \\ |h(p)| > u\beta(x)}} \frac{1}{p},$$

and assume that (for a certain $v > 0$) $x_1 < x_2 < \cdots$ is an unbounded sequence of values for which $\sigma(x_k, v) \to \infty$. Then the corresponding sequence of distribution functions $P_{x_j}(z)$ is compact; that is to say there is a subsequence of the x_j on which the $P_x(z)$ converge weakly.

Indeed, if $\varepsilon_1 > 0$ we choose z large enough that for all large x

$$F_x(z) - F_x(-z) \geq F(z) - F(-z) - \varepsilon_1 > 1 - 2\varepsilon_1.$$

We next choose ε so small that

$$\exp\left(-\frac{1}{8\varepsilon} \log \frac{1}{\varepsilon}\right) < \varepsilon_1,$$

and $u \, (> u_0)$ so large that

$$\sum_{\substack{x^\varepsilon < p \leq x \\ |h(p)| > u\beta(x)}} \frac{1}{p} < \varepsilon_1$$

for all sufficiently large values of x. With these choices lemma (21.2) enables us to assert that

$$P_x(z+2u) - P_x(-z-2u) \geq G_x(z) - G_x(-z) - 3c\varepsilon_1$$
$$\geq F_x(z) - F_x(-z) - (3c+1)\varepsilon_1$$
$$> 1 - (3c+3)\varepsilon_1,$$

where an appeal has been made to the first estimate of (5), and $\pm z$ assumed to be continuity points of $F(z)$, as they may be.

Without loss of generality we shall assume that the sequence of distribution functions $P_{x_j}(z)$ converges.

For any fixed $\varepsilon > 0$, and $t > 0$ the inequality

$$|h(p)| < \varepsilon \beta(x)$$

holds uniformly for $2 \leq p \leq t$ and all large x, thus

$$\limsup_{x \to \infty} \max_{p \leq x} P(\beta(x)^{-1}|z_p| > \varepsilon) < t^{-1}.$$

Since t may be chosen arbitrarily large the random variables $\beta(x)^{-1}Z_p$, $2 \leq p \leq x$, are infinitesimal. It follows from lemma (21.3) that

$$\lim_{k \to \infty} \sigma(x_k, w)$$

exists and is finite for almost all values of w. In particular

$$\limsup_{k \to \infty} \sigma(x_k, v)$$

must be finite.

This contradiction justifies our assertion that $s(u)$ is well defined for $u > 0$. Since $\beta(x)$ is measurable, and satisfies

$$\beta(x^y)/\beta(x) \to 1, \qquad x \to \infty,$$

for each fixed $y > 0$, there is a continuous increasing unbounded function $r(x)$, for which $(\log r(x))/\log x$ approaches zero as $x \to \infty$, and such that this asymptotic relation holds uniformly for $r(x) \leq x^y \leq x$. A detailed proof may be found as lemma (11.5) in Volume II of my book, Elliott [11].

Let ε be a real number, $0 < \varepsilon < 1$. Since $s(\varepsilon^2)$ is finite there is a number c_1 so that $\sigma(x, \varepsilon^2) < c_1$ for all sufficiently large values of x. Let d be a positive integer. Then, if x is sufficiently large

$$\sum_{m=0}^{d} \sum_{\substack{x^{\varepsilon^{m+1}} < p \leq x^{\varepsilon^m} \\ |h(p)| > \varepsilon^2 \beta(x)}} \frac{1}{p} \leq \sigma(x, \varepsilon^2) < c_1.$$

By fixing d at a value suitably large in terms of c_1, ε only, we can find an integer m in the range $0 \leq m \leq d$ for which

$$\sum_{\substack{x^{\varepsilon^{m+1}} < p \leq x^{\varepsilon^m} \\ |h(p)| > \varepsilon^2 \beta(x)}} \frac{1}{p} < \frac{c_1}{d+1} < \varepsilon.$$

Let us denote this integer, which is a function of x, by m.

Define an unbounded sequence of real numbers $w_1 < w_2 < \cdots < w_k < \cdots$ so that the above construction can be carried out with $\varepsilon = 2^{-k}$ for all $x \geq w_k$, and that if $x \geq w_k$ then $x^{\varepsilon^m} \geq r(x)$.

To each number x we attach two further numbers $l = l(x)$, $\gamma = \gamma(x)$, which are defined by

$$l(x) = x^{\varepsilon^m} \quad \text{with} \quad \varepsilon = 2^{-k},$$

$$\gamma(x) = 2^{-k},$$

over the range $w_k \leq x < w_{k+1}$. In particular $\gamma(x)$ is non-increasing, and approaches zero as $x \to \infty$.

In the notation of these functions we have

$$\sum_{\substack{l^\gamma < p \leq l \\ |h(p)| > \gamma^2 \beta(x)}} \frac{1}{p} \leq \gamma$$

for all sufficiently large values of x.

We now appeal to the hypothesis concerning the rate of growth of $\beta(x)$. This allows us to assert that

$$\beta(x) \leq 2\beta(l),$$

because $r \leq l \leq x$.

Therefore, for all large enough values of x

$$\sum_{\substack{l^\gamma < p \leq l \\ |h(p)| > 2\gamma^2 \beta(l)}} \frac{1}{p} \leq \gamma.$$

It follows from lemma (21.2), using (5) and the triangle law for Lévy's metric, that

$$\rho(F_x, P_x) \leq c \left(\sum_{\substack{x^\varepsilon < p \leq x \\ |h(p)| > u\beta(x)}} \frac{1}{p} + \frac{u}{\varepsilon} + \exp\left(-\frac{1}{8\varepsilon} \log \frac{1}{\varepsilon}\right) \right) + o(1) \qquad (6)$$

as $x \to \infty$, uniformly for $0 < \varepsilon < 1$, $x \geq 2$, $u > 0$ and all functions $h(n)$.

Replacing x, ε and u in the bound at (6) by l, γ, $2\gamma^2$ respectively, we obtain

$$\rho(F_l, P_l) \leq c\left(\gamma + 2\gamma + \exp\left(-\frac{1}{8\gamma}\log\frac{1}{\gamma}\right)\right) + o(1) = o(1)$$

as x, and so l, $\to \infty$. Since $F_l(z) \Rightarrow F(z)$ as $x \to \infty$ we see that $P_l(z) \Rightarrow F(z)$ as $x \to \infty$.

From lemma (21.3) we deduce that for a certain function $N(u)$, and almost all positive numbers u,

$$\sum_{\substack{p \leq l \\ |h(p)| > u\beta(x)}} \frac{1}{p} \to -N(u), \qquad l \to \infty,$$

Consider now any (sufficiently large) number x. If ε is a real number, $0 < \varepsilon < u$, and $u + \varepsilon$ is a continuity point of $N(u)$,

$$\sum_{\substack{p \leq x \\ |h(p)| > u\beta(x)}} \frac{1}{p} \geq \sum_{\substack{p \geq l \\ |h(p)| > (u+\varepsilon)\beta(l)}} \frac{1}{p} = -N(u+\varepsilon) + o(1), \qquad x \to \infty.$$

Since the function $r(x)$ is continuous and strictly increasing, for all sufficiently large values of x its inverse function $r^{-1}(x)$ is well defined. Set $w = r^{-1}(x)$ and $l_1 = l(w)$, so that $x \leq l_1 \leq w$. Then if $u - \varepsilon$ is a continuity point of $N(u)$,

$$\sum_{\substack{p \leq x \\ |h(p)| > u\beta(x)}} \frac{1}{p} \leq \sum_{\substack{p \leq l_1 \\ |h(p)| > (u-\varepsilon)\beta(l_1)}} \frac{1}{p} = -N(u-\varepsilon) + o(1), \qquad x \to \infty,$$

Since, as $x \to \infty$, $\beta(l_1)/\beta(w) \to 1$ and $\beta(w)/\beta(x) = \beta(w)/\beta(r(w)) \to 1$.

Putting these upper and lower bounds together gives

$$\limsup_{x \to \infty} \sum_{\substack{p \leq x \\ |h(p)| > u\beta(x)}} \frac{1}{p} - \liminf_{x \to \infty} \sum_{\substack{p \leq x \\ |h(p)| > u\beta(x)}} \frac{1}{p} \leq N(u+\varepsilon) - N(u-\varepsilon).$$

For continuity points u of $N(u)$, by allowing $\varepsilon \to 0+$ suitably we prove that

$$\lim_{x \to \infty} \sum_{\substack{p \leq x \\ |h(p)| > u\beta(x)}} \frac{1}{p} = -N(u).$$

Let u be a continuity point of $N(\ )$, and let ε and v be real numbers $0 < \varepsilon < 1$, $0 < v < u$, so that $v + u$ is also a continuity point of $N(\ )$. Then

$$\sum_{\substack{x^\varepsilon < p \leq x \\ |h(p)| > u\beta(x)}} \frac{1}{p} \leq \sum_{\substack{p \leq x \\ |h(p)| > u\beta(x)}} \frac{1}{p} - \sum_{\substack{p \leq x^\varepsilon \\ |h(p)| > u\beta(x)}} \frac{1}{p}$$

$$\leq -N(u) + o(1) - \sum_{\substack{p \leq x^\varepsilon \\ |h(p)| > (u+v)\beta(x^\varepsilon)}} \frac{1}{p} = N(u+v) - N(u) + o(1),$$

as $x \to \infty$, the second step being valid since $\beta(x^\varepsilon)/\beta(x) \to 1$. We let $v \to 0+$ suitably and deduce that for almost all $u > 0$, and therefore for all $u > 0$

(7) $$\sum_{\substack{x^\varepsilon < p \leq x \\ |h(p)| > u\beta(x)}} \frac{1}{p} \to 0, \qquad x \to \infty.$$

A further appeal to the bound (6), this time with $u = \varepsilon^2$, gives

$$\limsup_{x \to \infty} \rho(F_x, P_x) \leq c\left(\varepsilon + \exp\left(-\frac{1}{8\varepsilon} \log \frac{1}{\varepsilon}\right)\right).$$

Since ε may be chosen arbitrarily small, $\rho(F_x, P_x) \to 0$ as $x \to \infty$, or what is equivalent

$$P\left(\sum_{p \leq x} Z_p - \alpha(x) \leq z\beta(x)\right) \Rightarrow F(z), \qquad x \to \infty,$$

which is what we wished to prove.

In the other direction, if the distribution functions $P_x(z)$ converge weakly to a law $F(z)$ then the asymptotic estimate (7) will still be available to us. We can again deduce that $\rho(F_x, P_x) \to 0$ as $x \to \infty$. In particular, the $F_x(z)$ will also converge to $F(z)$.

The proof of lemma (21.4) is complete.

Remark. In the applications of lemma (21.3) which are made during the course of the proof of lemma (21.4), the functions $M(u)$ and $N(u)$ satisfy $M(-u) = -N(u)$ almost surely in $u > 0$.

We shall apply the following result concerning functional inequalities.

Lemma (21.5). *Let $\beta(x) > 0$ satisfy*

$$\beta(x^2) \le (2-\delta)\beta(x)$$

for some $\delta > 0$ and all sufficiently large values of x. Suppose further that $u(x)$ is a complex-valued function which for each fixed $y > 0$ satisfies

(8) $$|u(x^y) - yu(x)| \le c_1 \beta(x)$$

for all sufficiently large x. Here the constant c_1 may depend upon y. Then there are further constants U and c_2 so that

$$|u(x) - U \log x| \le c_2 \beta(x)$$

holds for all large x.

Proof. We begin by noting that if $\beta(x^2) \le (2-\delta)\beta(x)$ for all $x \ge x_1 > 1$, then over the same range of x-values

$$\beta(x^{2^m}) \le (2-\delta)^m \beta(x), \qquad m = 1, 2, \ldots.$$

We next show that for all large x

$$U(x) = \lim_{k \to \infty} 2^{-k} u(x^{2^k})$$

exists. Indeed, if with $y = 2$ the inequality (8) holds for $x \ge x_2 \ge x_1$, then arguing inductively

$$|u(x^{2^k}) - 2^{k-l} u(x^{2^l})| \le c_1 \sum_{j=1}^{k-l} 2^{j-1} \beta(x^{2^{k-j}})$$

$$\le c_1 \beta(x) \sum_{j=1}^{k-l} 2^{j-1} (2-\delta)^{k-j}$$

for $k \ge l \ge 0$. Dividing by 2^k gives

$$|2^{-k} u(x^{2^k}) - 2^{-l} u(x^{2^l})| \le \delta^{-1} c_1 \beta(x) (1 - \delta/2)^l.$$

The existence of the limit defining $U(x)$ now follows from Cauchy's criterion. Moreover, with $l = 0$, we obtain

(9) $$|U(x) - u(x)| \le \delta^{-1} c_1 \beta(x)$$

valid over the same range $x \ge x_2$.

For any given positive y the inequalities

$$|u(x^{y2^k}) - yu(x^{2^k})| \le c_1\beta(x^{2^k}) \le c_1(2-\delta)^k\beta(x)$$

certainly hold if $x \ge x_1$ and k is sufficiently large. Dividing by 2^k and letting $k \to \infty$ gives

$$U(x^y) = yU(x)$$

valid for $x \ge x_1$.

Every $x \ge x_1$ has a representation in the form x_1^y, and

$$U(x) = U \log x,$$

the constant U having the value $U(x_1)/\log x_1$. Together with (9) this completes the proof of lemma (21.5).

We come now to the proof of the main theorem.

Proof of theorem (21.1). Suppose first that the frequencies

$$\nu_x(n; f(n+1) - f(n) \le z\beta(x))$$

converge weakly to a law $F(z)$ with finite mean and variance, and that the means and variances of these frequencies are uniformly bounded. Then the sums

(10) $$(x\beta(x)^2)^{-1} \sum_{n \le x} |f(n+1) - f(n)|^2$$

are bounded uniformly for all large x. To begin with this last is all we shall use.

It follows directly from theorem (10.1), and this is the third ingredient of our proof of the theorem, that for a certain function $T(x)$

(11) $$\sum_{q \le x} q^{-1}(f(q) - T(x) \log q)^2 \le c_3 \sup_{x \le w \le x^c} \beta(w)^2$$

for some constant c. Here q denotes a prime-power. In the notation of that theorem $a = A = 1$, $b = 1$, $B = 0$ and $\Delta = 1$.

From the assumption on the growth of $\beta(x)$ which is made in the present theorem this upper bound does not exceed a constant multiple of $\beta(x)^2$.

Moreover, for each $y > 0$ there is a constant c_4 so that

$$|T(x^y) - T(x)| \log x \le c_4\beta(x)$$

for all sufficiently large x. This may be deduced by elimination between the inequality (11) and the similar inequality with x^y in place of x. An example of this argument is given towards the end of the proof of lemma (10.9).

Setting $u(x) = T(x) \log x$ we see that it satisfies the condition (8) of lemma (21.5) and we obtain a constant A so that

$$|u(x) - A \log x| \le c_2 \beta(x).$$

This together with (11) shows that

$$\sum_{q \le x} q^{-1}(f(q) - A \log q)^2 \le c_5 \beta(x)^2$$

for all large x.

With $h(n) = f(n) - A \log n$, the function $s(u)$ which is defined immediately preceding the statement of lemma (21.4) is seen to be defined for all positive u, and not to exceed c_5/u^2. Thus $s(u) \to 0$ as $u \to \infty$ and the hypotheses of that lemma are satisfied with $\alpha(x) = 0$. Note that

$$h(n+1) - h(n) = f(n+1) - f(n) - A \log(1 + n^{-1})$$

so that for each fixed $\varepsilon > 0$

$$\nu_x(n; |h(n+1) - h(n) - \{f(n+1) - f(n)\}| > \varepsilon \beta(x)) \to 0$$

as $x \to \infty$. Thus the frequencies

$$\nu_x(n; f(n+1) - f(n) \le z\beta(x)),$$

$$\nu_x(n; h(n+1) - h(n) \le z\beta(x))$$

converge (weakly) together.

We can now assert that the distribution functions $P_x(z)$ also converge to $F(z)$. Their characteristic functions are readily calculated to be

$$\phi_x(t) = \prod_{p \le x} \left(1 + \frac{1}{p} \{e^{ith(p)/\beta(x)} + e^{-ith(p)/\beta(x)} - 2\}\right).$$

If we write a typical term of this product in the form $1 + z_p$ then $p|z_p| \le 4$ always, and for each fixed p, $z_p \to 0$ as $x \to \infty$. Thus, with the principal value of the logarithm,

$$\log(1 + z_p) = z_p + O(|z_p|^2),$$

so that the above characteristic function can be put into the form

$$\exp\left(\sum_{p\leq x} z_p + O\left(\sum_{p\leq x} |z_p|^2\right)\right).$$

For any fixed $r>0$

$$\sum_{p\leq x} |z_p|^2 \leq o(1) + 16 \sum_{p>r} p^{-2}$$

as $x\to\infty$. By choosing the r increasingly large we see that as $x\to\infty$

$$\sum_{p\leq x} |z_p|^2 \to 0$$

and

$$\phi_x(t) = \exp\left(\sum_{p\leq x} p^{-1}\{e^{ith(p)/\beta(x)} + e^{-ith(p)/\beta(x)} - 2\} + o(1)\right).$$

In a similar manner we may replace the variable p (prime) in this sum by the variable q (prime-power).

Let

$$K_x(u) = \sum_{h(p)\leq u\beta(x)} \frac{1}{q}\left(\frac{h(q)}{\beta(x)}\right)^2 + \sum_{-h(q)\leq u\beta(x)} \frac{1}{q}\left(\frac{h(q)}{\beta(x)}\right)^2.$$

Then we have

$$\exp\left(\int_{-\infty}^{\infty} u^{-2}(e^{itu} - 1 - itu)\, dK_x(u)\right) \to \phi(t)$$

as $x\to\infty$, where $\phi(t)$ is the characteristic function of $F(z)$.

Here the exponential is the characteristic function of an infinitely divisible law of mean zero and variance

$$\int_{-\infty}^{\infty} 1\cdot dK_x(u) = \sum_{q\leq x} \frac{1}{q}\left(\frac{h(q)}{\beta(x)}\right)^2 \leq c_5$$

for all large x.

According to the theory of infinitely divisible distributions (see, for example, Gnedenko and Kolmogorov [1], Chapter 3, § 18) $\phi(t)$ must also be infinitely divisible, and have a representation

$$\exp\left(\int_{-\infty}^{\infty} u^{-2}(e^{itu} - 1 - itu)\, dK(u)\right).$$

Central Limit Theorem for Differences 371

Moreover, in the usual weak sense of measure theory, on bounded intervals

$$K_x(u) \Rightarrow K(u)$$

as $x \to \infty$.

In the other direction, if for some $K(u)$ and every bounded interval

$$K_x(u) \Rightarrow K(u)$$

as $x \to \infty$ then, in the above notation,

$$\phi_x(t) \to \exp\left(\int_{-\infty}^{\infty} u^{-2}(e^{itu} - 1 - itu)\, dK(u)\right)$$

as $x \to \infty$. Thus the distribution functions $P_x(z)$ (with $\alpha(x) = 0$) converge to a law $F(z)$ with finite mean and variance.

In view of lemma (21.4) the same is true for the frequencies

$$\nu_x(n; h(n+1) - h(n) \leq z\beta(x))$$

and so for the

$$\nu_x(n; f(n+1) - f(n) \leq z\beta(x)).$$

It is now straightforward to check that the means and variances of the frequencies (1) are uniformly bounded.

Remark. As one may readily check, the argument in the direction

$$P_x(z) \Rightarrow F(z) \quad \text{implies} \quad \nu_x(n; \ldots) \Rightarrow F(z)$$

does not require that the distribution functions $P_x(z)$ have bounded second moment. In this direction one can readily formulate and establish a result which allows the frequencies (1) to converge to limit laws which do not have a finite mean or variance. One might say that one can step readily from probability to number theory.

In the present situation the more severe difficulty is to reduce the number-theoretic problem to one of probability. At the moment the application of theorem (10.1), which implicitly makes demands upon the second moment of the frequencies (1), is all that is available.

In general one would expect a result along the lines of theorem (16.1) in my book, Elliott [11], Volume II.

Chapter 22

Density Theorems

In this chapter if $a_1 < a_2 < \cdots$ is an increasing sequence of positive integers, and $A(x)$ denotes the number of its members not exceeding a given real x, then its *upper density* is defined to be $\limsup x^{-1}A(x)$ as $x \to \infty$.

Theorem (22.1). *Let $a_1 < a_2 < \cdots$ be a sequence of positive integers of positive upper density d. Then there is a set of primes p for which $\sum p^{-1}$ converges to a sum bounded in terms of d, so that every positive integer n which is not divisible by one of these p has a representation*

$$n^k = \prod_i a_{m_i}^{\varepsilon_i},$$

where k is a positive integer not exceeding d^{-1} and each $\varepsilon_i = \pm 1$.

Thus a sufficiently dense sequence of integers will multiplicatively generate a power of almost every positive integer. In particular cases one may hope to decrease the value of k and remove the exceptional set of primes p by appealing to some arithmetic property of the sequence a_j. Examples of this are given in the following chapter.

If for some sequence of primes p_i the series $\sum p_i^{-1}$ converges, then those integers not divisible by any p_i have the asymptotic density $\prod (1 - p_i^{-1})$. This shows that the occurrence of an exceptional set of primes in theorem (22.1) is to be expected.

Theorem (22.1) is derived from the following result, which is of independent interest.

Theorem (22.2). *Let $a_1 < a_2 < \cdots$ be a sequence of positive integers of positive upper density d. Let $f(\ )$ be an additive arithmetic function, taking values in the group $\mathbb{R}/\mathbb{Z}$ of additive reals (mod 1), and which is constant on the a_j.*

Then there is an integer m, $1 \le m \le d^{-1}$, so that the series

$$\sum_{\|mf(p)\| \ne 0} \frac{1}{p}$$

converges. Moreover, the value of this sum is bounded in terms of d.

Here, as usual, $\|y\|$ denotes the distance of the real number y from the nearest integer. The value of this function depends only upon the value of y (mod 1), so that it is well defined on the group $\mathbb{R}/\mathbb{Z}$.

Let $L(\ )$ be an arithmetic function which takes values in an abelian group H. We shall write this group additively. Assume further that for coprime integers a and b, $L(ab) = L(a) + L(b)$.

Theorem (22.3) (Ruzsa). *The integers which L maps to a given element of H have an asymptotic density.*

Following the proofs of these theorems examples will be given in their application to problems considered earlier in this volume.

Lemma (22.4). *Let complex numbers $z_{n,\nu}$; $n = 1, \ldots, N$, $\nu = 1, \ldots, r$ be given. Then the inequality*

$$\sum_{\nu=1}^{r} \left| \sum_{n=1}^{N} d_n z_{n,\nu} \right|^2 \leq J \sum_{n=1}^{N} |d_n|^2,$$

with

$$J = \max_{1 \leq \lambda \leq r} \sum_{\nu=1}^{r} \left| \sum_{n=1}^{N} z_{n,\nu} \bar{z}_{n,\lambda} \right|$$

holds for all complex numbers d_n, $n = 1, \ldots, N$.

Proof. Consider the dual inequality

$$\sum_{n=1}^{N} \left| \sum_{\nu=1}^{r} b_\nu z_{n,\nu} \right|^2 \leq L \sum_{\nu=1}^{r} |b_\nu|^2.$$

Expanding the square summand in the expression on the left-hand side, and inverting the order of summation, we obtain an hermitian form

$$\sum_{\nu=1}^{r} \sum_{\lambda=1}^{r} b_\nu \bar{b}_\lambda \sum_{n=1}^{N} z_{n,\nu} \bar{z}_{n,\lambda}.$$

Since $|b_\nu \bar{b}_\lambda| \leq (|b_\nu|^2 + |b_\lambda|^2)/2$ for every ν, in absolute value this form does not exceed $J \sum |b_\nu|^2$. With $L = J$ the dual inequality holds for all complex b_ν, and lemma (22.4) is so proved.

We next note that for each real θ

$$1 - \operatorname{Re} e^{2\pi i \theta} = 2(\sin \pi \|\theta\|)^2 \begin{cases} \leq 2\pi \|\theta\|^2, \\ \geq 8\|\theta\|^2. \end{cases}$$

Lemma (22.5). *If for a real τ the series*

$$\sum \frac{\|\tau \log p\|^2}{p}$$

taken over the prime numbers converges, then $\tau = 0$.

Proof. Assume that $\tau \neq 0$. For any positive P

$$\sum_{p \leq P} \frac{1}{p} = \sum_{p \leq P} \frac{1 - \operatorname{Re} p^{-i\tau}}{p} + \operatorname{Re} \sum_{p \leq P} \frac{1}{p^{1+i\tau}}.$$

The first sum on the right-hand side is by hypothesis bounded, and after an integration by parts using the prime number theorem in the form

$$\sum_{p \leq y} 1 = \frac{y}{\log y} + O\left(\frac{y}{(\log y)^2}\right),$$

so is the second. Moreover, these bounds are uniform with respect to P. This contradicts the well-known divergence of the series of reciprocal primes.

Lemma (22.6). *For every multiplicative function $g(\ )$ which takes values in the unit circle of the complex numbers*

$$\beta = \lim_{x \to \infty} x^{-1} \left| \sum_{n \leq x} g(n) \right|$$

exists. There is an absolute constant c so that

$$\beta \leq c(T^{-1/4} + \exp(-m(T)/4))$$

with

$$m(T) = \inf_{|\tau| \leq T} \sum \frac{1}{p}(1 - \operatorname{Re} g(p) p^{-i\tau})$$

uniformly for $T \geq 2$. If the series involving the prime numbers p diverges for every real τ in the interval $|\tau| \leq T$ then the exponential term is understood to be zero.

Moreover, if $(g(p))^k = 1$ for some integer k and all primes p, then

$$\lim_{x \to \infty} x^{-1} \sum_{n \leq x} g(n) = \prod_p \left(1 - \frac{1}{p}\right)\left(1 + \frac{g(p)}{p} + \frac{g(p^2)}{p^2} + \cdots\right)$$

the limit existing finitely.

Proof. For a proof of the first two assertions, due to Halász, see Elliott [11], Volume I, Chapter 6, theorem (6.2) and lemma (6.10).

If $(g(p))^k = 1$ for every prime p then the series

$$\sum p^{-1}(1 - \operatorname{Re} g(p)p^{-i\tau})$$

diverges for every non-zero real τ, for otherwise define an additive function $f: Q^* \to \mathbb{R}/\mathbb{Z}$ by $\exp(2\pi i f(n)) = g(n)$. Then the series

$$\sum p^{-1} \|f(p) - \tau \log p\|^2$$

converges. But

$$\|k(f(p) - \tau \log p)\|^2 \le k \|f(p) - \tau \log p\|^2$$

and $\|kf(p)\| = 0$, giving the convergence of

$$\sum p^{-1} \|k\tau \log p\|^2.$$

According to lemma (22.5) this is not possible.

If now the series

$$\sum p^{-1}(1 - \operatorname{Re} g(p))$$

diverges then we may apply the inequality of lemma (22.6) with T arbitrarily large, to obtain $\beta = 0$. Otherwise the condition $(g(p))^k = 1$ will ensure the convergence of the series

$$\sum p^{-1}(1 - g(p))$$

and the final assertion of lemma (22.6) can be obtained from theorem (6.2) of the above reference. This part of lemma (22.6) was first established by Wirsing. This ends our indication of the proof of lemma (22.6).

The *lower density* of a sequence of positive integers $b_1 < b_2 < \cdots$ is defined to be $\liminf x^{-1} B(x)$ as $x \to \infty$. The highest common factor of such a sequence is the largest integer which divides every member of the sequence.

Lemma (22.7). *Let $b_1 < b_2 < \cdots$ be a sequence of positive integers, of positive lower density, and highest common factor 1. Then there are further integers s, t so that every positive integer $n \ge s$ has a representation as the sum of at most t of the integers b_i.*

Proof. A proof of this result may be found in Elliott [1] or Elliott [11]. It employs the notion of Schnirelmann density.

Proof of theorem (22.2). For the given additive function f define the multiplicative function $g(n) = \exp(2\pi i f(n))$. For each integer k let

$$\beta_k = \lim_{x \to \infty} x^{-1} \left| \sum_{n \leq x} g(n)^k \right|,$$

which exists by lemma (22.6). It is clear that $0 \leq \beta_k \leq 1$, $\beta_k = \beta_{-k}$, $\beta_0 = 1$.

We apply the inequality of lemma (22.4) with $d_n = 1$ if n belongs to the sequence A which appears in the statement of theorem (22.2), and $d_n = 0$ otherwise. For $z_{n,\nu}$ we take $g(n)^\nu$. Then dividing both sides of that inequality by x, and choosing x to approach infinity through a suitable sequence of values, we obtain

$$(r+1)d \leq 1 + 2 \sum_{k=1}^{r} \beta_k$$

for every positive integer r. Note that for each integer ν, $g(a_j)^\nu$ has a constant value for all j.

Choose a positive real number $\alpha > 2$ so that $c(\alpha^{-1/4} + \exp(-\alpha/4)) < d/4$, where c is the constant which appears in the statement of lemma (22.5). In the notation of that lemma we regard T as fixed at the value α from now on. Let K be the sequence of positive integers k for which

$$\inf_{|\tau| \leq \alpha} \sum \frac{1}{p} (1 - \operatorname{Re} g(p)^k p^{-i\tau}) \leq \alpha.$$

With our choice of α the number $K(r)$, of these not exceeding r, is bounded below by

$$K(r) \geq \sum_{\substack{k=1 \\ k \in K}}^{r} \beta_k \geq \sum_{k=1}^{r} \beta_k - \frac{rd}{4} \geq \frac{rd}{2} - 1.$$

In particular, K has positive lower density.

Let m be the highest common factor of the sequence K. Considerations of density show that m cannot exceed $2/d$. Then the sequence $m^{-1}k$, where k belongs to K, meets the requirements of lemma (22.7), so that with suitably chosen s, t, every multiple nm with $n \geq s$ has a representation as the sum of at most t integers in K, say

$$nm = \sum_{i=1}^{w} k_i, \quad w \leq t.$$

For each i, $1 \leq i \leq w$, there is a real τ_i, $|\tau_i| \leq \alpha$, so that

$$8 \sum \frac{1}{p} \| k_i f(p) - \tau_i \log p \|^2 \leq 2\alpha.$$

Since $\|u+v\| \le \|u\| + \|v\|$ for all real u, v, with $\lambda_n = \tau_1 + \cdots + \tau_w$ an application of the Cauchy–Schwarz inequality gives

$$8 \sum \frac{1}{p} \|nmf(p) - \lambda_n \log p\|^2 \le w2\alpha \le 2\alpha t.$$

This inequality holds for every integer $n \ge s$, and although λ_n may vary with n, it satisfies the uniform bound $|\lambda_n| \le \alpha t$.

Our next step is to prove that $\lambda_n = 0$ for all $n \ge s$. Indeed, for any two positive integers $n \; (\ge s)$ and d

$$(\lambda_{nd} - d\lambda_n) \log p = \lambda_{nd} \log p - ndmf(p) - d(\lambda_n \log p - nmf(p))$$

and after an application of the Cauchy–Schwarz inequality we obtain the convergence of the series

$$\sum \frac{1}{p} \|(\lambda_{nd} - d\lambda_n) \log p\|^2.$$

In view of lemma (22.5) $\lambda_{nd} = d\lambda_n$, and since the λ_{nd} are bounded uniformly for all d, $\lambda_n = 0$.

We have now that

$$\sum \frac{1}{p} \|nmf(p)\|^2 \le \alpha t$$

for all $n \ge s$.

For any real θ with $\|\theta\| \ne 0$,

$$\left| \frac{1}{M} \sum_{n=1}^{M} e^{2\pi i n \theta} \right| \le (M \sin \pi \|\theta\|)^{-1} \le (2M\|\theta\|)^{-1},$$

which bound approaches zero as $M \to \infty$. Thus for any positive P

$$\sum_{\substack{p \le P \\ \|mf(p)\| \ne 0}} \frac{1}{p} \le \sum_{p \le P} \frac{1}{p} \lim_{m \to \infty} \frac{1}{M} \sum_{n=1}^{M} (1 - \mathrm{Re} \exp(2\pi i n m f(p)))$$

$$\le \limsup_{M \to \infty} \frac{1}{M} \sum_{n=1}^{M} \sum_{p \le P} \frac{1}{p} (1 - \exp(2\pi i n m f(p)))$$

$$\le \limsup_{M \to \infty} \frac{1}{M} \sum_{n=s}^{M} \sum_{p \le P} 2\pi^2 \|nmf(p)\|^2 \le 2\pi^2 \alpha t.$$

This gives us the convergence of the sum

$$\sum_{\|mf(p)\|\neq 0} \frac{1}{p}$$

which is almost what we want, save that m is only restricted by $m \leq 2/d$.

Looking back to our earliest lower bound for the β_k we see that there is a positive integer k not exceeding $1/d$ for which β_k is not zero. From lemma (22.6) there will then be a real τ so that the series

$$\sum \frac{1}{p}(1 - \operatorname{Re} g(p)^k p^{-i\tau})$$

converges, or what is essentially the same, the series

$$\sum \frac{1}{p} \|kf(p) - \tau \log p\|^2$$

converges. Thus

$$\sum \frac{1}{p} \|smkf(p) -سm\tau \log p\|^2$$

also converges, and from what we have proved so far, so does the corresponding series where the term $-sm\tau \log p$ does not occur. Hence the series

$$\sum \frac{1}{p} \|sm\tau \log p\|^2$$

converges. Another appeal to lemma (22.5) shows that $sm\tau = 0$, so that $\tau = 0$ and

$$\sum \frac{1}{p} \|kf(p)\|^2$$

converges.

On those primes p for which $\|mf(p)\| = 0$, a non-zero value of $\|kf(p)\|$ will be at least as large as m^{-1}. Thus

$$\sum_{\|kf(p)\|\neq 0} \frac{1}{p} \leq m^2 \sum \frac{\|kf(p)\|^2}{p} + \sum_{\|mf(p)\|\neq 0} \frac{1}{p}$$

and the proof of theorem (22.2) is complete.

Groups of Bounded Order

For the proof of theorem (22.1) the following lemma is important. We recall that a group has *bounded order* if there is a positive integer n so that the nth-power of every element is the identity.

Lemma (22.8). *Let G be an abelian group with the property that every homomorphism of it into the group $\mathbb{R}/\mathbb{Z}$ has some power which makes it trivial. Then G has bounded order.*

Remark. Written additively the hypothesis states that for each homomorphism $\phi: G \to \mathbb{R}/\mathbb{Z}$ there is an integer m so that $m\phi(g) = 0$ for all g in G. The value of m may depend upon ϕ. The converse of lemma (22.8) is clear.

Proof. We begin by noting that the group $\mathbb{R}/\mathbb{Z}$ is divisible (see Chapter 15).

Suppose that G has an element g of infinite order. Then the map $ng \mapsto n\sqrt{2} \pmod{1}$, $n = 0, \pm 1, \pm 2, \ldots$ maps the cyclic subgroup generated by g isomorphically into $\mathbb{R}/\mathbb{Z}$. Since $\mathbb{R}/\mathbb{Z}$ is divisible we can extend this map to a homomorphism of the whole of G. Suppose the resulting map becomes trivial if it is raised to its mth-power. In particular, $m\sqrt{2} \equiv 0 \pmod{1}$, which is false. Thus G is a torsion group.

G can now be written as a direct sum of its primary groups (for a fixed p the subgroup of elements whose order is a power of p). Let Δ be one of these, with associated prime p. If Δ has elements of arbitrarily high order then it can be mapped homomorphically onto $\mathbb{Z}(p^\infty)$ the subgroup of $\mathbb{R}/\mathbb{Z}$ generated by the rational fractions whose denominators are powers of p. In this context note that if E, F are any two subgroups of $\mathbb{Z}(p^\infty)$ then either E is contained in F, or F in E. Suppose that ϕ maps Δ onto $\mathbb{Z}(p^\infty)$. Extending this map to the whole of G, as above, leads to an integer divisible by arbitrarily high powers of p, and another contradiction. Each primary group Δ is thus of bounded order. However, there may still be infinitely many such summands.

A primary group of bounded order is the direct sum of cyclic groups. Let H_j be a direct summand of the primary subgroup of G associated with the prime p_j. We shall assume that there are such groups for an infinity of p_j, and obtain another contradiction. Let the order of H_j be k_j. We map the direct sum of the H_j isomorphically into $\mathbb{R}/\mathbb{Z}$ by taking a generator of H_j onto $1/p_j^{k_j} \pmod{1}$. We can extend this map to the whole of G by making it trivial on the remaining cyclic summands of G. Let the mth-power of the resulting map be trivial, then $p_j^{k_j}$ divides m for every value of j, and a contradiction ensues.

Thus G has only a finite number of primary groups, each of which is of bounded order. The lemma is proved.

Measures on Dual Groups

Consider now a group G which is a direct sum of denumerably many cyclic groups H_j, of order k_j. For the purposes of the present chapter we shall regard its dual group $\hat{G}$ to be the group of all homomorphisms of G into the unit circle in the complex plane. Clearly any such homomorphism h will be determined by its values on the H_j. Typically h will map a generator of H_j onto one of the complex numbers $\exp(2\pi i m/k_j)$, where m is a positive integer not exceeding k_j. Thus $\hat{G}$ is isomorphic to the direct sum of the residue class groups $\mathbb{Z}/k_j\mathbb{Z}$, $j = 1, 2, \ldots$.

Regarding the individual classes (mod k_j) as forming a point set, let S be their cartesian product taken over all j. We can introduce a probability measure into S and so $\hat{G}$ by assigning non-negative weights to each residue class (mod k_j) so that their sum over all the classes (mod k_j) is one, and then in the usual way defining a product measure on S. There are clearly many such measures possible. We shall confine ourselves to that which assigns to each residue class (mod k_j) the weight $1/k_j$, and will denote it by μ. This measure has the advantage that if any set B which is μ measurable is translated by a member h of $\hat{G}$, then the translated set $h+B$ is also measurable, and with the same measure as B. We have constructed the Haar measure on $\hat{G}$.

It is now straightforward to obtain the orthogonality relation

$$E(h(g)) = \int_{\hat{G}} h(g)\, d\mu = \begin{cases} 1 & \text{if } g = 0, \\ 0 & \text{otherwise,} \end{cases}$$

where 0 denotes the identity of the group G, and h denotes a typical member of $\hat{G}$. Here the notation E(xpectation) emphasizes the fact that the various homomorphisms h of G into $\mathbb{R}/\mathbb{Z}$ may be thought of as random variables with respect to μ. To obtain this relation it suffices to note that when representing g in terms of the direct summands H_j only finitely many values of j need be considered. The integral then reduces to a finite exponential sum which is a product of finite geometric progressions.

Lemma (22.9). *Let g_j, $j = 1, \ldots, s$ be distinct elements of G. Then the identity*

$$E\left(\left|\sum_{j=1}^{s} \alpha_j h(g_j)\right|^2\right) = \sum_{j=1}^{s} |\alpha_j|^2$$

holds for all complex numbers α_j, $j = 1, \ldots, s$.

Proof. Expanding the square and inverting the order of summation in the expression on the left-hand side we obtain an hermitian form

$$\sum_{j=1}^{s}\sum_{k=1}^{s} \alpha_j \bar{\alpha}_k E(h(g_j - g_k))$$

and the asserted equality follows from the orthogonality relation.

We are now ready to establish theorem (22.1), and will obtain more.

Proof of theorem (22.1). Let Q^* be the group of positive rational fractions, and Γ its subgroup generated by the integers a_j. Let G be the quotient group Q^*/Γ. We begin by proving that G is essentially a torsion group.

The map $T: y \mapsto \log y$ takes Q^* and Γ isomorphically onto subgroups of the additive real numbers. These subgroups respectively generate vector spaces (say) V_1 and V_2 over the rational number field. Let $z_1, z_2, \ldots$ be a basis for the orthogonal complement of V_2 in V_1. Let $\omega_1, \omega_2, \ldots$ be a sequence of real numbers so that every finite subset of $\{1, \omega_1, \omega_2, \ldots\}$ is linearly independent over the rationals. We define a homomorphism of V_1 into the additive reals by $f(z_j) = \omega_j$, $f(y) = 0$ for all y in V_2. When restricted to the positive integers $h = f_\circ T$ becomes an additive arithmetic function with values in the additive group of the reals, moreover $h(a_j) = 0$ for every member a_j of a sequence of integers of positive upper density.

It follows from the *Surrealistic Continuity Theorem* (see Elliott [11], Theorem (7.3)) that the series $\sum p^{-1}$ taken over those primes p for which $h(p) \neq 0$, converges. Here it is important to apply the form of the theorem due to Elliott and Ryavec [1] in which no *a priori* restriction is placed upon the additive function involved. However, a more exact result can be obtained if we further compose h with the canonical homomorphism $\beta: \mathbb{R} \to \mathbb{R}/\mathbb{Z}$, which we illustrate by the following exact diagram

$$\begin{array}{c} 0 \\ \downarrow \\ 0 \to Q^* \xrightarrow{T} V_1 \to 0 \\ {\scriptstyle f} \downarrow \\ 0 \to \mathbb{Z} \to \mathbb{R} \xrightarrow[\beta]{} \mathbb{R}/\mathbb{Z} \to 0 \end{array}$$

We may then apply theorem (22.2) and obtain a positive rational integer m so that the series $\sum p^{-1}$ taken over those primes p for which $\|m\beta_\circ h(p)\| \neq 0$, converges. Moreover, an upper bound for this sum in terms of d can be given.

Let p be a prime for which $\|m\beta_\circ h(p)\| = 0$, and suppose that

$$\log p = \sum_{i=1}^{k} r_i z_i + y$$

for rational numbers r_i, and with y in V_2. Then

$$m \sum_{i=1}^{k} r_i \omega_i = mh(p) \equiv 0 \pmod{1},$$

and since $1, \omega_1, \ldots, \omega_k$ are linearly independent over the rationals, every $r_i = 0$. Hence $\log p$ is contained in V_2 or, what amounts to the same thing, for some integer t, p^t belongs to Γ. Note that this integer t is not necessarily m.

We call the above set of primes p for which $\beta_o h(p) \neq 0$ and which are such that the series $\sum p^{-1}$ converges *exceptional primes of the first kind*.

Let $x_1 < x_2 < \cdots$ be an unbounded sequence of real numbers for which $A(x_j) > dx_j/2$. We say that a prime q is an *exceptional prime of the second kind* if the bound

$$\sum_{\substack{a_i \leq x \\ a_i \equiv 0 \pmod{q}}} 1 \leq \frac{A(x)}{2q}$$

holds for all but finitely many of the choices $x = x_j$. These exceptional primes are also few in number. Let P be a positive integer. We apply the inequality (4) of Chapter 6 with $Q = P$, $M = 1$, $N = x$, setting the a_n there to be 1 if n belongs to the present sequence A, and to be zero otherwise. We obtain

$$\sum_{q \leq P} \frac{1}{q} \left(\frac{A(x)}{2}\right)^2 \leq c_1(x + P^2) A(x)$$

for an absolute constant c_1, and if we choose the $x = x_j$ to become unbounded suitably

$$\sum_{q \leq P} \frac{1}{p} \leq \frac{16 c_1}{d}.$$

Since P may be chosen arbitrarily large the exceptional primes of the second kind also have a convergent sum of reciprocals.

Let Q_1 be the subgroup of Q^* generated by those integers which are not divisible by any of the exceptional primes, and let G_1 be the subgroup of G generated by the cosets $n \pmod{\Gamma}$ with n in Q_1. Thus G_1 makes the following exact diagram commutative

$$\begin{array}{ccccccc} 0 \to \Gamma \to & Q^* & \to & G & \to 0 \\ & \uparrow & & \uparrow & \\ & Q_1 & \to & G_1 & \to 0 \\ & \uparrow & & \uparrow & \\ & 0 & & 0 & \end{array}$$

We next prove that G_1 has bounded order. To do this we employ lemma (22.8).

Let $f: G_1 \to \mathbb{R}/\mathbb{Z}$ be a homomorphism. Since $\mathbb{R}/\mathbb{Z}$ is divisible we can extend f to the whole of G, and then by composing it with the canonical map $Q^* \to Q^*/\Gamma$ to the whole of Q^*.

$$\begin{array}{c} 0 \to \Gamma \to Q^* \to G \to 0 \\ \uparrow \quad \searrow \\ G_1 \to \mathbb{R}/\mathbb{Z} \\ \uparrow \\ 0 \end{array}$$

Considered restricted to the positive integers f (we use the same symbol!) is an additive arithmetic function, with values in $\mathbb{R}/\mathbb{Z}$, which vanishes on the sequence of integers $a_1 < a_2 < \cdots$. From theorem (22.2) there is a positive integer m so that the series

$$\sum_{\|mf(p)\| \neq 0} \frac{1}{p}$$

converges. Let us label the primes for which $\|mf(p)\| \neq 0$, p_j, $j = 1, 2, \ldots$. Let s be chosen so that

$$\sum_{j=s+1}^{\infty} \frac{1}{p_j} < \frac{d}{16}$$

where d is the upper density of the a_i. Let the further integer v be chosen so that

$$\sum_{j=1}^{s} \frac{1}{p_j^v} < \frac{d}{16}.$$

Consider now a prime p which belongs to the group Q_1. Since it is not exceptional

$$\sum_{\substack{a_i \leq x \\ a_i \equiv 0 \pmod{p}}} 1 \geq \frac{A(x)}{2p} > \frac{dx}{4p}$$

for an unbounded sequence $x = x_j$. The number of a_i not exceeding x which are divisible by some p_j with $j > s$ is certainly not more than

$$\sum_{j=s+1}^{\infty} \left[\frac{x}{p_j}\right] < \frac{dx}{16}.$$

Similarly not more than $dx/16$ of these $a_i \le x$ can be divisible by any of the p_j, $1 \le j \le s$, to a power greater than the vth. We can thus find an a_i of the form $a_i = ptr$, where $\text{nq } p_j$ divides r, and t is bounded, independently of p. In fact the product of the p_j^v with $j = 1, \ldots, s$ is a permissible bound for t. Hence $mf(pt) = mf(a_i) - mf(r) = 0$.

Since no p is one of the first exceptional primes, there is an integer w so that for all permissible t, $wf(t) = 0$. Hence $wmf(p) = -wf(t) = 0$ for all p in Q_1.

The wm^{th}-power of the map f is trivial, and by lemma (22.8) G_1 has bounded order.

To continue we shall prove that G_1 is essentially bounded, and that its order does not exceed a function of d.

Let D be the homomorphism $Q^* \to G_1$ given by projecting Q^* onto Q_1 and then employing the map $Q_1 \to G_1$ in the above diagram. For each element g_j of G_1 define

$$\alpha_j = \min\left(\frac{d}{5}, \sum_{D(p)=g_j} p^{-1}\right).$$

Without loss of generality we can assume that $\alpha_1 \ge \alpha_2 \ge \cdots$, so that α_1 is positive. Let k be a positive integer and set

$$\theta = \sum_{j=1}^{k} \alpha_j^2 \left(\sum_{j=1}^{k} \alpha_j\right)^{-2}.$$

Since G_1 is of bounded order we can apply to it our above remarks on the construction of a Haar measure μ on the dual group. In terms of this measure

$$P\left(\left|\sum_{j=1}^{k} \alpha_j h(g_j)\right| > \frac{1}{2} \sum_{j=1}^{k} \alpha_j\right) < 4\theta.$$

We now try to construct a collection of t elements h_r in the dual group $\hat{G}_1$ so that

(1) $$\left|\sum_{j=1}^{k} \alpha_j h_r(g_j) \bar{h}_s(g_j)\right| \le \tfrac{1}{2} \sum_{j=1}^{k} \alpha_j$$

holds for all pairs $1 \le r$, $s \le t$, with $r \ne s$.

We can certainly find at least one map h if $4\theta < 1$. Supposing $t-1$ already obtained we see that the probability of a new map h_t failing to satisfy the inequality (1) for any $\bar{h}_s$ with $s < t$ is at most $(t-1)4\theta$. Here the translation invariant property of the measure μ is applied many times. Thus a tth map will be available so long as $2\theta(2 + t(t-1)) < 1$. Let us temporarily assume that $4\theta < 1$, which allows the choice $t = [\theta^{-1/2}]$.

Since the unit circle is divisible we can extend the maps h_r to be homomorphisms of G into the unit circle. In terms of these extended maps let ρ be the canonical map $Q^* \to G$ and consider the sum

$$U(x) = \sum_{r=1}^{t} \left| \sum_{a_j \leq x} h_r(\rho(a_j)) \right|^2.$$

We obtain a bound for this sum by applying lemma (22.4) with $z_{n,\nu} = h_\nu(\rho(n))$, setting $d_n = 1$ if n belongs to A, and $d_n = 0$ otherwise. In our present circumstances if $r \neq s$

$$\lim_{x \to \infty} x^{-1} \left| \sum_{n \leq x} z_{n,r} \bar{z}_{n,s} \right| = \lim_{x \to \infty} x^{-1} \left| \sum_{n \leq x} h_r(\rho(n)) \bar{h}_s(\rho(n)) \right|,$$

which by lemma (22.6) does not exceed

$$c(T^{-1/4} + \exp(-m(T)/4)),$$

where

$$m(T) = \inf_{|\tau| \leq T} \sum_p \frac{1}{p}(1 - \operatorname{Re} h_r(\rho(p)) \bar{h}_s(\rho(p)) p^{-i\tau}).$$

In view of lemma (22.5) this infimum (if it exists finitely at all) may be taken at $\tau = 0$. Here we apply that G_1 has bounded order. Thus whatever the value of T

$$m(T) \geq \sum_{j=1}^{k} (1 - \operatorname{Re} h_r(g_j) \bar{h}_s(g_j)) \sum_{D(p) = g_j} \frac{1}{p},$$

$$\geq \sum_{j=1}^{k} \alpha_j (1 - \operatorname{Re} h_r(g_j) \bar{h}_s(g_j)) \geq \tfrac{1}{2} \sum_{j=1}^{k} \alpha_j,$$

and the factor corresponding to J in lemma (22.4) presently does not exceed

$$(1 + o(1)) cx \exp\left(-\sum_{j=1}^{k} \alpha_j / 8\right), \qquad x \to \infty.$$

This gives for $U(x)$ the upper bound

$$xA(x)\left(1 + (1 + o(1))(t-1)c \exp\left(-\sum_{j=1}^{k} \alpha_j / 8\right)\right), \qquad x \to \infty.$$

On the other hand each map h_r is defined on G so that every $h_r(\rho(n)) = 1$, and $U(x) = tA(x)^2$.

Combining these two estimates and choosing x to become unbounded suitably gives

(2) $$td \leq 1 + (t-1)c \exp\left(-\sum_{j=1}^{k} \alpha_j/8\right).$$

We shall now prove that

(3) $$\sum_{j=1}^{k} \alpha_j \leq \max(2/d, 8\log 2c/d).$$

For if this fails then the sum of the α_j certainly exceeds 1, so that $\theta \leq \frac{1}{5}$. This permits the choice $t = [\theta^{-1/2}]$, and since

$$c \exp\left(-\sum_{j=1}^{k} \alpha_j/8\right) \leq d/2,$$

we see from inequality (2) that $[\theta^{-1/2}] \leq 2/d$. Thus $\theta \geq d^2/9$ and

$$\left(\sum_{j=1}^{k} \alpha_j\right)^2 \leq 9d^{-2} \sum_{j=1}^{k} \alpha_j^2 < 2d^{-1} \sum_{j=1}^{k} \alpha_j,$$

leading to a contradiction. The inequality (3) holds, and uniformly in all positive integers k.

There can be at most $\max(10d^{-2}, 40d^{-1} \log 2cd^{-1})$ elements g_i, for which $\alpha_i = d/5$; and the sum of the reciprocals of the primes which map under D into one of the remaining elements does not exceed the upper bound at (3). By considering these primes p as being exceptional, and restricting ourselves to the subgroup G_2 of G_1, and so G, which is generated by the $n \pmod{\Gamma}$ with integers n not divisible by any of these exceptional primes, we obtain a group G_2 with at most c_1/d^2 generators. In particular G_2 is finite.

To complete the proof of theorem (22.1) we shall show that by the loss of a few more primes we can arrange for a subgroup of G_2 every element of which has order at most d^{-1}.

Suppose that G_2 has an element g of order $l > 1/d$. Bearing in mind that the group $\mathbb{R}/\mathbb{Z}$ is divisible we may begin with the homomorphism given by $g \mapsto 1/l \pmod 1$ and extend it first to a map of G_2 then of G, and finally, by a composition with the canonical $Q^* \to Q^*/\Gamma$, to a map of Q^*. This is illustrated by the following commutative exact diagram,

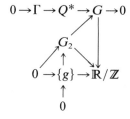

where $\{g\}$ denotes the cyclic subgroup of G_2 which is generated by g. Restricting this map to the positive integers and denoting it by f we see that $f(a_j) = 0$ for every j. By theorem (22.2) there is an integer m, $1 \le m \le d^{-1}$, for which the series

$$\sum_{\|mf(p)\| \ne 0} \frac{1}{p}$$

converges, and with a sum which is bounded in terms of d. Those primes p which under ρ are taken to the element g satisfy $mf(p) \equiv m/l \pmod 1$ and, since $0 < m/l < 1$, must be counted in this series.

Considering the elements of G_2 in turn we obtain a subgroup G_3, of order at most c_1/d^3, every element of which has order at most $1/d$, with the property that those primes p for which $\rho(p)$ does not belong to G_3 have a series of reciprocals whose sum can be explicitly bounded in terms of d.

The proof of theorem (22.1) is complete.

It is not clear if the bound c_1/d^3 for the order of G_3 is of the correct size.

Proof of theorem (22.3). Assume first that H is finite. If h denotes a typical character on H (here a homomorphism of H into the unit circle), then

$$\frac{1}{|G|} \sum_{h \in \hat{H}} h(\delta) \bar{h}(\gamma) = \begin{cases} 1 & \text{if } \delta = \gamma, \\ 0 & \text{otherwise}. \end{cases}$$

Thus the asymptotic density of those integers n which L maps to γ is given by

$$\frac{1}{|G|} \sum_{h \in \hat{H}} \bar{h}(\gamma) \lim_{x \to \infty} x^{-1} \sum_{h \le x} h_\circ L(n).$$

Note that if m is the order of $|G|$ then $(h_\circ L(n))^m = 1$, so that the inner limits exist by lemma (22.6).

If we do not assume H to be finite we may at any rate assume the existence of an element γ of H, and a sequence of integers a_j of positive upper density for which $L(a_j) = \gamma$. Otherwise all the desired densities exist and have the value zero.

Let Γ be the subgroup of Q^* which is generated by those a_j, and define $G = Q^*/\Gamma$. Let ρ be the canonical map $Q^* \to G$. According to the result obtained in the proof of theorem (22.1), there is a finite subgroup G_3 of G, and a sequence of primes p whose reciprocals converge, so that any integer n which is not divisible by such a p is mapped by ρ into a member of G_3. Thus there is a representation

$$n = r \prod a_{j_i}^{\varepsilon_i}$$

with r belonging to a certain finite collection of rationals, and every $\varepsilon_i = \pm 1$. In particular $L(n)$ will have the form $L(r) + s\gamma$. The group H is essentially finitely generated.

Suppose now that γ has infinite order. Then the map given by $\gamma \mapsto \sqrt{2} \pmod{1}$ can be lifted to a homomorphism $H \to \mathbb{R}/\mathbb{Z}$, and then by composing it with L, to a map $f: Q^* \to \mathbb{R}/\mathbb{Z}$. This map satisfies $f(a_j) = f(\gamma)$ for a sequence a_j of positive upper density, hence there is a positive integer m so that the series

$$\sum_{\|mf(p)\| \neq 0} \frac{1}{p}$$

converges. Let t be the order of the group G_3. If a prime p is taken under L to $L(r) + s\gamma$ with $s \neq 0$ then $tmf(p) = tms\sqrt{2} \pmod{1}$ and $mf(p) \neq 0 \pmod{1}$, so that it will be counted in this last sum.

It is now clear that we can find a finite subgroup K of H such that the series $\sum p^{-1}$ taken over the primes p with $L(p)$ not in K, converges. We are essentially back to the first case.

Every positive integer n may be uniquely represented as a product $n_1 n_2$ where n_2 is made up of the primes p with $L(p)$ in K, and n_1 of those primes remaining. Then

$$x^{-1} \sum_{\substack{n \leq x \\ L(n) = \gamma}} 1 = x^{-1} \sum_{\substack{n_1 n_2 \leq x \\ L(n_1) + L(n_2) = \gamma}} 1$$

(4)
$$= \sum_{\alpha \in K} x^{-1} \sum_{\substack{n_1 n_2 \leq x \\ L(n_1) = \gamma - \alpha \\ L(n_2) = \alpha}} 1.$$

Note that the series $\sum n_1^{-1}$ taken over those n_1 with $L(n_1) = \beta$ (say), converges, since its terms are selected from those of the series $\sum n_1^{-1}$ with no side condition, whose sum is $\prod (1 - p^{-1})^{-1}$ taken over those primes p with $L(p)$ not in K. Moreover, this convergence is uniform for all β in H.

The contribution towards the sum at (4) which comes from those products $n_1 n_2$ with $n_1 > \sqrt{x}$ does not exceed

$$\sum_{n_1 > \sqrt{x}} n_1^{-1} = o(1),$$

as $x \to \infty$. From the remaining products we obtain

$$\sum_{\alpha \in K} \sum_{\substack{n_1 \leq \sqrt{x} \\ L(n_1) = \gamma - \alpha}} (1 + o(1)) \psi(\alpha) n_1^{-1}, \qquad x \to \infty,$$

where $\psi(\alpha)$ is the density of those integers n_2 for which $L(n_2) = \alpha$. The existence of this density is obtained by considering the map $Q^* \to K$ obtained by composing L restricted to Q_1, the positive rationals generated by primes p with $L(p)$ in K, with the projection $Q^* \to Q_1$.

Thus the integers n with $L(n) = \gamma$ have the asymptotic density

$$\sum_{\alpha \in K} \psi(\alpha) \sum_{L(n_1) = \gamma - \alpha} n_1^{-1}.$$

This completes the proof of theorem (22.3).

EXAMPLES. (i) Let G be a finite abelian group and L a map of the positive integers into G which satisfies $L(ab) = L(a) + L(b)$ when a and b are coprime. Ruzsa's theorem (theorem (22.3)) asserts that those integers n for which L has a given value in G have an asymptotic density. When will the value of this density be $|G|^{-1}$ where $|G|$ denotes the order of G?

For each homomorphism h in the dual group $\hat{G}$ the limit

$$\delta(h) = \lim_{x \to \infty} x^{-1} \sum_{n \leq x} h_{\circ} L(n)$$

exists, and the density $\Delta(g)$ of those integers with $L(n) = g$ is given by

$$\Delta(g) = \frac{1}{|G|} \sum_{h \in \hat{H}} \bar{h}(g) \delta(h).$$

It follows from a theorem of Artin (see Lang [1], Theorem 7, p. 209) that the characters h are linearly independent over the complex numbers. In particular the determinant $\det(\bar{h}(g))$, where h runs through $\hat{G}$ and g through G, is non-zero. The densities $\Delta(g)$ thus uniquely determine the limits $\delta(h)$.

Define $\varepsilon(h)$ to be 1 if h is the principal (trivial) character, and to be zero for the remaining characters. Since

$$\frac{1}{|G|} = \frac{1}{|G|} \sum_{h \in \hat{G}} \bar{h}(g) \varepsilon(h)$$

we see that in order for every density $\Delta(g)$ to have the value $|G|^{-1}$ it is necessary and sufficient that for non-principal characters h the limit $\delta(h)$ be zero.

An application of lemma (22.6) shows that $\delta(h)$ will be zero if and only if either the series

$$\sum p^{-1}(1 - \operatorname{Re} h_{\circ} L(p))$$

diverges, or this series converges and for every positive integer k, $h_{\circ} L(2^k) = -1$. This last is not possible when k is even.

A sufficient condition for every $\Delta(g)$ to be $|G|^{-1}$ is that for every element g of G which is not the identity the series $\sum p^{-1}$, taken over those primes for which $L(p) = g$, diverges.

(ii) Let $a > 0$, b, $A > 0$, B be integers for which $(a, b) = (A, B) = (a, A) = 1$. Let k be a further positive integer. Let Γ be the subgroup of Q generated by the positive fractions $(an+b)/(An+B)$ with $n > k$, and let G be the quotient group Q^*/Γ. According to the Second Intermezzo this group is finite.

Since the First Intermezzo gives a bound for a set of generators for G, the remarks on algorithms made in Chapter 15 show that an upper bound can be given for the size of $|G|$ itself. Besides being rather large, this bound will depend upon k. It seems very likely that G does not depend upon k.

Let ρ denote the canonical map $Q^* \to G$. I shall now show that there must be a set of generators g of G for each of which the series $\sum p^{-1}$, taken over the primes p with $\rho(p) = g$, diverges.

Suppose that for some element g of G there is an unbounded sequence of x-values so that, in the notation of lemma (12.4),

$$ \sum_{\substack{x^c < p \le x^{1-c/4} \\ \rho(p) = g}} \frac{1}{p} < \tfrac{1}{4}. $$

Then according to lemmas (12.1) and (12.4) (see the remark following the proof of lemma (12.4)) for each positive integer D we can find arbitrarily large integers n so that

$$ an + b = Dr_1, \qquad An + B = r_2, $$

where every prime divisor of the r_j can be taken larger than $(n/2)^c$, and so than D. Moreover, neither of $an+b$ or $An+B$ is divisible by a prime p for which $\rho(p) = g$. Our conditions on the integers a, b, A and B ensure that in this case the fixed-divisor pair is $(1; 1)$ (see Chapter 3). Thus

$$ D = \frac{an+b}{An+B} \cdot \frac{r_2}{r_1}, $$

and $\rho(D)$ is generated by the elements of G with g removed.

Proceeding inductively we arrive at a set of (say) s generators of G, for each of which the estimate

$$ \sum_{\substack{x^\alpha < p \le x \\ \rho(p) = g}} \frac{1}{p} > c_0 $$

holds for some positive constants α, c_0, and all sufficiently large x. Moreover, the same values of c_0 and α can be taken for all generators. Putting together these lower bounds gives

$$\sum_{\substack{p \leq x \\ \rho(p)=g}} \frac{1}{p} \geq c_1 \log\log x + O(1), \qquad x \to \infty.$$

for some positive constant c_1. We can now compare these lower bounds with the well-known asymptotic estimate for $\sum p^{-1}$, $p \leq x$, and obtain $sc_1 \leq 1$. This gives an upper bound for s which does not involve k.

(iii) Let S be the subgroup of Q^* generated by the integers prime to 3. Form the direct sum $S \oplus S$ of two copies of S, and let Γ be its subgroup generated by the elements of the form $(3n-17) \oplus (3n+19)$, with $n = 7, 8, \ldots$. Let G be the quotient group $(S \oplus S)/\Gamma$.

According to theorem (19.2), a pair of integers m_1, m_2, neither of which is divisible by 3, will have $m_1 \oplus m_2$ in Γ if and only if $m_1 \equiv m_2 \equiv 1$, 4 or 7 (mod 9). If $(x, 3) = 1$ then one of x, $x/2$ will be $\equiv 1$, 4 or 7 (mod 9). Each element $x \oplus y$ of $S \oplus S$ will thus be congruent (mod Γ) to an element in the list $1 \oplus j$, $2 \oplus j$, $1 \leq j \leq 9$, $(j, 3) = 1$. It is straightforward to check that there is an isomorphism

$$G \simeq (\mathbb{Z}/6\mathbb{Z}) \oplus (\mathbb{Z}/2\mathbb{Z}).$$

We can now define a "two-dimensional" density, analogous to those considered in theorem (22.3), by

$$\Delta(G) = \lim_{x \to \infty} \frac{9}{4x^2} \sum_{\substack{m_1 \leq x \\ \sigma(m_1 \oplus m_2) = g}} \sum_{m_2 \leq x} 1,$$

where it is to be understood that the integers m_j are prime to 3, and σ is the canonical map $S \oplus S \to G$. It is an easy corollary of theorem (19.2) that every $\Delta(g)$ with g in G exists.

Consider the case $g = 2 \oplus 1 \pmod{\Gamma}$. Let W denote the multiplicative group of reduced residue classes (mod 9). Theorem (19.2) asserts that the value of $\sigma(m_1 \oplus m_2)$ depends only upon the image $\tau(m_1 \oplus m_2)$ in $W \oplus W$ under the map $\tau: m_j \mapsto m_j \pmod 9$. In particular $\sigma(m_1 \oplus m_2) = g$ if and only if $\tau(m_1 \oplus m_2)$ has one of the values $2 \oplus 1$, $8 \oplus 4$, $5 \oplus 7$, these last three direct sums being considered (in an obvious notation) in $W \oplus W$. In this case $\Delta(g) = \frac{1}{12}$.

A simple modification of this argument, together with a suitably strong form of Dirichlet's theorem on primes in arithmetic progression, gives

$$\lim_{x \to \infty} \left(\frac{\log x}{x}\right)^2 \sum_{\substack{p_1 \leq x \\ \sigma(p_1 \oplus p_2) = g}} \sum_{p_2 \leq x} 1 = \frac{1}{12},$$

where p_1, p_2 denote primes. This last result is not derivable from the quantitatively weak general argument given in example (i). In the present circumstances it is available since theorem (19.2) asserts the existence of a map $\psi\colon W\oplus W\to G$ which makes the following exact diagram commutative:

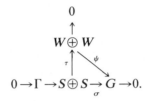

One would expect an analogue of this situation to hold in (ii), and indeed in many similar situations when one considers a subgroup Γ of (say) Q^* or $Q^*\oplus Q^*,\ldots$, generated by fractions with pronounced ring theoretic arithmetic properties.

Arithmic Groups

We say that a group G is *arithmic* if it is the homomorphic image of a direct sum of finitely many groups of reduced residue classes derived from $\mathbb{Z}$.

Let

$$Y = (\mathbb{Z}/D_1\mathbb{Z})^* \oplus \cdots \oplus (\mathbb{Z}/D_k\mathbb{Z})^*$$

be a group of the latter type. For each positive integer d let Q_d^* be the subgroup of Q^* generated by the integers m which are prime to d, and M_d its subgroup generated by those $m \equiv 1 \pmod{d}$. Let H be the direct sum of the $Q_{D_i}^*$ with the above D_i, and let H_0 be the direct sum of the corresponding M_{D_i}.

If the arithmic group G is a homomorphic image of Y under a map ψ then we have the following commutative exact diagram.

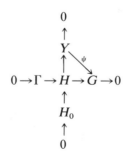

Here H is mapped canonically onto Y by taking $m_1 \oplus \cdots \oplus m_k$ onto the direct sum of the corresponding residue classes m_j (mod D_j). Composition then gives a map $H \to G$, which defines the kernel Γ.

Since each of the groups Q_{D_i} is free, so is H, and we have exhibited G as a quotient of a free group by a group of generators. This is easy. More difficult, and part of which forms the subject matter of the present volume, is to obtain the converse and decide whether a given denumerable abelian group which is presented in terms of a free group and a set of generators is arithmic and can be fitted into a diagram of the above type.

Concluding Remarks

For his proof of theorem (22.3), see Ruzsa [1]. As lemma (9.2) of his paper he asserts that by a theorem of Wirsing a multiplicative function $g(n)$ which takes values in the (complex) unit circle with an arc removed possesses a mean-value. As the function $g(p^k) = \exp(2\pi i(\log \log p^k)^{-1})$ shows this need not be true. It will suffice if there is a positive constant c so that whenever $g(p) \neq 1$ for a prime p then $|g(p)-1| \geq c$, and since this is true in the case he has to hand, no harm is done.

In the present chapter I have given a proof of theorem (22.3) which runs along quite different lines, more in accord with the philosophy of the earlier chapters of this volume. The greatest burden is thrown upon the study of arithmetic functions with values in the group $\mathbb{R}/\mathbb{Z}$. From Ruzsa's proof I took the idea of a random character, except that by introducing more group theory and reducing the problem to a study of denumerable abelian groups of bounded order I was able to avoid an appeal to the theory of Haar measure on (locally) compact groups. This was possible since the relevant dual groups had a readily identifiable structure.

Taking matters further my present direct method allows quantitative bounds to be obtained for both the sums and the orders of the groups involved. Indeed, as I indicate in some of the problems of the following chapter, it is possible to localize the argument to the study of integers in a given bounded interval.

For another application and discussion of Ruzsa's density theorem, with emphasis on the multiplicative properties of the integers $p+1$, p prime, see Meyer [2].

Chapter 23

Problems

This chapter is in two parts. The first part contains exercises that can be solved using ideas related to those which appear in the present volume or my two-volume work on Probabilistic Number Theory, denoted by PNT. They are not organized as to difficulty. The second part contains unsolved problems.

Exercises

1. Let $k \geq 1$ be given. Prove that every positive integer m which is not divisible by 3 has a representation

$$m = \prod_i \left(\frac{3n_i + 1}{3n_i - 1} \right)^{\varepsilon_i},$$

where every $n_i > k$.

2. Let $f(\ )$ be a completely additive arithmetic function, with values in a $\mathbb{Z}$-module, which satisfies

$$3f(5n+1) + 2f(5n+3) = \text{constant}$$

for all $n \geq 1$.

Obtain from this a simpler relation. Examples are $3^4 7 f(10n+1) = $ constant, and $5f(5n+1) = $ constant.

Prove that every integer $m \geq 1$ which is prime to 5 has a representation

$$m^v = \prod_i \left(\frac{(5n_i+1)^3}{(5n_i+3)^2} \right)^{\varepsilon_i}$$

with positive integers n_i, and that the best universal value for v (valid for all permissible m) is $v = 2$.

Give an algorithm to obtain such a representation.

Exercises

3. Let $t \geq 1$ be given. Prove that every positive integer m with $(m, 5) = 1$ has a representation

$$m^k = \prod_i \left(\frac{(5n_i + 1)^3 (5n_i - 9)^7}{(5n_i + 11)^2} \right)^{\varepsilon_i}$$

with $n_i > t$.

Prove that there is a universal value of k. Determine the smallest value of k for each permissible m.

4. Let $t \geq 1$ be given. Prove that every positive integer m with $(m, 7) = 1$ has a representation

$$m^k = \prod_i \left(\frac{(7n_i + 2)^3 (7n_i - 12)^5}{(7n_i - 5)^2} \right)^{\varepsilon_i}.$$

Prove that there is a universal k, and find the smallest such.

5. Let $t \geq 2$ be given. Let $f_1(\ )$, $f_2(\ )$ be completely additive arithmetic functions with values in a $\mathbb{Z}$-module, which satisfy

$$f_1(3n + 2) - f_2(n + 1) = \text{constant}$$

for all $n \geq t$.

Prove that there are constants α, β_1 and β_2 so that

$$f_1(3n + 1) = \alpha n + \beta_1,$$

$$f_2(n) = \alpha n + \beta_2$$

for all $n \geq t$.

Deduce that $f_2(n) = 0$ for all positive n, and $f_1(n) = 0$ for all positive n prime to 3.

Prove that every pair of positive integers (m_1, m_2) with m_1 not divisible by 3, has a simultaneous representation

$$m_1 = \prod_i (3n_i + 2)^{\varepsilon_i},$$

$$m_2 = \prod_i (n_i + 1)^{\varepsilon_i},$$

where every $n_i \geq t$.

6. Let the prime p satisfy $p \equiv 1 \pmod{3}$.

Prove that there is a positive solution to the congruence

$$n^2 + n + 1 \equiv 0 \pmod{p}$$

which does not exceed $(p-1)/2$.

7. Prove that every prime which satisfies $p \equiv 1 \pmod{3}$ has a representation

$$p = \prod (n_i^2 + n_i + 1)^{\varepsilon_i}$$

with positive integers n_i.

8. Let $f(\ )$ be a complex-valued completely additive arithmetic function which satisfies

$$f(n^2 + n + 1) = \sum_{j=1}^{k} P_j(n) \omega_j^n$$

for some polynomials $P_j(x)$ and distinct ω_j.

By means of the identity

$$n^4 + n^2 + 1 = [n^2 + n + 1][(n-1)^2 + (n-1) + 1]$$

or otherwise, prove that every $|\omega_j| \leq 1$, and that if some $|\omega_j| = 1$ then the corresponding polynomial $P_j(x)$ is a constant.

9. Let t be given. Prove that every prime $p \equiv 1 \pmod{3}$ has a representation

$$p^v = \prod_i \left(\frac{(n_i^2 + 5n_i + 7)^5}{(n_i^2 - 3n_i + 3)^7} \right)^{\varepsilon_i}$$

with integers $v > 0$, $\varepsilon_i = \pm 1$ and $n_i > t$.

10. Let θ_j, $|\theta_j| = 1$, $j = 1, \ldots, s$ be distinct complex numbers which satisfy

$$\sum_{j=1}^{s} \alpha_j (\theta_j^{n^2} - \theta_j^n - \theta_j^{n-1}) = O(c^{-n})$$

for some $c > 1$ and all positive integers n.

Prove that

$$\sum_{j=1}^{s}{}' \alpha_j (\theta_j^{m^2} - \theta_j^m - \theta_j^{m-1}) = 0$$

for all positive m, where $'$ indicates summation over those θ_j which are roots of unity. Prove that every θ_j is a root of unity.

Exercises

11. In the notation of question 8 assume that no ω_j with $|\omega_j|=1$ is a root of unity. Then there is a constant $c>1$ so that

$$|f(n^2+n+1)|\le c^{-n}.$$

Prove that $f(p)$ is uniformly bounded on the primes $p\equiv 1\pmod 3$.

12. Prove there is a positive number δ with the following property: Given any positive integer m which is a product of primes $p\equiv 1\pmod 3$, there are infinitely many integers n so that

$$n^2+n+1=dm,$$

where every prime divisor q of d satisfies $q>n^\delta$.

13. In the notation of question 11 prove that $f(p)=0$ for all primes $p\equiv 1\pmod 3$.

14. Let t be given. Prove that every prime $p\equiv 1\pmod 3$ has a representation

$$p^v=\prod_i\left(\frac{[n_i^2+9n_i+21]^4[n_i^2-n_i+1]}{[n_i^2+5n_i+7]^2}\right)^{\varepsilon_i}$$

with integers $v>0$, $\varepsilon_i=\pm 1$, $n_i>t$.

15. Let α, β be integers and let p be a prime, $p>3\alpha^2/4+|\beta|$.
Prove that if the congruence

$$n^2+\alpha n+\beta\equiv 0\pmod p$$

is soluble, then it has a positive solution not exceeding $(p+\alpha)/2$.

16. Let $t\ge 1$ be given. In the notation of question 15 let $\alpha^2\ne 4\beta$. Let Q_1 be the group of positive rationals generated by the primes p for which the Legendre symbol satisfies

$$\left(\frac{\alpha^2-4\beta}{p}\right)=1,0$$

and let Γ be its subgroup generated by the integers of the form $n^2+\alpha n+\beta$ with $n>t$.

Prove that the group Q_1/Γ is finitely generated.

17. Let $h(x)=x^k+\alpha x^{k-1}+\cdots+\delta$ be a polynomial with integer coefficients.
Prove that there is a number c, depending upon the coefficients of this polynomial, so that if the congruence $h(n)\equiv 0\pmod p$ is solvable for some prime p, then it has a positive solution not exceeding $p(k-1)/k+c$.

18. Let $a<0$ and b be integers. Let D be an odd positive integer.

Prove that if the congruence $x^2+ax+b\equiv 0 \pmod{D}$ has a root, then it has one with a representative x in the interval $0\leq x\leq (D+|a|)/2$.

19. In the notation of question 16 prove that the group Q_1/Γ is finite.

20. Let Γ be the subgroup of the positive rationals which is generated by the positive integers of the form $n^2+\alpha n+\beta$ where $n>t$ and the integers α, β satisfy $\alpha^2\neq 4\beta$.

Prove that the quotient group Q^*/Γ is the direct sum of a free group and a finite group.

21. Let Γ be the subgroup of Q^* generated by the n^2+1 with positive integers n.

Prove that Q^*/Γ is free.

22. Let $a>0$, b, $A>0$, B be integers for which $aB\neq Ab$. Let f_1 and f_2 be real-valued additive functions so that $f_1(an+b)+f_2(An+B)$ is bounded uniformly for all large n.

Prove that there is a constant c so that

$$|f_i(n)|\leq (\log n)^c$$

uniformly for all $n\geq 2$ which is prime to aA, and for $i=1,2$.

23. Obtain an analogue of the inequality of theorem (10.1) with $f(an+b)-f(An+B)$ replaced by $f_1(an+b)+f_2(An+B)$.

24. Let $a>0$, b, $A>0$, B be integers for which $aB\neq Ab$.

Characterize all pairs f_1, f_2 of real-valued additive arithmetic functions for which

$$\lim_{n\to\infty}(f_1(an+b)+f_2(An+B))$$

exists and is finite.

25. Let $a_1, a_2, \ldots$ be a sequence of positive integers, and suppose that every positive integer n has a representation

$$n^k = a_{j_1}^{\varepsilon_1}\cdots a_{j_r}^{\varepsilon_r},$$

where each $\varepsilon_i=\pm 1$, $a_{j_i}\leq n^c$, and the positive integral exponent k does not exceed k_0.

Prove that if χ is a non-principal Dirichlet character $\pmod{p}$ p prime, of order greater than k_0, then for each fixed $\delta>0$ there is a constant c_1, depending only upon δ, and an integer $a_j\leq c_1 p^{(c/4\sqrt{e})+\delta}$ so that $\chi(a_j)\neq 0, 1$.

Exercises

26. Define a rational function $R(x) = \prod_{j=1}^{h} (x + a_j)^{b_j}$ where the a_j are distinct non-negative integers, and the integral exponents b_j have highest common factor 1.

Prove that for each fixed $\delta > 0$ there is a constant c, depending at most upon δ and the a_j, b_j, so that for any non-principal Dirichlet character χ defined to a prime modulus p there will be an integer n in the interval $1 \leq n \leq p^{1/(4\sqrt{e}) + \delta}$ for which $\chi(R(n)) \neq 0, 1$ (cf. Burgess [1]).

27. Let positive integers a_j, b_j, $j = 1, \ldots, k$ be given.

Prove that there is a constant c so that the congruence

$$y^2 \equiv (x + a_1)^{b_1} \cdots (x + a_k)^{b_k} \pmod{p}$$

with a prime p, has a solution with $1 \leq y \leq c$. This result was conjectured by Chowla and Chowla [1] and proved by Stephens [1]; see also Elliott [13].

28. Let f_1, f_2, f_3 be real completely additive functions which satisfy

$$f_1(n) + f_2(n+1) + f_3(n+3) = O(1)$$

for all positive integers n.

Prove that

$$f_1(2n-1) + f_2(n) + f_3(n+1) = O(1)$$

holds for all positive n.

Using the parity of n, and also its value (mod 3) construct an algorithm to prove that

$$|f_i(n)| \leq (\log n)^c$$

for some constant c, all $n \geq 2$ and $i = 1, 2, 3$.

29. Characterize those real completely additive functions f_i, $i = 1, 2, 3$, which satisfy

$$f_1(n) + f_2(n+1) + f_3(n+3) \to 0$$

as $n \to \infty$.

30. Let the complex-valued additive functions f_j, $j = 1, 2, 3$, satisfy

$$f_1(n) + f_2(n+1) + f_3(n+3) = \sum_{j=1}^{k} P_j(n) \omega_j^n$$

for polynomials $P_j(x)$ and distinct ω_j. Let

$$\omega = |\omega_1| = |\omega_2| = \cdots = |\omega_r| > |\omega_{r+1}| \geq \cdots \geq |\omega_k|$$

and

$$d = \deg P_1(x) \geq \deg P_2(x) \geq \cdots \geq \deg P_r(x).$$

Prove that if $\omega > 1$ or $d \geq 1$ then there is a constant λ so that

$$|f_i(n)| \leq \lambda n^d \max(\omega, 1)^{2n}$$

uniformly for all $n \geq 1$ and $1 \leq i \leq 3$.

31. Let $0 < \delta \leq \frac{1}{4}$. Prove that the number of integers not exceeding a given positive x for which at least one of n, $n+1$, $n+3$ does not have a prime factor q in the range $x^\delta < q \leq x$, is at most $C\delta x$. Here we may take for C an absolute constant provided that x is sufficiently large depending upon δ.

32. In the notation of question 30 let

$$V(n) = \sum_{j=1}^r \alpha_j (\omega_j \omega_1^{-1})^n,$$

where for $j = 1, \ldots, r$, α_j is the coefficient of x^d in $P_j(x)$. Suppose that $\omega > 1$. Let $0 < \varepsilon < 1$.

Prove that there are constants λ and $\beta > 1$, possibly depending upon ε, so that for all but $C\varepsilon x$ of the integers n not exceeding x

$$|V(n)| \leq \lambda \beta^{-n}.$$

Deduce that

$$\lim_{x \to \infty} x^{-1} \sum_{n \leq x} V(n) = 0,$$

and so that $\alpha_1 = 0$.

In this way prove that $\omega \leq 1$, and if any $|\omega_j| = 1$ then the corresponding polynomial $P_j(x)$ is a constant.

33. Let $a_1, \ldots, a_k$; $b_1, \ldots, b_k$ be integers, not all the b_j being zero. Define the rational function

$$R(x) = \prod_{j=1}^k (x + a_j)^{b_j},$$

and let $t \geq 1$ be given.

Exercises

Determine which triplets (m_1, m_2, m_3) of positive integers have a simultaneous representation

$$m_1^v = \prod_i R(n_i)^{\varepsilon_i},$$

$$m_2^v = \prod_i R(n_i+1)^{\varepsilon_i},$$

$$m_3^v = \prod_i R(n_i+3)^{\varepsilon_i},$$

where each $\varepsilon_i = \pm 1$ and v and the n_i are positive integers, $n_i > t$.

34. In the notation of question 33 prove that a value can be given for v which holds for all the permissible triplets (m_1, m_2, m_3).

Until now we have obtained the product representation of integers by appealing to the algebraic properties of the representing sequences. We now consider sequences which are assumed to have many members rather than a simple algebraic structure.

If $A: a_1 < a_2 < \cdots$ is a sequence of positive integers, $A(x)$ will denote the numbers of its members not exceeding x.

The *lower density* of A is defined to be

$$\underline{d}(A) = \liminf_{x \to \infty} x^{-1} A(x).$$

The *upper density* of A is likewise

$$\bar{d}(A) = \limsup_{x \to \infty} x^{-1} A(x).$$

35. Let A be a sequence of positive integers with lower density $d > \frac{1}{2}$.
Prove that some positive power of 2 can be expressed in the form a_i/a_j.

36. Let A and B be sequences of positive integers for which $\underline{d}(A) + \underline{d}(B) > 1$.
Prove that for those primes p which do not have a representation of the form $p = a_i/b_j$ the series $\sum (1/p)$ converges.
Hint: Apply lemma (15.2) to prove that for most primes p a typical interval $[1, p^{-1}x]$ contains at least about $p^{-1}x\underline{d}(A)$ integers $p^{-1}a_i$ with $a_i \equiv 0 \pmod{p}$, and at least about $p^{-1}x\underline{d}(B)$ integers b_j.

37. In the notation of the previous question prove that the primes p which do not have infinitely many representations of the form $p = a_i/b_j$ are so few that the series $\sum (1/p)$ converges.

38. Let $q_1, q_2,$ be a sequence of primes for which the series $\sum q_i^{-1}$ converges.

Prove that the sequence of integers $a_1, a_2, \ldots$ which are not divisible by any of these primes q_j satisfies

$$\lim_{x \to \infty} x^{-1} A(x) = \prod_{i=1}^{\infty} \left(1 - \frac{1}{q_i}\right).$$

Prove that if the series $\sum q_i^{-1}$ diverges, but the product is interpreted to mean *zero*, then this result remains true.

39. Let A be a sequence of positive integers with $\bar{d}(A) > \frac{1}{2}$.

Prove that the primes p which do not have a representation of the form

$$p = \prod_i a_{n_i}^{\varepsilon_i}$$

are such that the series $\sum (1/p)$ converges.

This result may be compared with that of question 36 with $A = B$.

40. Let q_i, $i = 1, 2, \ldots,$ be a sequence of primes for which the series $\sum q_i^{-1}$ converges.

Prove that for every positive integer m there are infinitely many primes p for which

$$p + 5 = mt,$$

where the integer t is prime to m and is not divisible by any of the primes q_i.

41. Prove that there is a positive integer k, which can be determined, so that every positive integer m has a representation of the form

$$m^k = \prod_j (p_j + n_j^7)^{\varepsilon_j},$$

where the p_j are primes, the n_j positive integers, and $\varepsilon_j = \pm 1$.

Hint: Romanoff.

42. Prove that there is a positive integer v, whose value can be computed, so that every positive integer m has a representation of the form

$$m^v = \prod_i (p_i + 5^{n_i})^{\varepsilon_i},$$

where the p_i are primes, each n_i is a positive integer, and $\varepsilon_i = \pm 1$.

43. Let k be a positive integer. A theorem of Miech [1] asserts that those integers which are not expressible in the form $p + m^2$ for some prime p and positive integer m have asymptotic density zero. Use the Hardy–Littlewood circle method to establish this.

Hint: See Vaughan [2].

44. Let k be a positive integer. Prove that every positive integer D has infinitely many representations of the form

$$D = \frac{p_1 + m_1^k}{p_2 + m_2^k}$$

with primes p_i and integers m_j.

45. Let n be a positive integer in an interval $1 < n < x^{4/5}$.

Prove that the number $r(n)$ of solutions to the equation

$$n = \frac{p+1}{q+1}$$

in primes $p \leq x$ and q, does not exceed

$$\frac{c_1}{n} \prod_{\substack{p \mid n(n-1) \\ p > 2}} \left(\frac{p-1}{p-2}\right) \frac{x}{(\log x)^2},$$

where c_1 may be given an absolute value for all (absolutely) large x.

Deduce that

$$\sum_{n \leq x^{4/5}} nr(n)^2 \leq c_2 \frac{x^2}{(\log x)^3}.$$

46. Prove that

$$\sum_{n < x^{4/5}} r(n) > c_1 \frac{x}{\log x}$$

for some constant c_1 and all large x.

47. Prove that the positive integers which are representable in the form $(p+1)/(q+1)$ with primes p, q, have a positive upper density.

48. Prove that there is a positive integer v so that every integer $m > 0$ has a representation

$$m^v = \prod_i (p_i + 1)^{\varepsilon_i}.$$

Prove that the same result holds even if it is required that the primes p_i exceed a given value t.

49. Let $a_1 < a_2 < \cdots < a_k \leq x$ be a finite sequence of positive integers with $kx^{-1} \geq d > 0$.

Prove that there is a positive integer s, and real numbers T, m bounded in terms of d only, so that any additive function $f: Q^* \to \mathbb{R}/\mathbb{Z}$ which vanishes on these integers a_j also satisfies

$$\sum_{p \leq x} \frac{\|srf(p) - \tau_r \log p\|^2}{p} \leq m$$

with $|\tau_r| \leq T$ for all positive integers r uniformly for all $x \geq 2$.

50. Let T, m be given positive reals. Let

$$\sum_{p \leq x} \frac{\|\tau \log p\|^2}{p} \leq m$$

for a real τ, $|\tau| \leq T$.

Prove that there is a constant c, depending upon m and T only, so that

$$|\tau| < c(\log x)^{-1}$$

for all x large enough in terms of m, T.

51. In the notation of question 49 prove that

$$|\tau_{r+s} - \tau_r - \tau_s| < c_1(\log x)^{-1}$$

for some constant c_1, all positive integers r, s and all x sufficiently large in terms of d.

52. A real-valued arithmetic function $h(n)$ satisfies

$$|h(r+s) - h(r) - h(s)| \leq y$$

for all positive integers r, s.

Prove that

$$\lim_{k \to \infty} 2^{-k} h(2^k m)$$

exists for each positive m.

Prove that there is a constant A so that

$$|h(r) - Ar| \leq y$$

for all integers $r > 0$.

53. Let $a_1 < a_2 < \cdots < a_k \leq x$ be $k > dx$ positive integers. Let $f: Q^* \to \mathbb{R}/\mathbb{Z}$ be an additive function which satisfies $f(a_j) = 0, j = 1, \ldots, k$.

Prove that there is a positive integer s, not exceeding d^{-1}, and a real number m which depends only upon d, so that

$$\sum_{\substack{p \leq x \\ \|sf(p)\| \neq 0}} \frac{1}{p} \leq m$$

holds for all positive reals x.

54. In the notation of question 48, prove that there is a constant c so that one may restrict the primes p_i occurring in the product representation by $p_i < m^c$.

Let A be a sequence of positive integers and define Γ to be the subgroup of Q^* generated by these integers, and G to be the quotient group Q^*/Γ. If A has a positive upper density, then there is a sequence of primes p for which $\sum p^{-1}$ converges, and a finite subgroup H of G so that if Q_1 denotes the subgroup of Q^* generated by the primes distinct from these p the canonical map $Q^* \to G$ leads readily to the commutative diagram

$$\begin{array}{ccccccccc} 0 & \to & \Gamma & \to & Q^* & \to & G & \to & 0 \\ & & & & \uparrow & & \uparrow & & \\ & & & & Q_1 & \to & H & & \\ & & & & & & \uparrow & & \\ & & & & & & 0 & & \end{array}$$

Define the *multiplicative density* of such a sequence A to be $\delta(A) = |H|^{-1}$, the reciprocal of the order of an H chosen so that $|H|$ is minimal.

If A has upper density zero, then define $\delta(A) = 0$.

Thus the positive integers have multiplicative density 1, the integers of the form $4k + 3$ have multiplicative density $\frac{1}{2}$.

55. Prove that if A is a subsequence of the sequence of integers B, then $\delta(B)$ is an integral multiple of $\delta(A)$.

56. Prove that there is an absolute constant c so that $\delta(A)\bar{d}(A)^3 \leq c_0$ holds for every sequence A of positive integers.

57. For sequences of rationals A, B let $A \wedge B$ denote the sequence of positive integers which are generated by the members of A and B using multiplication and division. Note that $A \wedge A = A$ need not be true, but $(A \wedge A) \wedge A = A \wedge A$ will be. Let T_p be the set of pth-powers of the positive rational integers, for some prime p.

Prove that if $\delta(A)$ is positive, then $\delta(A \wedge T_p) = p^{-t}$ for some non-negative integer t.

58. Let $a_1, a_2, \ldots$ be a sequence A of positive integers of upper density $\delta > 0$. Suppose further that for every positive integer D and sequence of primes p_j for which $\sum p_j^{-1}$ converges there is a member a_i of A which has the form $a_i = Dr$ where the rational number r is not divisible by any of the p_j.

Prove that a finite collection of s ($\ll \delta^{-3}$) primes q_t can be found so that every positive integer n has a representation in the form

$$n = q_1^{\alpha_1} \cdots q_s^{\alpha_s} \prod_{i=1}^{k} a_{j_i}^{\varepsilon_i},$$

where the α_t are integers and each $\varepsilon_i = 1$ or -1.

And now for a paper chase around the theory of sieves. For general background references see the books of Halberstam and Richert [1], Davenport [1], Prachar [1].

For the next 28 problems, up until number 82, P will denote the subgroup of the positive rationals generated by integers of the form $p+1$ with p prime, G will denote the quotient group Q^*/P, and ρ the canonical map $Q^* \to G$.

59. Let D be an even positive integer, and define

$$w(D) = \frac{1}{D} \prod_{\substack{p \mid D \\ p > 2}} \left(\frac{p-1}{p-2}\right) \prod_{p > 2} \left(1 - \frac{1}{(p-1)^2}\right).$$

Let δ be a real number, $0 < \delta < \frac{1}{2}$. Let x be a positive real number and q a prime which exceeds $x^{1-\delta}$.

Prove that the number of primes p not exceeding x which satisfy $p \equiv -1 \pmod{Dq}$ are are such that $(p+1)/Dq$ is not divisible by any odd prime up to x^δ is as $x \to \infty$ not more than

$$(8 + o(1)) w(D) x (\log x)^{-2}.$$

Here the function $o(1)$ may depend upon D.

60. In the same notation prove that if $\delta < \frac{1}{4}$, then the number of primes p not exceeding x which satisfy $p \equiv -1 \pmod{D}$ and for which $(p+1)/D$ is not divisible by any prime up to x^δ is at least

$$(1 + o(1)) w(D) f\left(\frac{1}{2\delta}\right) \frac{2x}{\delta e^\gamma (\log x)^2}, \qquad x \to \infty,$$

where

$$u f(u) = 2 e^\gamma \log(u - 1)$$

for $2 \le u \le 4$.

Exercises

61. Let μ be a further positive real number, not exceeding $\delta/2$. Let q be a prime in the range $x^\delta < q \le x^{1-\delta}$.

Prove that the number of primes $p \le x$ for which $p \equiv -1 \pmod{Dq}$ and $(p+1)/Dq$ is not divisible by any prime up to x^μ, is not more than

$$\frac{2+o(1)}{\mu^2} \cdot \frac{(q-1)w(D)}{q(q-2)} \cdot \frac{x}{(\log x)^2}$$

as $x \to \infty$.

62. Let σ be a (possibly infinite) collection of primes.

Prove that the number of primes p up to x which satisfy $p \equiv -1 \pmod{Dq}$ for some prime q belonging to σ and lying in the range $x^\delta < q \le x^{1-\delta}$ is as $x \to \infty$ not more than

$$\frac{8+o(1)}{\delta^2} \cdot \frac{xw(D)}{(\log x)^2} \sum_{\substack{x^\delta < q \le x^{1-\delta} \\ q \in \sigma}} \frac{1}{q}.$$

63. In the notation of the previous question prove that if

$$8\delta^{-2} \sum_{\substack{x^\delta < q \le x^{1-\delta} \\ q \in \sigma}} q^{-1} < 2\, e^{-\gamma} \delta^{-1} f((2\delta)^{-1}) - 8$$

for an unbounded sequence of x-values, then there is a prime p so that $p+1 = Dt$ with none of the prime divisors of t belonging to σ.

For $0 < \delta < \frac{1}{4}$ define

$$\beta(\delta) = (\delta\, e^{-\gamma} f(1/2\delta) - 4\delta^2)/4 \log(1-\delta)/\delta.$$

64. Prove that there are s generators g_i of G for each of which

$$\sideset{}{'}\sum_{q \le z} \frac{1}{q} \ge \beta(\delta) \log \log z + O(1), \qquad z \to \infty,$$

where $'$ indicates that summation is confined to those primes for which $\rho(q) = g_i$.

65. Prove that G has at most 325 generators.

66. Let M and $d \ge 2$ be positive integers, M regarded as "fixed". For each $x \ge 1$, $v > 0$ let $G_d(x, v)$ denote the number of pairs of primes p, q for which $p+1 = Md(q+1)$, with $q \le x^v - 1$.

Prove that

$$G_d(x, v) \le (8+o(1))\alpha g(Md) \frac{x^v}{(v \log x)^2}$$

as $x \to \infty$, where g is the function defined on the integers $n > 1$ by

$$g(n) = \prod_{\substack{p \mid n(n-1) \\ p > 2}} \frac{p-1}{p-2},$$

and

$$\alpha = \prod_{p > 2} \left(1 - \frac{1}{(p-1)^2}\right).$$

67. Prove that for each fixed $v < 1$

$$\sum_{2 \leq d \leq x/M} G_d(x, v) \geq \frac{1}{(1+v) \log x} \sum_{q \leq x^v - 1} \frac{q+1}{\phi(M(q+1))} + O\left(\frac{x^{1+v}}{(\log x)^3}\right)$$

as $x \to \infty$.

68. Prove that the sum

$$\sideset{}{''}\sum_{2 \leq d \leq x/M} g(Md)$$

taken over those integers d for which $G_d(x, v) > 0$ is at least

$$\frac{(v + o(1))}{8\alpha(1+v)} \frac{x}{\phi(M)} \prod_{p \nmid M} \left(1 + \frac{1}{(p-1)^2}\right), \qquad x \to \infty.$$

69. For integers $n \geq 2$ define

$$g_M(n) = \prod_{\substack{p \mid n(Mn-1) \\ p \nmid M, \, p > 2}} \left(\frac{p-1}{p-2}\right).$$

Prove that for all large x

$$\sideset{}{''}\sum_{2 \leq d \leq x/M} g_M(d) \geq \frac{(v + o(1))}{8\alpha(1+v)} \frac{x}{\varphi(M)} \prod_{\substack{p \mid M \\ p > 2}} \left(\frac{p-2}{p-1}\right) \prod_{p \nmid M} \left(1 + \frac{1}{(p-1)^2}\right).$$

70. For real $t \geq \tfrac{1}{2}$ define

$$a_M(t) = \prod_{p \nmid 2M} \left(1 + \frac{\left(\frac{p-1}{p-2}\right)^{2t} - 1}{p}\right).$$

Prove that for each $\theta > 1$

$$\sum_{2 \le \alpha \le x/m} g_M(d)^\theta \le (1+o(1))\frac{x}{M}a_M(\theta), \qquad x \to \infty.$$

For each positive integer m let B_m denote the set of positive integers which can be represented in the form $(p+1)(m(q+1))^{-1}$ with primes p and q.

71. Define

$$c_M = \prod_{p \nmid 2M} \left(1 + \frac{1}{p(p-2)}\right).$$

By applying Hölder's inequality prove that for any $u > 1$, $1/u + 1/w = 1$, the lower density of B_M is at least

$$\left(\frac{c_M}{8a_M(w)^{1/w}}\right)^u.$$

Questions 66 through 71 I have derived from the paper [3] of Meyer. They may be compared with questions 45 through 47 which I derived from my paper [4]. Between these two papers lie the further papers Elliott [7] and Wirsing [4].

The conclusion of question (71) shows that

$$\underline{d}B_1 \ge \left(\frac{c_1}{8a_1(w)^{1/w}}\right)^u.$$

Meyer [3] gives the approximate value $c_1 = 2.14069$ and states that considered as a function of u the maximum of this lower bound occurs with u near to 1.156, leading to $\underline{d}B_1 \ge 1/39.57$. We continue until with question 74 we reach his main conclusion.

72. Prove that if in the notation of question 71 we replace M by any multiple, then the expression which is given as a lower bound for $\underline{d}B_M$ does not decrease.

73. Let r be twice the product of the first eight primes.
Prove that $\underline{d}B_r \ge 1/8.88$.
Hint: According to Meyer [3] u approximately 1.031 will secure this.

74. (Meyer [3]). Prove that every positive integer n has a representation

$$n^\nu = (p_1 + 1)(p_2 + 1)(p_3 + 1)^{-1}(p_4 + 1)^{-1}$$

with primes p_j and a positive integer ν not exceeding 8.

In his paper [2] Meyer announces that this upper bound on ν can be reduced to 4.

75. Prove that amongst any collection of nine positive integers m_i there must be a pair m_i, m_j with $1 \le i < j$ so that

$$m_j = m_i(p_1+1)(p_2+1)(p_3+1)^{-1}(p_4+1)^{-1}$$

with primes p_s.

76. Prove that for each element g of G there are infinitely many integers m so that $\rho(m) = g$.

77. Prove that the order of G is at most 8.

For each Dirichlet character χ let

$$\psi(x,\chi) = \sum_{n \le x} \Lambda(n)\chi(n)$$

where $\Lambda(n)$ denotes von Mangoldt's function.

78. Prove that

$$\sum_{\substack{q \le Q \\ q \text{ prime}}} \frac{q}{\varphi(q)} \sum_{\substack{\chi \\ \text{non-principal}}} \max_{y \le x} |\psi(y,\chi)| \le 1.93 Q^2 x (\log x)^4$$

uniformly for $x^{1/3} \le Q \le x^{1/2}$, $x \ge (202)^6$.

For integers $D > 0$, l and real x, let

$$\psi(x) = \sum_{n \le x} \Lambda(n),$$

$$\psi(x, D, l) = \sum_{\substack{n \le x \\ n \equiv l \pmod{D}}} \Lambda(n).$$

79. Prove that

$$\sum_{\substack{q \le Q \\ q \text{ prime}}} \frac{q}{\varphi(q)} \max_{(l,q)=1} \max_{y \le x} \left| \psi(y, q, l) - \frac{\psi(y)}{\phi(q)} \right| < 2Q^2 x^{1/2} (\log x)^4$$

uniformly for $x^{1/3} \le Q \le x^{1/2}$, $x \ge (202)^6$.

80. Let D be a positive even integer. Let $g(n)$ be a multiplicative function defined on squarefree integers by $g(p) = \omega(p)(p - \omega(p))^{-1}$ where $\omega(p) = 1$ if $p | D$, and $= 2$ otherwise.

Prove that

$$\sum_{n \le z} \mu^2(n) g(n) \ge \tfrac{1}{4}\left(\log \frac{z}{D}\right)^2 \prod_{\substack{p | D \\ p \ge 3}} \left(\frac{p-2}{p-1}\right)$$

uniformly for $2 \le D < z$.

Exercises

81. In the notation of the previous question prove that if D has the form qm where q is an odd prime, $(q, m) = 1$, then the above lower bound may for $m < z$ be replaced by

$$\frac{1}{4}\left(1-\frac{2}{q}\right)\left(\log\frac{z}{m}\right)^2 \prod_{\substack{p\mid m \\ p\geq 3}} \left(\frac{p-2}{p-1}\right).$$

Let N be a positive integer. Inductively define sets of primes S_1, $S_2, \ldots$, with S_i in the interval $(2^{i-1}N, 2^iN]$ as follows:

S_1 is the empty set.

For $i > 1$ S_i consists of those primes r which do not have a representation $p+1 = rt$ where p is a prime, and each prime divisor of t is less than r and does not belong to S_j for any $j < i$.

82. Let $\alpha > 1$ be given. Prove that if N is chosen sufficiently large in terms of α, then S_i contains at most $2^i N (\log 2^i N)^{-\alpha}$ members for every i.

83. In the notation of the previous question prove that $\alpha = 2$, $N = 10^{2042}$ is possible.

84. Prove that every positive integer n has a representation

$$n = t \prod_{i=1}^{s} (p_i + 1)^{\varepsilon_i}$$

with the p_i primes, each ε_i has a value 1 or -1, and every prime factor of the rational number t is at most 10^{2042}. Moreover, indicate an algorithm which will obtain such a representation.

A result of this type is implicit in the paper of Katái [2], except that in his argument he applied the Bombieri–Vinogradov theorem on primes in arithmetic progression and since that inequality contains a constant which cannot be determined he was not able to give a bound for N, nor to construct an algorithm to obtain a representation (cf. Elliott [26]).

85. In the notation of the previous question, prove that such a representation for n exists with t a positive integer possessing at most eight distinct prime factors; so that t does not exceed 10^{15000}.

86. Prove that if G is trivial, then there exists a k such that every positive integer n has a representation of the form

$$n = \prod_{i=1}^{s} (p_i + 1)^{\varepsilon_i}$$

with primes p_i, each $\varepsilon_i = \pm 1$, and with at most k factors $p_i + 1$ occurring in the product.

We say that a polynomial $f(x)$, defined over a commutative ring R without divisors of zero, has *persistence of form* if there are distinct polynomials $P_i(x)$ and integers d_i not all zero, $i = 1, \ldots, k$, so that

$$\prod_{i=1}^{k} f(P_i(x))^{d_i}$$

is identically a constant.

Such identities have interesting applications to the representation of integers by products, as was shown in some of the earlier questions.

87. Let $f(x)$ be a polynomial, with complex coefficients, which for some $a \neq 0$ satisfies the identity

$$f(x^2) = f(x)f(x+a).$$

Prove that all the roots of $f(x)$ lie on the circle $|z| = 1$, and there are at most two of them.

88. The polynomial $f(x)$ has integer coefficients, and for some non-zero integer a satisfies the identity

$$f(x^2) = f(x)f(x+a).$$

Prove that it has the form

$$(x-1)^m, \; m \geq 1 \quad \text{with } a = 2,$$

or

$$(x^2 + x + 1)^m, \; m \geq 1 \quad \text{with } a = -1.$$

89. Let $\omega_1, \omega_2, \ldots$ be a sequence of real numbers which satisfies

$$\omega_{n+1} \geq \omega_n^2 - \delta$$

for a real $\delta \geq 2$.

Prove that if $\omega_1 > \delta$ then $\omega_n \to \infty$ as $n \to \infty$.

90. Let a complex number y, $|y| \geq 2$, be given. Let S be a bounded set of points in the complex plane which is taken into (but not necessarily onto) itself under the map $z \mapsto (z-y)^2$.

Prove that S is contained in the disc $|z - y| \leq |y|$; and that if y is real and positive then also in the half-disc $|z - y| \leq \sqrt{2}|y|$, $\operatorname{Re} z \leq y$.

91. Let $f(x)$ be a polynomial with complex coefficients which satisfies the identity

$$f(x^2) = f(x+a)f(x+b)$$

for some fixed real numbers $b > a \geq 2$.

Prove that all the roots of $f(x)$ lie on the intersection of the circles $|z| = a$ and $|z - b| = \sqrt{a}$. Deduce that $f(x)$ is identically zero.

92. Let $f(x)$ be a polynomial with complex coefficients which satisfies the identity

$$f(x^2) = f(x+a)f(x+b)$$

for some non-zero rational integers a and b.

Prove that $|a| \leq 9$, $|b| \leq 9$.

Determine all possible polynomials $f(x)$.

93. Let t be a real number. Prove that every positive integer m has a representation

$$m = \prod_{i=1}^{k} (n_i(n_i^2 - 2))^{\varepsilon_i},$$

where the n_i are positive integers exceeding t, and each $\varepsilon_i = \pm 1$.

Hint (see Burgess [2]): Let c be an integer and $f(x)$ the polynomial $x(x^2 - c(c+1))$. Consider

$$\frac{f(x^2 + x - c(c+1))}{f(x)f(x+1)}.$$

We next recall the (Fejér kernel) representation:

$$\frac{1}{n}\left(\frac{\sin \frac{1}{2}nt}{\sin \frac{1}{2}t}\right)^2 = \sum_{\nu=-(n-1)}^{n-1} \left(1 - \frac{|\nu|}{n}\right) e^{i\nu t}$$

valid for real t, $0 < |t| < \pi$, integers $n \geq 2$.

94. A real-valued additive arithmetic function $f(n)$ is said to be *strongly-additive* if $f(p^m) = f(p)$ for every power m of a prime p. We extend its definition to all integers by $f(0) = 0$, $f(-n) = f(n)$ if $n > 0$. Let $x \geq 2$ be real.

Prove that for integers n in the range $1 \leq n \leq x$, such a function has a representation

$$f(n) = \sum_{r=0}^{p-1} \sum_{p \leq x} \frac{f(p)}{p} e^{2\pi i r n / p}.$$

95. For any strongly-additive function $f(n)$ and integer $N \geq 2$ define

$$V(f, N) = \sum_{m=-N}^{N} \left(1 - \frac{|m|}{N+1}\right) |f(m) - E(N)|^2,$$

where

$$E(N) = \sum_{p \leq N} \frac{f(p)}{p}.$$

Let f_0 denote the function obtained from $f(n)$ by replacing $f(p)$ with $|f(p)|$. Prove that

$$V(f, N) \leq V(f_0, N).$$

96. In the notation of question 95 prove that

$$V(f_0, N) \leq \left(N + O\left(N \left(\frac{\log \log N}{\log N}\right)^{1/2}\right)\right) \sum_{p \leq N} \frac{|f(p)|^2}{p}.$$

Deduce the following form of the Turán–Kubilius inequality:

$$\sum_{n \leq x} |f(n) - E(x)|^2 \leq \left(2x + O\left(x \left(\frac{\log \log x}{\log x}\right)^{1/2}\right)\right) \sum_{p \leq x} \frac{|f(p)|^2}{p}.$$

97. Prove that the inequality

$$\sum_{p < x} p \left| \sum_{\substack{m < x \\ m \equiv 0 \pmod{p}}} \left(1 - \frac{m}{x}\right) a_m - \frac{1}{p} \sum_{m < x} \left(1 - \frac{m}{x}\right) a_m \right|^2$$

$$< \frac{1}{2} \left(x + O\left(x \left(\frac{\log \log x}{\log x}\right)^{1/2}\right)\right) \sum_{m < x} \left(1 - \frac{m}{x}\right) |a_m|^2$$

holds for all complex numbers a_m, $m = 1, 2, \ldots, [x]$, for all $x > 2$.

98. For odd primes p let $\alpha(p)$ denote the smallest positive integer n so that both n and $n+1$ are quadratic non-residues $\pmod{p}$. In PNT, Chapter 4, by using the dual of a Turán–Kubilius inequality, I proved that for each $\varepsilon > 0$ there is a constant c, depending upon ε, so that $\alpha(p) < cp^{(1-\delta)/4+\varepsilon}$ with $\delta = e^{-10}/2$. Improve that result.

99. Let A be a sequence of positive integers with positive upper density. Prove that those primes p for which one cannot solve each congruence $a_i + a_j \equiv r \pmod{p}$, $r = 0, 1, \ldots, p-1$, with suitable members a_i, a_j of the sequence A, are finite in number.

Exercises

100. Let A be a sequence of integers with positive lower density. Prove that the primes p for which one cannot solve the congruence $a_i^m \equiv r \pmod{p}$ when $1 \leq r \leq p-1$, and the a_i is to be chosen from A, are such that the series $\sum \varphi(p-1)/(p-1)$ converges.

Let $\nu_x(n; \ldots)$ denote the frequency $N/[x]$ where N denotes the number of integers n, not exceeding x, which have the property....

101. Let $f(n)$ be a real-valued additive arithmetic function which for real z satisfies

$$\nu_x(n; f(n) + f(n+1) \leq z) \Rightarrow F(z), \qquad x \to \infty.$$

Here $\Rightarrow$ denotes the usual weak convergence of probability/measure theory.

Let ε be given, $0 < \varepsilon < 1$. Prove that there is a constant c_1 so that the integers n for which $|f(n) + f(n+1)| < c_1$ have a lower density of at least $1 - \varepsilon$.

Prove that the integers for which $|f(n) - f(n+2)| < 2c_1$ have a lower density of at least $1 - 2\varepsilon$.

Prove that there is a constant c_2 so that those integers for which $|f(n)| \leq c_2$ have a lower density of at least $\frac{1}{4} - 3\varepsilon$.

Give necessary and sufficient conditions for the above limiting relation to hold.

102. Let $h(x)$ be a polynomial with integer coefficients, positive for all sufficiently large integer values of x. For each prime p let $\rho(p)$ denote the number of incongruent solutions $n \pmod{p}$ to $h(n) \equiv 0 \pmod{p}$.

Let $f(\ )$ be a non-negative real valued additive arithmetic function which for real z satisfies

$$\nu_x(n; f(h(n)) \leq z) \Rightarrow F(z), \qquad x \to \infty.$$

Prove that for a suitable positive constant c the series

$$\sum_{f(p) > c} \frac{\rho(p)}{p}$$

converges.

Prove that for each $\varepsilon > 0$ the series

$$\sum_{f(p) > \varepsilon} \frac{\rho(p)}{p}, \qquad \sum_{f(p) \leq \varepsilon} \frac{\rho(p) f(p)^2}{p}$$

converge.

Let $\nu_x(p; \ldots)$ denote the frequency $N/\pi(x)$ where N denotes the number of primes not exceeding x for which property ... is satisfied, and $\pi(x)$ denotes the total number of primes not exceeding x.

103. Let $h(x)$ be a polynomial with integer coefficients, positive for all large positive integers x. Let $\psi(p)$ denote the number of solutions $n \pmod{p}$ with $(n, p) = 1$, to the congruence $h(n) \equiv 0 \pmod{p}$.

Let $f(\ )$ be a non-negative real additive arithmetic function which satisfies

$$\nu_x(p; f(h(p)) \leq z) \Rightarrow F(z), \qquad x \to \infty.$$

Prove that for each $\varepsilon > 0$ the series

$$\sum_{f(p) > \varepsilon} \frac{\psi(p)}{p}, \qquad \sum_{f(p) \leq \varepsilon} \frac{\psi(p) f(p)^2}{p}$$

converge.

104. Prove that the frequencies

$$\nu_x(p; f(p+1) \leq z)$$

converge weakly for a non-negative real additive arithmetic function if and only if the series

$$\sum_{f(p) > 1} \frac{1}{p}, \qquad \sum_{f(p) \leq 1} \frac{f(p)}{p}$$

converge.

105. Prove that there is a constant c so that

$$\limsup_{x \to \infty} \frac{1}{x} \left| \sum_{n < x} \mu(n(n+1)) \right| < c < 1.$$

Here $\mu(n)$ denotes the usual Möbius function.

For a survey discussion of the word problem for groups see Stilwell [1]. As lemma (2.31) of this paper [1] Britten gives the following result:

There is a one-one mapping ϕ of the set of positive integers into itself such that the image of any given integer can be calculated but there exists no finite process for deciding whether or not any given integer is an image under ϕ.

106. (Britten). Let f_n, g_n, h_n, be the generators of a group with relations $f_{\phi(n)} h_n = 1$, $g_{\phi(n)} h_n^{-1} = 1$; $n = 1, 2, 3, \ldots$. Prove that no finite decision process can be given to decide if $f_m g_m = 1$.

107. The foregoing question defines a denumerable group with a recursively unsolvable word problem. This group need not be commutative. Define a commutative denumerable group with an unsolvable word problem.

108. Prove that there is a sequence of positive rationals $r_1, r_2, \ldots$ so that no algorithm can be given to decide whether an arbitrary positive integer n has a representation of the form

$$n = r_{t_1}^{\varepsilon_1} \cdots r_{t_k}^{\varepsilon_k},$$

where each ε_i has a value $+1$ or -1.

Unsolved Problems

1. Given polynomials $P_j(x)$, with integer coefficients and positive for all large integers x, and complex-valued multiplicative functions $g_j(n), |g_j(n)| \leq 1$ for all n, $j = 1, \ldots, k$, when can

$$\lim_{x \to \infty} x^{-1} \sum_{n \leq x} g_1(P_1(n)) \cdots g_k(P_k(n))$$

exist? Even a simple case such as

$$\lim_{x \to \infty} x^{-1} \sum_{n \leq x} g_1(n) g_2(n+1)$$

is at present apparently beyond reach.

2. Decide if

$$\lim_{x \to \infty} x^{-1} \sum_{n \leq x} \mu(n(n+1)) = 0.$$

The answer must surely be "yes"!

Is there a connection between sums of this type and prime-pairs?

3. Decide whether

$$\lim_{x \to \infty} x^{-1} \sum_{n \leq x} \mu(n^2+1) = 0.$$

Once again the answer must surely be "yes".

4. Develop a theory for additive functions which take values in a finite field.

5. Let the polynomial $P(x)$ with integer coefficients be positive for all large x.

What is the necessary and sufficient condition that a real additive function $f(\ )$ (extended by $f(n) = 0$ if $n \leq 0$) should have the sums

$$x^{-1} \sum_{n \leq x} |f(P(n))|^2, \qquad x \geq 2,$$

uniformly bounded? We do not know of any general result with degree $P(x) \geq 2$.

6. Let $R_j(x)$, $j = 1, 2$, be rational functions of x, with rational coefficients. Let χ_j be a character mod D_j, $j = 1, 2$. Let $R_j(x)$ be defined mod D_j; that is to say the coefficients in $R_j(x)$ are to be prime to D_j.

How large is the least positive integer n for which

$$\chi_1(R_1(n))\chi_2(R_2(n)) \neq 0, 1?$$

Is it in general $O((D_1 D_2)^\varepsilon)$ for each fixed $\varepsilon > 0$? For example, let

$$R(x) = \prod_{j=1}^{h} (x + a_j)^{b_j}$$

where the a_j are non-negative integers, $(b_1, \ldots, b_h) = 1$. There is a simultaneous representation

$$m_1 = \prod_i R(n_i)^{\varepsilon_i},$$

$$m_2 = \prod_i R(n_i + 1)^{\varepsilon_i},$$

for each pair of positive integers m_1 and m_2. In particular

$$\chi_1(m_1)\chi_2(m_2) = \prod_i (\chi_1(R(n_i))\chi_2(R(n_i + 1)))^{\varepsilon_i}.$$

Thus if every $\chi_1(R(n_i))\chi_2(R(n_i + 1)) = 1$ we shall have $\chi_1(m_1)\chi_2(m_2) = 1$. This forces the characters χ_1, χ_2 to be principal. This argument can be made exact, and in this context it is desirable to localize the integers n_i in terms of m_1 and m_2.

7. An additive arithmetic function $f(\ )$ takes values in a finite field and satisfies $f(p + 1) = 0$ for every prime p.

Prove that it is identically zero.

8. Prove that every pair of positive integers m_1, m_2 has a (simultaneous) representation

$$m_1 = \prod_i (p_i + 1)^{\varepsilon_i},$$

$$m_2 = \prod_i (p_i + 3)^{\varepsilon_i},$$

with primes p_i and integers $\varepsilon_i = \pm 1$.

Unsolved Problems

9. Prove that every prime q which satisfies $q \equiv 1 \pmod{4}$ has a representation

$$q = \prod_i (p_i^2 + 1)^{\varepsilon_i}$$

with primes p_i and integers $\varepsilon_i = \pm 1$.

10. For what polynomials $P_j(x)$ in $\mathbb{Z}[x]$, $j = 1, 2$, are simultaneous representations

$$m_1 = \prod P_1(n_i)^{\varepsilon_i},$$

$$m_2 = \prod P_2(n_i)^{\varepsilon_i}, \qquad n_i > 0, \ \varepsilon_i = \pm 1,$$

possible for every pair of positive integers m_1 and m_2?

11. If $a > 0$, b, $A > 0$, B are integers satisfying $aB \neq Ab$, then the group Q^*/Γ, where Γ is the subgroup of Q^* generated by fractions $(an + b)/(An + B)$, has the form $G_1 \oplus G_2$ where G_1 is free, G_2 is finite. It is known which elements belong to G_2. How can one theoretically determine their precise order?

12. Let $F(x)$ be a rational function with rational coefficients which is not the square or higher power of another such function. Let $F(x)$ be positive for all sufficiently large positive reals and let $r_n = F(n)$ for $n = 1, 2, \ldots$. Let Γ be the subgroup of Q^* which is generated by the positive r_n.

I conjecture that the quotient group $G = Q^*/\Gamma$ is the direct sum of a free group and a finite group.

This can be presently obtained when $F(x)$ has one of the forms

$$\prod_{i=1}^{k} (x + a_i)^{b_i}, \quad a_i, b_i \text{ integers}, (b_1, \ldots, b_k) = 1,$$

$$\frac{ax + b}{Ax + B}, \qquad a > 0, A > 0, b, B \text{ integers with } aB \neq Ab,$$

$$x^2 + ax + b, \quad a, b \text{ integers with } a^2 \neq 4b.$$

In general, one would expect G to be determined by the splitting fields over the rationals of the irreducible polynomials which appear in the numerator and denominator of $F(x)$. Perhaps the ideal class groups of these fields also play a rôle, although some evidence for the case when $F(x)$ is an irreducible cubic would be desirable.

I conjecture that when $F(x) = (ax + b)/(Ax + B)$, $aB \neq Ab$, the group G is arithmic.

Surely some analogue of this property holds in the general case?

13. Let $0 < \beta < 1$. Let $f(\ )$ be a real additive arithmetic function. What is the necessary and sufficient condition that

$$\lim_{x \to \infty} x^{-\beta} \sum_{x < n \leq x + x^\beta} |f(n)|^2$$

exist and be finite?

14. In the notation of question 12, when does the frequency

$$x^{-\beta} \sum_{\substack{x < n \leq x + x^\beta \\ f(n) \leq z}} 1$$

converge weakly as $x \to \infty$?

One may formulate other such analogues of problems in Probabilistic Number Theory.

15. If $0 < \beta < 1$ and $g(n)$ is a complex-valued multiplicative function which satisfies $|g(n)| \leq 1$ for all positive n, give necessary and sufficient conditions that

$$\lim_{x \to \infty} x^{-\beta} \sum_{x < n \leq x + x^\beta} g(n)$$

exists.

What is the smallest value of β presently available when $g(n) = \mu(n)$, the Möbius function?

16. There is a positive integer v so that every positive integer m which is not divisible by 3 has a representation

$$m^v = \prod_{i=1}^{k} \left(\frac{3n_i + 1}{3n_i + 2} \right)^{\varepsilon_i},$$

where $n_i > 0$ and $\varepsilon_i = \pm 1$. The arguments of the present volume allow one to obtain such a representation with $k < c_0 (\log m)^{c_1}$ for certain constants c_0 and c_1.

What is the smallest possible value for k in terms of m?

17. Develop a theory which allows one to establish product representations

$$n = \prod_{j=1}^{k} a_j^{d_j},$$

where emphasis lies more on the number of fractions a_j occurring (here $|d_1| + \cdots + |d_k|$), rather than on their position (size), as is presently the case.

18. In the notation of theorem (7.1) it was proved (in particular) that

$$\sum_{\substack{\log x < p \le Q \\ (p,c)=1}} (p-1) \max_{(r,p)=1} \max_{y \le x} |E(y, p, r)|^2 \ll \left(\frac{x^{2(1-\sigma)}}{\log x} + Q^c\right) \sum_{q \le x} \frac{|f(q)|^2}{q}$$

for some value of $c > 0$.
   What is the best possible value for c?

Supplement

Progress in Probabilistic Number Theory

I present here remarks concerning my work *Probabilistic Number Theory*, published in 1979/80 as Volumes 239, 240 of the present series. Particular attention is paid to recent developments. References in this supplement that are to those volumes are prefaced by PNT.

PNT Chapter 1. In his review of *Probabilistic Number Theory*, I. Z. Ruzsa [4] pointed out the PNT lemma (1.6), which I attributed to Halász, is in fact due to Raikov [1].

The proof which I gave was adapted from Halász' paper [2], and is completely different from that of Raikov. Indeed, Raikov reduces the proof to an application of a theorem of Davenport, itself a rediscovery of a result of Cauchy. The Cauchy–Davenport theorem represents an early analogue of Mann's (α, β) theorem (see Halberstam and Roth [1]), so that the analogy between that result and PNT lemma (1.6) is now perfectly natural.

PNT Chapter 4. For a real-valued additive function $f(n)$ let

$$M(n) = \sum_{p^k \leq n} \frac{f(p^k)}{p^k}\left(1 - \frac{1}{p}\right),$$

$$D(n) = \left(\sum_{p^k \leq n} \frac{|f(p^k)|^2}{p^k}\right)^{1/2} \geq 0.$$

The Turán–Kubilius inequality asserts that

$$c = \sup(nD(n)^2)^{-1} \sum_{m \leq n} |f(m) - M(n)|^2,$$

where the supremum is taken over all additive functions not identically zero on the interval $1 \leq m \leq n$, is bounded by an absolute constant, uniformly for all $n \geq 2$. In PNT lemma (4.1) it was shown that $c \leq 32$, with an argument yielding lim sup $c \leq 2$ as $n \to \infty$.

The best asymptotic value of c has now been determined, with proofs given by Kubilius [4], [5] (apparently announced in a conference on Analytic Number Theory, Budapest, 1981) and by Hildebrand [2]: as $n \to \infty$, $c \to \frac{3}{2}$.

For convenience of exposition let us consider the ratio

$$\tau = (nB^2)^{-1} \sum_{m=1}^{n} \left| f(m) - \sum_{p \leq n} f(p) p^{-1} \right|^2,$$

with

$$B^2 = \sum_{p \leq n} p^{-1} f(p)^2,$$

which is appropriate to a strongly additive function $f(n)$. Expanding the square and inverting the order of summation gives for τ the representation $Q_1 + Q_2 - Q$ where

$$Q_1 = n^{-1} \sum_{p \leq n} [n/p] p x_p^2 - \sum_{p \leq n^{1/2}} x_p^2/p,$$

$$Q_2 = \sum_{p \leq n} \sum_{l \leq n} (\alpha(p) p^{1/2} l^{-1/2} + \alpha(q) p^{-1/2} l^{1/2}) x_p x_l$$

$$- \sum_{\substack{pl \leq n \\ p \neq l}} \alpha(pl) p^{1/2} l^{1/2} x_p x_l,$$

with $\alpha(v) = v^{-1} - n^{-1}[n/v]$,

$$Q = \sum_{\substack{p \leq n \, l \leq n \\ pl > n}} p^{-1/2} l^{-1/2} x_p x_l,$$

and for each prime p (or l)

$$x_p = f(p) p^{-1/2} B^{-1}$$

so that

$$\sum_{l \leq n} x_l^2 = 1.$$

A straightforward application of the Cauchy–Schwarz inequality shows that $Q_2 \ll (\log n)^{-1/2}$. Moreover, it is clear that the bound $Q_1 \leq 1$ is essentially best possible. The main interest lies in Q.

Let **A** be the coefficient matrix of the quadratic form Q. It is symmetric, and after the remarks in Chapter 5 concerning self-adjoint maps, we need an estimate for its lowest eigenvalue. In fact almost its entire spectrum can be obtained.

To obtain suitable candidates for eigenvectors of **A**, Kubilius considers the problems of estimating

(1) $$V(\psi) = \frac{\iint \psi(u)\psi(v) u^{-1} v^{-1} \, du \, dv}{\int_0^1 (\psi(u))^2 u^{-1} \, du},$$

where the double integral is taken over the triangle $0 \le u \le 1$, $0 \le v \le 1$, $u + v > 1$. This can be viewed as an extremal problem involving functions of the class $L^2[0, 1]$. That it is an appropriate analogue is perhaps not immediately clear, and I shall discuss this presently.

By considering $V(\psi + yh)$ for a (temporarily fixed) function h, and varying real numbers y, Kubilius shows that any function ψ which gives rise to an extreme value λ of $V(\psi)$ must satisfy

$$\int_{1-u}^{1} \psi(v) v^{-1} \, du = \lambda \psi(u)$$

for all u, $0 \le u \le 1$. It must be an eigenfunction of the operator

(2) $$\psi(u) \mapsto \int_{1-u}^{1} \psi(v) v^{-1} \, dv.$$

Such eigenfunctions will satisfy the relation

(3) $$\lambda (1-u) \psi'(u) = \psi(1-u),$$

where $'$ denotes differentiation with respect to u, and hence the second order differential equation

(4) $$u(1-u)\phi''(u) - u\phi'(u) + \lambda^{-2} \phi(u) = 0.$$

This is an equation of hypergeometric type and solutions

$$\phi_r(u) = u \sum_{k=0}^{r-1} (-1)^{r-1-k} \binom{r}{k} \binom{r-1}{k} u^{r-k-1}(1-u)^k$$

with $\lambda = \lambda_r = (-1)^{r-1} r^{-1}$ are exhibited, one for each positive integer r. In terms of the standard notation for Jacobi polynomials,

$$\psi_r(u) = u P_{r-1}^{(1,0)}(1 - 2u).$$

See, for example, Rainville [1], Chapter 16. In particular

$$\int_0^1 \psi_r(v)\psi_s(v)v^{-1}\,dv = \begin{cases} 0 & \text{if } r \neq s, \\ (2r)^{-1} & \text{if } r = s. \end{cases}$$

Remark. The simplest procedure is perhaps to check that the polynomials $\psi_r(u)$ indeed satisfy the relation (3), and so the differential equation (4). Using this equation in the form

$$((1-u)\psi'(u))' + u^{-1}\lambda^{-2}\psi(u) = 0$$

one obtains the first of the orthogonality relations. To obtain the second such relation, involving $\psi_r(u)^2$ say, we begin by noting that $\psi_r(u)$ may be written as a linear combination of u^r and the $\psi_s(u)$ with $s < r$. By what we have already established we need only consider

$$\int_0^1 \rho v^{r-1}\psi_r(v)\,dv,$$

where ρ, the coefficient of v^r in $\psi_r(v)$, is the constant coefficient in the Laurent series expansion of

$$([1+z]^r - z^r)[1+z^{-1}]^{r-1}(-1)^{r-1},$$

and so has the value $(-1)^{r-1}\binom{2r-1}{r}$. We now employ the representation

$$\psi_r(u) = \frac{1}{(r-1)!}\frac{d^{r-1}}{du^{r-1}}(u^r(1-u)^{r-1})$$

and integrate by parts $r-1$ times.
This ends the remark.

Let E be the vector space $\mathbb{C}^{\pi(n)}$, and denote by x_p a typical component (one for each prime p not exceeding n) of a typical vector $\mathbf{x}$ in E. Let $\mathbf{z}_r$ be the vector with p-component $p^{-1/2}\psi_r((\log p)/\log n)$. A direct computation shows that

$$|\mathbf{A}\mathbf{z}_r - \lambda_r \mathbf{z}_r| \ll (\log n)^{-1/2}.$$

We can now apply lemma (5.6) and deduce the existence of an eigenvalue ν_r of $\mathbf{A}$ which satisfies $\nu_r = \lambda_r + O((\log n)^{-1/2})$. For $r \leq c_1(\log n)^{1/4}$ and a suitably chosen positive constant c_1, the eigenvalues for differing r will be distinct. We do not as yet know whether any occur multiply for $\mathbf{A}$.

Let ν_r, $r = 1, \ldots, t$, with $t = \pi(n)$, denote the entire collection of eigenvalues of **A**. Then

$$\sum_{r=1}^{t} \nu_r^2 = \operatorname{tr} \mathbf{A}^2$$

and this trace Kubilius estimates, in a straightforward manner, to be $\zeta(2) + O(\log \log n / \log n)$, in terms of the Riemann zeta function. Since demonstrably

$$\sum_{r=1}^{\infty} \lambda_r^2 = \zeta(2),$$

we see that all eigenvalues with $r > c_1 (\log n)^{1/4}$ are $O((\log n)^{-1/4})$ in size. Their effect in **A** is small.

The highest eigenvalue of $-\mathbf{A}$ is thus $-\nu_2$, giving

$$\tau \leq \tfrac{3}{2} + O((\log n)^{-1/2}).$$

Moreover, with $\mathbf{x} = \mathbf{z}_2$ we can obtain for τ a similar estimate in the other direction. This gives for the maximal value of τ the estimate $\tfrac{3}{2} + O((\log n)^{-1/2})$, which is the main result in Kubilius' paper.

A second method is sketched. Direct computation shows that for each positive integer k

$$\operatorname{tr} \mathbf{A}^k = \sum (p_1 p_2 \cdots p_k)^{-1},$$

where the summation is taken over those k-tuples of primes that satisfy $p_j \leq n$ for every j, $p_i p_{i+1} > n$ for $1 \leq i \leq k-1$, and $p_k p_1 > n$. This sum in turn is closely approximated by

$$\int (u_1 \cdots u_k)^{-1} du_1 \cdots du_k$$

taken over the region $0 \leq u_j \leq 1$ for every j, $u_i + u_{i+1} > 1$ for $1 \leq i \leq k-1$ and $u_k + u_1 > 1$.

Defining

$$\theta(u_k, u_1) = \begin{cases} 1 & \text{if } u_k > 1 - u_1, \\ 0 & \text{otherwise,} \end{cases}$$

this last integral may be written in the form

$$\int_0^1 u_1^{-1} du_1 \int_{1-u_1}^1 u_2^{-1} du_2 \cdots \int_{1-u_{k-1}}^1 \theta(u_k, u_1) u_k^{-1} du_k.$$

Expanding the function θ in terms of the eigenfunctions ψ_r, say

$$\theta(u, u_1) \sim \sum_{r=1}^{\infty} a_r \psi_r(u),$$

we see that

$$a_r = 2r \int_{1-u_1}^{1} \psi_r(u) u^{-1} \, du = 2r\lambda_r \psi_r(u_1),$$

and readily obtain the estimate

$$\operatorname{tr} \mathbf{A}^k = \sum_{r=1}^{\infty} (-1)^{(r-1)k} r^{-k} + o(1),$$

valid as $n \to \infty$.

Since $\mathbf{A}$ is symmetric, a result from lemma (5.5) shows that

$$\sum_{r=1}^{t} \nu_r^k = \zeta(k) + o(1)$$

for *even* integers k. Assuming that $|\nu_r|$ does not increase with r, we deduce the estimate

$$\limsup_{n \to \infty} |\nu_1| \leq \liminf_{k \to \infty} \zeta(k)^{1/k} = 1.$$

Indeed, for $\mathbf{x} = \mathbf{z}_1$, Q has the value $1 + O((\log n)^{-1/2})$, so that $\nu_1 = 1 + o(1)$. Removing ν_1^k from the trace gives

$$\nu_2^k \leq \zeta(k) - 1 + o(1)$$

for every *even* k, and so

$$\limsup_{n \to \infty} |\nu_2| \leq \liminf_{k \to \infty} (\zeta(k) - 1)^{1/k} = \tfrac{1}{2}.$$

Since for $\mathbf{x} = \mathbf{z}_2$, Q has the value $-\tfrac{1}{2} + O((\log n)^{-1/2})$, we get $\nu_2 = -\tfrac{1}{2} + o(1)$, and so on.

At the end of his first paper Kubilius acknowledges several helpful suggestions that were made by H. L. Montgomery.

Remarks. Results from the probabilistic theory of numbers show that τ is small only when $f(p)$ is nearly a multiple of $\log p$. Then in some sense

$$\sum_{p\leq z}\frac{f(p)}{p} \sim \sum_{p\leq z}\frac{c\log p}{p} \sim c\log z \sim \int_1^z \frac{c}{v}\,dv,$$

which suggests the consideration of the function (1) if one applies the transformation $v = z^u$ to reduce the range of integration to $0 \leq u \leq 1$.

Let W be the space of complex-valued functions $h(u) = ug(u)$ where $g(u)$ is continuous on the interval $0 \leq u \leq 1$, and define

$$\|h\| = \sup_{0\leq u\leq 1} |g(u)|.$$

This space is complete. It is mapped into itself by the transformation Φ which takes h to the function

$$u \mapsto \int_{1-u}^1 \frac{h(v)}{v}\,dv, \qquad 0 \leq u \leq 1.$$

This transformation is bounded on W, and of norm 1.

Let H be the Hilbert space of complex-valued Lebesgue measurable functions $k(u)$, defined on the unit interval, with

$$\int_0^1 \frac{|k(u)|^2}{u}\,du < \infty.$$

We make the usual identification of functions differing on a set of measure zero. The inner product on H is given by

$$[k, t] = \int_0^1 \frac{k(u)\overline{t(u)}}{u}\,du.$$

The points of W are everywhere dense in H (in the topology of H), and an application of Fubini's theorem shows that Φ is a self-adjoint operator in H.

Applying a well-known theorem of Weierstrass we see that the polynomials $\psi_r(u)$ span W, and therefore H. In particular, for every non-zero complex number ρ, and member x of H,

$$\|(\Phi - \rho I)x\| \geq \|x\| \min_r |\lambda_r - \rho|.$$

This shows that the spectrum of Φ contains only the eigenvalues λ_n and possibly the real number zero.

If $\Phi g = 0$ in H then by differentiation of Φg, g is seen to be zero also. Thus zero is not an eigenvalue of Φ. However the range of Φ is not the whole of H, since there are functions in H which are not absolutely continuous. Thus zero does belong to the spectrum of Φ in H.

In particular, Φ has norm 1 on H, which is not immediately obvious. This ends the remarks.

In his paper, Hildebrand treats at once an arbitrary additive function, and gives a kind of asymptotic expansion for the analogue of $B^2\tau$. He reduces its estimation to that of a quadratic form, introducing the polynomials $p_k(u)$ which are orthogonal over the interval $0 \le u \le 1$ with weight u. Thus $up_k(u) = \psi_k(u)$.

In order to illustrate his results within the earlier format, let F be the space $\mathbb{C}^s$ where s is the number of prime-powers p^k not exceeding n.

Let

$$[\mathbf{x}, \mathbf{y}] = \bar{\mathbf{y}}^T G \mathbf{x}$$

be an inner product on the space F, defined in terms of an hermitian matrix G. Let H be a further hermitian matrix. Then it is readily checked that the map T given by

$$\mathbf{x} \mapsto G^{-1} H \mathbf{x}$$

is self-adjoint with respect to this inner product, and that

$$[G^{-1} H \mathbf{x}, \mathbf{x}] = \bar{\mathbf{x}}^T H \mathbf{x}.$$

To specialize let

$$\omega(p^k) = [np^{-k}] - [np^{-k-1}],$$

so that

$$\tfrac{1}{4} \le p^k \omega(p^k) \le 1$$

holds uniformly for all prime powers not exceeding n. If we take G to be the diagonal matrix with $\omega(p^k)$ in the p^k-th position, then in particular

$$\tfrac{1}{4} \le \|\mathbf{x}\|^2 \left(\sum_{p^k \le n} p^{-k} |x_{p^k}|^2 \right)^{-1} \le 1.$$

A standard argument now shows that with $x_{p^k} = f(p^k)$

(5) $$n^{-1} \sum_{m=1}^{n} \left| f(m) - n^{-1} \sum_{m=1}^{n} f(m) \right|^2 = \|\mathbf{x}\|^2 - \bar{\mathbf{x}}^T H \mathbf{x}$$

where the hermitian matrix $\mathbf{H}$ satisfies

$$(\mathbf{Hx})_{p^k} - p^{-k} \sum_{n/p^k < q \le n} q^{-1} x_q \ll \|\mathbf{x}\| p^{-k} \{\max(1, \log np^{-k})\}^{-1/2},$$

where q denotes a prime-power.

The spectral theorem for T provides a basis $\mathbf{b}_1, \ldots, \mathbf{b}_s$ for F, made up of eigenvectors of T, which are orthonormal with respect to the inner product [,]. Let μ_r be the eigenvalue corresponding to $\mathbf{b}_r$, $r = 1, \ldots, s$. Then for distinct integers i and j, not exceeding s,

$$[T\mathbf{b}_i, \mathbf{b}_j] = [\mu_i \mathbf{b}_i, \mathbf{b}_j] = \mu_i [\mathbf{b}_i, \mathbf{b}_j] = 0,$$

and the basis vectors are, so-to-speak, orthogonal with respect to $[T\mathbf{x}, \mathbf{y}]$.

We expand $\mathbf{x}$ in terms of these $\mathbf{b}_j$ and separate off the contribution corresponding to the first v:

$$\mathbf{x} = \sum_{j=1}^{s} c_j \mathbf{b}_j = \sum_{j=1}^{v} c_j \mathbf{b}_j + \mathbf{w}.$$

To gain an asymptotic expansion for $S(x)$, the expression which occurs on the left-hand side of equation (5), we note that

(6) $$S(\mathbf{x}) = \sum_{j=1}^{v} (1 - \mu_j) |c_j|^2 + \|\mathbf{w}\|^2 - [T\mathbf{w}, \mathbf{w}],$$

and that

$$|[T\mathbf{w}, \mathbf{w}]| \le \|\mathbf{w}\|^2 \max_{j > v} |\mu_j|.$$

As $S(\mathbf{x}) = 0$ can only hold with $\mathbf{x} = \mathbf{0}$, every $\mu_j < 1$. However, we expect μ_1 to be very near to 1.

The eigenvalues of T may be estimated by showing that if $\mathbf{y}_r$ is the vector with p^k-coordinate $\psi_r((\log p^k)/\log n)$, then

$$\|T\mathbf{y}_r - \lambda_r \mathbf{y}_r\| = \|\mathbf{G}^{-1}(\mathbf{H} - \lambda_r \mathbf{G})\mathbf{y}_r\| = o(1).$$

In this connection it is helpful that $\mathbf{G}^{-1}$ is a diagonal matrix with typical diagonal element $\omega(p^k)^{-1}$. Another application of lemma (5.6) gives, without loss of generality,

$$\mu_r = (-1)^{r-1} r^{-1} + o(1)$$

and as before, a trace computation shows that this estimate is essentially uniform in r. Although they are close, ν_r and μ_r need not be equal.

Since

$$\max_{j>v}|\mu_j| \leq (v+1)^{-1} + o(1),$$

the term in the expansion (6) which involves $T\mathbf{w}$ may be considered small in comparison with $\|\mathbf{w}\|^2$.

To round out the practicality of the asymptotic expansion we apply lemma (5.7) to obtain, without loss of generality, the estimate

$$\|\mathbf{b}_r - \mathbf{y}_r\| \|\mathbf{y}_r\|^{-1}\| = o(1).$$

This enables us to replace the $\mathbf{b}_r$ by the rescaled $\mathbf{y}_r$ for $1 \leq r \leq v$, and a typical coefficient c_j, by

$$(2j)^{1/2}[\mathbf{x}, \mathbf{y}_j].$$

We can now readily deduce lemma 1 and almost all of lemma 2 in Hildebrand's paper, and so deduce his main results. What remains to be shown, and this can be done directly, as in his paper, is that for any λ

$$(7) \qquad S(\mathbf{x} + \lambda \mathbf{y}_1) = S(\mathbf{x}) + \frac{\lambda^2(1+o(1))}{(\log n)^2} + o\left(\frac{|\lambda|\|\mathbf{x}\|}{\log n}\right).$$

Here it seems essential that one considers a general additive function rather than a strongly additive one, for the very good distribution of $\log n$ over the residue classes $0 \pmod{p^k}$ has to be employed.

As an example, let us apply (6) with $v=1$ and $\mathbf{x}$ replaced by $\mathbf{x} - \lambda \mathbf{y}_1$ where λ is chosen so that $\lambda[\mathbf{y}_1, \mathbf{b}_1] = [\mathbf{x}, \mathbf{b}_1]$. Then

$$S(\mathbf{x} - \lambda \mathbf{y}_1) \geq (\tfrac{2}{3}) \sum_{j=2}^{s} \|[\mathbf{x} - \lambda \mathbf{y}_1, \mathbf{b}_j]\|^2$$

$$= (\tfrac{2}{3}) \|\mathbf{x} - \lambda \mathbf{y}_1\|^2$$

for all sufficiently (absolutely) large n. Moreover, from (7) we obtain

$$S(\mathbf{x}) \geq S(\mathbf{x} - \lambda \mathbf{y}_1) + \frac{\lambda^2(1+o(1))}{(\log n)^2} + o(\|\mathbf{x} - \lambda \mathbf{y}_1\|^2).$$

Together these give

$$7S(\mathbf{x}) \geq \min_{\beta}\left(|\beta|^2 + \sum_{p^k \leq n} p^{-k}|f(p^k) - \beta \log p^k|^2\right),$$

which, except for a different constant in place of the 7, was first obtained by Ruzsa [3], using ideas from the theory of probability together with an elaboration of the method of Halász expounded in PNT Chapter 6.

An interesting corollary of the above arguments is the estimate

$$\mu_1 = 1 - 2(\log n)^{-2} + O((\log n)^{-5/2}).$$

Perhaps an asymptotic series expansion can be given for each (fixed) eigenvalue μ_j.

Hildebrand's paper also contains a result of independent interest in connection with the value distribution of additive functions. I consider it in the notes on PNT Chapter 12.

As a last application of the ideas in this section let us return to the space E, and introduce the inner product

$$[\mathbf{x}, \mathbf{y}] = \sum_{p \leq n} p^{-1} x_p \bar{y}_p.$$

It is readily checked that the linear map K which takes x_p to

$$\sum_{n/p < h \leq n} h^{-1} x_h, \quad h \text{ prime},$$

is self-adjoint with respect to this inner product. The eigenvalues of K and of the map $\mathbf{y} \mapsto A\mathbf{y}$ are the same, and if $\mathbf{w}$ is an eigenvector of the second map which satisfies $|\mathbf{w}| = 1$ in terms of the standard inner product, then the vector $\mathbf{x}$ with $x_p = p^{1/2} w_p$ is an eigenvector of K and satisfies $\|\mathbf{x}\| = 1$.

Consider now the expression

$$\sum_{p \leq n} \frac{1}{p} \left| f(p) - \sum_{n/p < h \leq n} \frac{f(h)}{h} \right|^2, \quad h \text{ prime},$$

which I shall denote by ε_n^2, and regard as "small". With $x_p = f(p)$ we can write

$$\|K\mathbf{x} - \mathbf{x}\|^2 = \varepsilon_n^2,$$

showing that $\mathbf{x}$ is an approximate eigenfunction for the eigenvalue ν_1.

Assume for the moment that $\|\mathbf{x}\| = 1$, and expand $\mathbf{x}$ in terms of a basis $\mathbf{d}_j$, $1 \leq j \leq t$, of E consisting of eigenvectors of K:

$$\mathbf{x} = \sum_{j=1}^{t} \mathbf{a}_j \mathbf{d}_j.$$

Then if ν_j is the eigenvalue corresponding to $\mathbf{d}_j$, we have

$$\|K\mathbf{x}-\mathbf{x}\|^2 = \sum_{j=1}^{t} (\nu_j-1)^2 |a_j|^2$$

$$\geq (\tfrac{1}{3}) \sum_{j=2}^{t} |a_j|^2 = (\tfrac{1}{3})\|\mathbf{x}-a_1\mathbf{d}_1\|^2.$$

In particular

$$|\|\mathbf{x}\|-|a_1|| \leq \|\mathbf{x}-a_1\mathbf{d}_1\| \leq \varepsilon_n\sqrt{3},$$

so that

$$\|\mathbf{x}-\mathbf{d}_1\| < 4\varepsilon_n.$$

This inequality is interesting in its own right, and with $\|\mathbf{x}\|\mathbf{d}_1$ in place of $\mathbf{d}_1$ it holds whether $\|\mathbf{x}\| = 1$ or not.

To continue, a direct computation with $\mathbf{1}$, the vector whose p-coordinate is $\sqrt{2}\psi_1((\log p)/\log n) = \sqrt{2}(\log p)/\log n$, shows that

$$\|K\mathbf{1}-\mathbf{1}\|^2 \ll ((\log\log n)/\log n)^3.$$

Applications of lemmas (5.6) and (5.7) give

$$\|\mathbf{1}-c\mathbf{d}_1\| \ll ((\log\log n)/\log n)^{3/2}$$

for some constant c. Moreover, since the same upper bound serves for $\|\mathbf{1}\|-c$, the value of c is $1+O((\log n)^{-3/2+\delta})$ for any fixed positive number δ.

Together, therefore,

(8) $$\|\mathbf{x}-\|\mathbf{x}\|\mathbf{1}\| \ll \varepsilon_n + \|\mathbf{x}\|(\log n)^{-3/2+\delta},$$

whether $\mathbf{x}$ is a unit vector or not.

The inductive argument given in the example at the end of Chapter 8 shows that for a suitable absolute constant α,

$$\|\mathbf{x}\| \ll \max_{p\leq\alpha}|f(p)|(\log n)^{1+\delta} + \max_{m\leq n} \varepsilon_m(\log\log n)^3,$$

which combines with our estimate (8) to give (say)

$$\|\mathbf{x}-\|\mathbf{x}\|\mathbf{1}\| \ll \varepsilon_n + \max_{p\leq\alpha}|f(p)|(\log n)^{-1/3} + \max_{m\leq n}\varepsilon_m(\log n)^{-1}.$$

This result may be compared with the solution of the approximate functional equation considered in Chapter 9, theorem (9.11). In that chapter, we were concerned with the consequences of

$$\sum_{p \leq n^\theta} \frac{1}{p} \left| f(p) - \sum_{n/p < h \leq n} \frac{f(h)}{h} \right|^2, \quad h \text{ prime,}$$

being small, for some $\theta < 1$. Since this last hypothesis gives a control over only the first $\pi(n^\theta)$ coordinates of $K\mathbf{x} - \mathbf{x}$, a different and more complicated treatment was required.

Analogues of the Turán–Kubilius Inequality

Let $\alpha > 1$, and let $\| \ \|_1$ denote the standard norm

$$\|\mathbf{a}\|_1 = \left(\sum_{j=1}^{n} |a_j|^\alpha \right)^{1/\alpha}.$$

If s (once again) denotes the number of prime-powers q not exceeding n, then we introduce into $\mathbb{C}^s$ the norm

$$\|\mathbf{x}\|_2 = \left(\sum_{q \leq n} q^{-1} |x_q|^2 \right)^{1/2} + \left(\sum_{q \leq n} q^{-1} |x_q|^\alpha \right)^{1/\alpha}.$$

In terms of these norms the linear map T_α from $\mathbb{C}^s$ to $\mathbb{C}^n$ which is given by

$$x_q \mapsto n^{-1/\alpha} \left(\sum_{q \| n} x_q - \sum_{q \leq n} q^{-1} x_q \right)$$

has norm bounded by a constant depending at most upon α. This is a high-moment analogue of the Turán–Kubilius inequality, to which it reduces when $\alpha = 2$. For a proof of this and related inequalities see Elliott [19].

A method for estimating general moments of additive arithmetic functions in terms of the corresponding moments of appropriately defined independent random variables was given by Ruzsa [6]. As he shows in Ruzsa [7], this method is sufficiently powerful that it will also yield the appropriate analogues for additive functions of Kolmogorov's inequality and inequalities related to it in the theory of probability. A standard proof of Kolmogorov's inequality is given in PNT lemma (1.21).

It was mentioned earlier that for real additive arithmetic functions f, Ruzsa proved that

$$n^{-1} \sum_{m=1}^{n} \left| f(m) - n^{-1} \sum_{m=1}^{n} f(m) \right|^2$$

lies between absolute constant multiples of

$$\min_{\beta} \left(\beta^2 + \sum_{q \leq n} q^{-1} |f(q) - \beta \log q|^2 \right).$$

It is natural to expect an analogue of this result to hold for moments other than the second. For moments of every positive order this was obtained by Hildebrand [1]. According to a remark in Ruzsa [7], Manstavičius also obtained the appropriate analogues for moments of order greater than 2.

It is easy to check that the minimum which appears in Ruzsa's result is attained with

$$\beta = L^{-1} \sum_{q \leq n} q^{-1} f(q) \log q, \qquad L = 1 + \sum_{q \leq n} q^{-1} (\log q)^2.$$

We introduce into $\mathbb{C}^s$ yet another norm:

$$\|x\|_3 = |\beta_0| + \left(\sum_{q \leq n} q^{-1} |x_q - \beta_0 \log q|^2 \right)^{1/2},$$

where β_0 is obtained from β by replacing $f(q)$ with x_q. His result can now be viewed as an assertion that between the spaces $\mathbb{C}^s$ with $\| \ \|_3$ and $\mathbb{C}^n$ with $\| \ \|_1$ and $\alpha = \frac{1}{2}$, the map $T_{1/2}$ is an approximate isometry.

Surely underneath all this there is somewhere a genuine isometry?

It seems appropriate at this point to mention the maximal form of the large sieve obtained by Montgomery [2]. Employing results of Carleson and Hunt from the theory of Fourier series, he proves that

$$\sum_j \max_{k \leq N} \left| \sum_{n=1}^{k} a_n \exp(2\pi i n \delta_j) \right|^2 \ll (N + \delta^{-1}) \sum_{n=1}^{N} |a_n|^2$$

is valid for all complex a_n, and sets of points δ_j which satisfy

$$\min_{m \in \mathbb{Z}} |\delta_j - \delta_k - m| \geq \delta, \qquad j \neq k.$$

As an application of this theorem let Q be a positive real number, and specialize the δ_j to be the rationals sp^{-1}, $1 \leq s \leq p-1$, with p running through the primes not exceeding Q. Then in particular we may deduce

$$\sum_{p\leq Q} \max_{k\leq N} p \sum_{r=0}^{p-1} \left| \sum_{\substack{n=1 \\ n\equiv r(\bmod p)}}^{k} a_n - p^{-1} \sum_{n=1}^{k} a_n \right|^2 \ll (N+Q^2) \sum_{n=1}^{N} |a_n|^2.$$

Let $k(p)$ be a sequence of real numbers, one for each prime p, and lying in the interval $1 \leq k(p) \leq N$. We set $Q = N^{1/2}$, restrict ourselves to the zero class (mod p) for each p, and then take the inner sums over the integers n not exceeding $k(p)$. Dualizing gives

$$\sum_{n=1}^{N} \left| \sum_{\substack{p|n, p\leq N^{1/2} \\ k(p)\geq n}} f(p) - \sum_{\substack{p\leq N^{1/2} \\ k(p)\geq n}} p^{-1} f(p) \right|^2 \ll N \sum_{p\leq N^{1/2}} p^{-1} |f(p)|^2.$$

Moreover, a similar inequality holds with the opposite summation condition(s) $k(p) < n$. Straightforward applications of the Cauchy–Schwarz inequality enable us to replace the restrictions $p \leq N^{1/2}$ by $p \leq N$.

As a further specialization, let $1 \leq w(1) \leq w(2) \leq \cdots \leq w(N) \leq N$, and choose $k(p)$ to the largest integer s for which $w(s) \leq p$. Then

$$\sum_{n=1}^{N} \left| \sum_{\substack{p|n \\ p>w(n)}} f(p) - \sum_{w(n)<p\leq N} p^{-1} f(p) \right|^2 \ll N \sum_{p\leq N} p^{-1} |f(p)|^2,$$

where the implied constant is absolute.

This and the similar inequality with prime-powers q in place of the primes p, and the condition $q \| n$ in place of $p|n$, are new analogues of the Turán–Kubilius inequality.

PNT Chapter 5. Let the real-valued additive function $f(n)$ have a limiting distribution whose characteristic function is $v(t)$. If the series $\sum p^{-1}$ taken over those primes for which $f(p) \neq 0$ diverges, then the limiting distribution is continuous. In PNT Chapter 5 I gave two proofs of this last result, mentioning a short proof of Szüsz [1] which I felt was incomplete. His proof began with

$$J(T) = \frac{1}{2T} \int_{-T}^{T} |v(t)|^2 \, dt = \tfrac{1}{2} \int_{-1}^{1} |v(Ty)|^2 \, dy$$

$$\leq 2c \int_{0}^{1} \exp\left(-\sum_{p} p^{-1}(1 - \cos f(p) Ty)\right) dy,$$

and treated this last integral under the assumption that the $f(p)$ took distinct values. He remarked, without details, that this extra condition could be dispensed with.

In a letter dated February 22, 1982, he sent me a further argument which, in response to his request, I include here.

Denote by a_l, $l = 1, 2, \ldots$, the possible non-zero values of $f(p)$, and define

$$b_l = \sum_{f(p)=a_l} \frac{1}{p}.$$

Then we have to consider the sum

$$-\sum_{l \leq N} b_l(1 - \cos a_l Ty).$$

Since this is smaller than

$$-b_\lambda(1 - \cos a_\lambda Ty)$$

with b_λ maximal amongst the b_l, and

(9) $$\int_0^1 \exp(-(1 - \cos a_\lambda Ty)b_\lambda) \, dy \to 0$$

as $b_\lambda \to \infty$, we may without loss of generality assume that b_λ is $O(1)$. Consider now the integral

$$\int_0^1 \exp\left(-\sum_{l \leq N} b_l(1 - \cos a_l Ty)\right) dy$$

which corresponds to formula (1.5) of his paper. Let y be a random variable uniformly distributed over the interval $(0, 1)$. Then

$$E\left(\sum_{l \leq N} b_l \cos a_l Ty\right) = o(1)$$

and

$$D^2\left(\sum_{l \leq N} b_l \cos a_l Ty\right) = \frac{1}{2} \sum_{l \leq N} b_l^2 + o(1)$$

as $T \to \infty$. Therefore Tchebyshev's inequality yields

$$P\left(\sum_{l \leq N} b_l \cos a_l Ty > \sum_{l \leq N} b_l - K\right) \leq (\tfrac{1}{2} + o(1))\left(\sum_{l \leq N} b_l - K\right)^{-2} \sum_{l \leq N} b_l^2.$$

For a given K and large enough N this bound does not exceed

$$\left(\sum_{l \le N} b_l\right)^{-2} \sum_{l \le N} b_l^2,$$

since the series $\sum b_l$ diverges. Now if b_λ is $O(1)$, this last bound is $o(1)$ as $N \to \infty$.

Hence for a given ε and N large enough (in the above notation)

$$\limsup_{T \to \infty} J(T) \le \varepsilon + e^{-K},$$

and the limit is clearly zero.

In view of a well-known criterion of Wiener this ensures the continuity of the limiting distribution.

Remarks. The success of the above argument depends upon the b_l being uniformly bounded. For a fixed value of T the truth of the estimate (9) is almost immediate, but to obtain a result essentially uniform in T perhaps needs some further argument.

For a positive integer k and real $x \ge 2$ let

$$b = \sum_{\substack{p \le x \\ f(p) = a_k}} \frac{1}{p} > 1.$$

Define β by $2\pi\beta = |a_k|T$, and let $\delta < 1$ be a positive real number, to be thought of as small.

Let B denote the subset of the interval $0 \le y \le 1$ for which the inequality

$$1 - \cos a_k T y < \delta$$

is satisfied. For this inequality to hold we must have

$$2(\sin \pi \|\beta y\|)^2 < \delta,$$

where $\|z\|$ in the present context denotes the distance from z to the nearest integer. Employing the lower bound $\sin \pi\theta \ge 2\theta$ which is valid for $0 \le \theta \le \tfrac{1}{2}$, we deduce that $\|\beta y\|^2 < \delta/8$ and so obtain a representation

$$\beta y = m + \psi$$

with $|\psi| < (\delta/8)^{1/2}$ and m an integer. There are at most $[\beta] + 2$ possibilities for this integer. It follows readily that B has measure not more than

$$(\beta+2)\frac{2}{\beta}\left(\frac{\delta}{8}\right)^{1/2} < \left(1+\frac{4\pi}{|a_k|T}\right)\delta^{1/2}.$$

By integrating first over the set B we obtain the upper bound

$$\int_0^1 \exp(-(1-\cos a_k Ty)b)\, dy < \left(1+\frac{4\pi}{|a_k|T}\right)\delta^{1/2} + \exp(-\delta b),$$

and choosing $\delta = b^{-1/2}$ deduce that

$$\limsup_{T\to\infty} J(T) \leq 4cb^{-1/4}.$$

It is now clear that we may assume the b_l to be uniformly bounded.

Altogether this gives another treatment of the continuity property of the limit law, interesting more for its probabilistic nature than for its length.

PNT Chapter 7. The proof of PNT theorem (7.10) is a little compressed at the end. If

$$h(p) = \text{Sign } g(p)(|g(p)|p^{-ic})^t,$$

then it helps to note that the series

$$\sum p^{-1}(1 - \text{Re } h(p)p^{-i\tau})$$

diverges for real non-zero values of τ. Indeed, when $g(p)$ is not zero $h(p)$ has the form $\exp(i\theta_p)$ say. Without loss of generality we may assume that

$$||g(p)p^{-c}| - 1| \leq \varepsilon < 1$$

for all primes p. If the above series converges, then so does

$$\sum \frac{1}{p} \left\| \frac{\theta_p}{2\pi} - \tau \log p \right\|^2,$$

and from PNT lemma (8.7) the $\theta_p/(2\pi)$ are well distributed (mod 1). Since their values are clearly concentrated at 0 and $\frac{1}{2}$ (mod 1), this last possibility cannot arise.

PNT Chapter 10. For each real $\alpha > 0$ let L^α denote the complex linear space of complex-valued arithmetic functions f for which

$$\|f\| = \limsup_{n \to \infty} n^{-1} \sum_{m \leq n} |f(m)|^\alpha$$

is finite. Then the following result is valid:

Theorem A. *In order for a multiplicative function g to belong to L^α, $\alpha > 1$, and possess a non-zero mean-value*

$$A = \lim_{n \to \infty} n^{-1} \sum_{m=1}^{n} g(m),$$

it is both necessary and sufficient that the series

$$\sum_{|g(p)| \leq 3/2} p^{-1} |g(p) - 1|^2, \qquad \sum_{|g(p)| > 3/2} p^{-1} |g(p)|^\alpha, \qquad \sum_{p, k \geq 2} p^{-k} |g(p^k)|^\alpha$$

converge, and that for each prime p

$$\sum_{k=1}^{\infty} p^{-k} g(p^k) \neq -1.$$

For $\alpha = 2$ this is contained in PNT theorem (10.1), see also Elliott [6]. A modified approach to the necessity of the conditions was given by Daboussi and Delange [1]. An alternative proof of their sufficiency was given by Schwarz and Spilker [1]. For a general $\alpha > 1$ this theorem was proved independently and with different methods by Daboussi [2] and Elliott [20].

In Elliott [20] a form of the above theorem which considers zero mean-values is obtained.

Theorem B. *Let the multiplicative function $g(n)$ belong to L^α, $\alpha > 1$. In order that $g(n)$ possess a zero mean-value*

$$\lim_{n \to \infty} n^{-1} \sum_{m=1}^{n} g(m) = 0$$

it is necessary and sufficient that one of the following four conditions be satisfied:

(i) *One of the series*

$$\sum_{||g(p)|-1| \leq 1/2} p^{-1} |1 - |g(p)||^2, \qquad \sum_{||g(p)|-1| > 1/2} p^{-1} |1 - |g(p)||^\alpha$$

diverges.

(ii) *The condition* (i) *fails, but for each real value of t the series*

(10) $$\sum p^{-1}(|g(p)| - \operatorname{Re} g(p)p^{-it})$$

diverges.

(iii) *The conditions* (i) *and* (ii) *fail, but there is a real t so that the series* (10) *converges and*

$$\sum_{k=1}^{\infty} g(p^k) p^{-k(1+it)} = -1$$

for some prime p.

(iv) *The conditions* (i), (ii) *and* (iii) *fail, but*

$$\operatorname{Re}\left(\sum_{p \leq x} \frac{1 - g(p)}{p}\right) \to \infty$$

as $x \to \infty$.

Let N_α be the subspace of functions f in L^α for which $\|f\| = 0$. Then $\| \ \|$ induces a norm on the quotient space L^α / N_α, and the machinery of functional analysis becomes available. There is an extensive literature concerning additive and multiplicative functions which belong to L^α and related spaces; for examples, see Daboussi [1], Schwarz [1], Schwarz and Spilker [1], the last two of these containing many further references.

For real additive functions which belong to L^α, $\alpha > 0$, a complete classification in terms of their values on the prime-powers has been given; independently by Elliott [19], and Hildebrand and Spilker [1]. Such a classification can also be deduced from PNT lemma (7.8), using the methods of probabilistic number theory. In this connection I note the following result.

Let l_i run through a sequence of primes for which the series $\sum l_i^{-1}$ converges. If now a strongly additive function f satisfies $f(p) = 0$ when p is not an l_i, and for some $\alpha > 0$

$$\liminf_{x \to \infty} x^{-1} \sum_{n \leq x} |f(n)|^\alpha$$

is finite, *then the series*

$$\sum_{j=1}^{\infty} |f(l_j)|^\alpha l_j^{-1}$$

converges. Indeed, those integers m whch are not divisible by any of the primes l_j have a positive asymptotic density

$$\prod_{j=1}^{\infty} (1 - l_j^{-1}).$$

Then for a suitable unbounded sequence of x-values

$$\sum_{l_i \le x} \sum_{m \le x l_i^{-1}} |f(l_i m)|^\alpha \ll x,$$

and since $f(m) = 0$, the asserted result follows almost at once.

For multiplicative functions a complete classification has not yet been obtained. The results of Elliott [20], [21] give a classification of those multiplicative functions which are the representatives of the non-zero elements of the quotient spaces L^α/L^β, $0 < \beta < \alpha < \infty$. The arguments given there for the cases with $\alpha > 1$ may be extended to the general case by noting that a function g belongs to L^α/L^β (so-to-speak) if and only if $|g|^{\alpha/2}$ belongs to $L^2/L^{2\beta/\alpha}$.

In establishing results of these kinds, high-moment analogues of the Turán–Kubilius inequality, and their related duals, can be applied to advantage.

Suppose now that g is a non-negative multiplicative function. Then in Elliott [21] the following is established

Theorem C. *If g has a finite mean-value, then so does g^δ when $0 < \delta < 1$. Moreover, if any of these latter mean-values is non-zero, then the series*

(11) $$\sum p^{-1}(g(p)^{1/2} - 1)^2$$

taken over the prime numbers, converges.

As an application of these general considerations to a particular arithmetic function, let $\tau(n)$ be Ramanujan's function defined by

$$\sum_{n=1}^\infty \tau(n) x^n = x \prod_{j=1}^\infty (1 - x^j)^{24}.$$

Then for $0 < \delta \le 2$

$$\lim_{x \to \infty} x^{-1} \sum_{n \le x} \left(\frac{|\tau(n)|}{n^{11/2}}\right)^\delta$$

exists. For $\delta = 2$ this is due to Rankin [1], and since Mordell [1] proved that $\tau(n)$ is multiplicative, for $0 < \delta < 2$ it follows from an application of Theorem C. In this case the convergence of the series (11) would not be consistent with a well-known conjecture of Sato and Tate, and I conjectured that for $\delta < 2$ these limits were zero.

This is indeed the case, and in Elliott, Moreno and Shahidi [1] it is proved that the sum featured in the above limit is $O((\log x)^{-w(\delta)})$ where

$$w(\delta) = \inf_{-1 \le y \le 3} y^{-2}(1 + \delta y - (1 + y)^\delta).$$

In particular

$$\sum_{n\leq x} |\tau(n)| \ll x^{13/2}(\log x)^{-1/18}.$$

The lower bound

$$\sum_{n\leq x} |\tau(n)| \geq cx^{13/2}(\log x)^{-1/2}, \qquad x \geq 2,$$

with $c > 0$ has been obtained by Rankin [2]. For related topics see also Ram Murty [1].

PNT Chapter 12. As pointed out by Delange in his April 1984 letter to me, the statement of theorem (12.1) in PNT is not quite correct.

In fact PNT lemma (1.28) should be reformulated. One may require at the outset that the mean and variance of the $F_n(z)$ converge to those of the limit law $F(z)$. The weak convergence of the $K_n(u)$ on the whole real line is then appropriate, since $K_n(\infty)$ and $K(\infty)$ are the variances of $F_n(z)$ and $F(z)$, respectively. This is Theorem 3 of §9 in Gnedenko and Kolmogorov [1].

Alternatively one can omit the assertion that the variance of $F_n(z)$ converges to that of $F(z)$, and interpret $K_n(u) \Rightarrow K(u)$ as weak convergence on the compact sets of the real line. This requires that

$$K_n(z) - K_n(y) \to K(z) - K(y),$$

whenever y and z are continuity points of $K(u)$, $-\infty < y < z < \infty$, and does not include $K_n(\infty) \to K(\infty)$.

Let $f(m)$ be a real-valued strongly additive function and define

$$A(x) = \sum_{p\leq x} p^{-1} f(p),$$

$$B(x) = \left(\sum_{p\leq x} p^{-1} f(p)^2 \right)^{1/2} \geq 0.$$

The statement of PNT theorem (12.1) may be corrected in two different ways. The first is to require that the limiting distribution have variance 1 (PNT Volume II, p. 13, line 1). With this formulation theorem (12.1) would then match Theorem 4.1 of Kubilius' monograph [3].

A better change is to require only that the convergence

$$\frac{1}{B(x)^2} \sum_{\substack{p\leq x \\ f(p) \leq uB(x)}} \frac{f(p)^2}{p} \to K(u), \qquad x \to \infty,$$

be weak convergence on the compact sets of the real line. In this form theorem (12.1) would then match Theorem 2 in Kubilius' report [2].

Likewise, appropriate changes should be made in the papers of Kubilius [1], Delange [1], Misevičius [1], Elliott [25].

In an early draft of PNT I had incorporated a form of theorem (12.1) involving the weak convergence of measures. Feeling that this would demand further explanation, I sought to simplify matters by regarding $K(u)$ as a distribution function, and fell into a trap that had held others before me. For the total variation of $K(u)$ is at most 1, but as Delange remarks it is not difficult to give examples where it is strictly less than 1.

Suppose now that $B(n)$ is unbounded with n, and that as $n \to \infty$

(12) $$n^{-1} \sum_{m=1}^{n} (f(m) - A(n))^2 = (1 + o(1)) B(n)^2.$$

A problem of this kind is considered by Hildebrand [2], in his paper on the Turán–Kubilius inequality. We can give a slightly modified form of his argument here.

It follows from (12), bearing in mind the expansion for τ given in our comments on PNT Chapter 4, that

$$\sum_{\substack{p,l \leq n \\ pl > n}} \frac{f(p)f(l)}{pl} \leq o(B(n)), \qquad n \to \infty.$$

Here p and l denote prime numbers. Replacing n by n^t and integrating over the interval $0 \leq t \leq 1$ gives

$$\int_0^1 \sum_{\substack{p,l \leq n^t \\ pl > n^t}} \frac{f(p)f(l)}{pl} \, dt \leq o(B(n)), \qquad n \to \infty.$$

If we set $\alpha_p = (\log p)/\log n$ and change the order of integration and summation in the expression here bounded above, we obtain

$$\sum_{p,l \leq n} \frac{f(p)f(l)}{pl} \int_{\max(\alpha_p, \alpha_l)}^{\alpha_p + \alpha_l} 1 \, dt = \sum_{p,l \leq n} \frac{f(p)f(l)}{pl} \int_0^{\min(\alpha_p, \alpha_l)} 1 \, dt$$

$$= \int_0^1 \sum_{n^t < p, l \leq n} \frac{f(p)f(l)}{pl} \, dt = \int_0^1 h(t)^2 \, dt$$

with

$$h(t) = \sum_{n^t < p \leq n} \frac{f(p)}{p}.$$

Let y be a (fixed) number in the interval $0<y<1$. An application of the Cauchy–Schwarz inequality shows that

$$|h(t) - h(y)| \le B(n)\left(\sum_{n^y < p \le n^t} \frac{1}{p}\right)^{1/2}$$

$$= B(n)\left(\log\frac{t}{y} + O((\log n)^{-1})\right)^{1/2}$$

uniformly for $y \le t \le 1$. Hence for any z, $0 < z \le 1-y$,

$$z|h(y)|^2 = \int_y^{y+z} |h(y)|^2 \, dt$$

$$\le \int_y^{y+z} |h(t)|^2 \, dt + O\left(B(n)^2\left\{z\log\frac{y+z}{y} + o(1)\right\}\right)$$

$$\ll B(n)^2\left(z\log\frac{y+z}{y} + o(1)\right),$$

and therefore

$$\limsup_{n\to\infty} |h(y)| B(n)^{-1} \ll \left(\log\frac{y+z}{y}\right)^{1/2}.$$

We can now let z approach zero, and deduce that

(13) $$B(n)^{-1} \sum_{n^y < p < n} p^{-1} f(p) \to 0, \qquad n \to \infty,$$

for each fixed y, $0 < y < 1$. This brings us to essentially the end of his argument.

Let $F(z)$ be a distribution function with mean zero and variance 1, and suppose now that

(14) $$\nu_n(n; f(m) - A(n) \le zB(n)) \Rightarrow F(z), \qquad n \to \infty,$$

with the mean and variance of the frequency converging to those of the limit law. Then (12) holds, and so the conclusion (13).

Suppose further, that $f(p) \ge 0$ for all p, an undesirable assumption. Then for every positive ε

$$\sum_{\substack{n^y < p \le n \\ |f(p)| > \varepsilon B(n)}} \frac{1}{p} \to 0, \qquad n \to \infty,$$

and we may apply PNT theorem (12.5) to show that

$$P\left(\sum_{p\le n} X_p - A(n) \le zB(n)\right) \Rightarrow F(z),$$

where the independent random variables X_p are distributed according to

$$X_p = \begin{cases} f(p) & \text{with probability } 1/p, \\ 0 & \text{with probability } 1 - 1/p. \end{cases}$$

Note that the function f need not be of Kubilius class H.

Since the variance of the sum of the $X_p B(n)^{-1}$, $2 \le p \le n$, is bounded uniformly in n, arguments from the theory of probability enable necessary and sufficient conditions to be given in order for the weak convergence (14) to occur. In particular, if $F(z)$ is the normal law, then the condition is that for each fixed $\varepsilon > 0$,

$$\sum_{\substack{p \le n \\ |f(p)| > \varepsilon B(n)}} \frac{f(p)^2}{pB(n)^2} \to 0, \qquad n \to \infty.$$

In other words, for *non-negative* strongly additive functions f, Shapiro's conjecture [1] is valid. His full conjecture, that this condition is also necessary when f need not be non-negative, remains open.

PNT Chapters 13 and 14. The necessary and sufficient condition for a real additive function $f_x(n)$, which may depend upon x, to have a renormalization $\alpha(x)$ so that as $x \to \infty$, the frequency

$$\nu_x(n; f_x(n) - \alpha(x) \le z)$$

approaches the improper law with jump at the origin, has been obtained by Ruzsa [5]. It is that

$$\min_\lambda \left(\lambda^2 + \sum_{q \le x} q^{-1} \min(1, |f(q) - \lambda \log q|^2) \right) \to 0$$

as $x \to \infty$, the sum being taken over the prime-powers q. Moreover, if the function here being minimized asymptotically approaches zero with the choice $\lambda = \lambda(x)$, then one may take

$$\alpha(x) = \lambda(x) \log x + \sum_{\substack{p \le x \\ |f(p) - \lambda(x) \log p| \le 1}} p^{-1}(f(p) - \lambda(x) \log p).$$

The essential ingredient in his proof of this result, and of his treatment of the Turán-Kubilius inequality that was noted in the section on PNT Chapter 4, is an estimate for the concentration function

$$Q(x) = \sup_a x^{-1} \sum_{\substack{n \leq x \\ a \leq f(n) < a+1}} 1.$$

For this he applies a quantitative version of the method of Halász [1] together with ideas from the theory of probability to obtain

$$Q(x) \ll (\min_\lambda U(x, \lambda))^{-1/2}$$

with

$$U(x, \lambda) = \lambda^2 + \sum_{p \leq x} p^{-1} \min(1, |f(p) - \lambda \log p|^2).$$

A detailed proof, together with a number of applications, can be found in Ruzsa [5].

In PNT Chapter 14 the case $f_x(n) = \beta(x)^{-1} f(n)$ is treated for an additive function $f(n)$ which does not depend upon x, and a positive function $\beta(x)$ which satisfies $\beta(x^y) \ll \beta(x)$ for each fixed positive y. See, also, Manstavičius [1].

PNT Chapter 17. For a further treatment of the possible limit laws for renormalized real additive arithmetic functions see Timofeev [1].

PNT Chapter 19. In his review of PNT, Heath-Brown [1] gave a result which makes precise and extends PNT theorem (19.9) in the case of cubic characters. For such a character defined to the prime modulus p, let

$$L(s, \chi) = \sum_{n=1}^{\infty} \chi(n) n^{-s}, \qquad s = \sigma + i\tau$$

be the corresponding L-series. Then his explicit bounds are

$$L(1 + i\tau, \chi) \ll \begin{cases} (\log p|\tau|)^{26/27}, & |\tau| \geq 1, \\ (\log p)^{16/17} |\tau|^{-1/17}, & 0 < |\tau| \leq 1, \end{cases}$$

where τ is real.

References

Aczél, J., Daróczy, Z.
  1. *On Measures of Information and their Characterizations.*
     Academic Press, New York, London, 1975.

Akilov, G. P., Kantorovich, L. V.
  1. *Functional Analysis in Normed Spaces.*
     Translated from the Russian, Pergamon, Macmillan, New York, 1964.

Besicovitch, A. S.
  1. On additive functions of a positive integer.
     Studies in Mathematical Analysis and Related Topics. Essays in honor of G. Polya.
     Stanford University Press, Stanford, Calif., 1962, pp. 38–41.

Bombieri, E.
  1. *Le grand crible dans la théorie analytique des nombres.*
     Astérisque **18** (1974).

Britten, J. L.
  1. Solution of the word problem for certain types of groups, I.
     Proc. Glasgow Math. Soc. **3** (1956–58), 45–56.

Burgess, D. A.
  1. On Dirichlet characters of polynomials.
     Proc. London Math. Soc. **13** (1963), 537–548.
  2. Dirichlet characters and polynomials.
     Proc. Int. Conf. in Number Theory, Moscow, 14-18 Sept. 1971. *Trud. Mat. Inst. Steklov*, CXXXII, Moscow, 1973.

Chowla, P., Chowla, S.
  1. On kth power residues.
     J. Number Theory **10** (1978), 351–353.

Chowla, S., Chowla, P.
  1. On kth power residues.
     J. Number Theory **10** (1978), 351–353.

Conway, J. H.
  1. Unpredictable Iterations.
     Proceedings of the Number Theory Conference, University of Colorado, Boulder, 1972, pp. 49-52.

Csazszár, A.
  1. Solution of problem 28.
     Mathematikai Lapok **3** (1952), 92.

Daboussi, H.
  1. Caractérisation des fonctions multiplicatives ppB^λ à spe tre non vide.
     Ann. Inst. Fourier **30** (1980), 141–166.
  2. Sur les fonctions multiplicatives ayant une valeur moyenne non nulle.
     Bull. Soc. Math. France **109** (1981), 183–205.

Daboussi, H., Delange, H.
1. On a theorem of P. D. T. A. Elliott on multiplicative functions.
   J. London Math. Soc. (2) **14** (1976), 345–356.

Daróczy, Z., Aczél, J.
1. *On Measures of Information and their Characterizations.*
   Academic Press, New York, London, 1975.

Davenport, H.
1. *Multiplicative Number Theory*, second edition.
   Graduate Texts in Mathematics, Vol. 74. Springer-Verlag, New York, Berlin, Heidelberg, 1980.

Davenport, H., Halberstam, H.
1. The values of a trigonometrical polynomial at well-spaced points.
   Mathematika **13** (1966), 91–96. See also, Corrigendum and addendum, *Mathematika* **14** (1967), 229–232.

Delange, H.
1. Sur les fonctions arithmétiques fortement additives.
   C.R. Acad. Sci. Paris, Sér. A **244** (1957), 2122–2124.

Delange, H., Daboussi, H.
1. On a theorem of P. D. T. A. Elliott on multiplicative functions.
   J. London Math. Soc. (2) **14** (1976), 345–356.

Dress, F., Volkmann, B.
1. Ensembles d'unicité pour les fonctions arithmétiques additives ou multiplicatives.
   C.R. Acad. Sci. Paris, Sér. A **287** (1978), 43–46.

Elliott, P. D. T. A.
1. On sequences of integers.
   Quart. J. Math. (Oxford) (2) **16** (1965), 35–45.
2. On inequalities of large sieve type.
   Acta Arith. **18** (1971), 405–422.
3. On the distribution of the values of Dirichlet L-series in the half-plane $\sigma > \frac{1}{2}$.
   Proc. Koninkl. Nederl. Akad. van Wetenschappen (Series A), **74**, No. 3 = *Indag. Math* **33** (1971), 222–234.
4. A conjecture of Kátai.
   Acta Arith. **26** (1974), 11–20.
5. The law of large numbers for additive arithmetic functions.
   Math. Proc. Camb. Philos. Soc. **78** (1975), 33–71.
6. A mean value theorem for multiplicative functions.
   Proc. London Math. Soc. (3) **31** (1975), 418–438.
7. On two conjectures of Kátai.
   Acta Arith. **30** (1976), 35–39.
8. General asymptotic distributions for additive arithmetic functions.
   Math. Proc. Camb. Philos. Soc. **79** (1976), 43–54.
9. On the difference of additive functions.
   Mathematika **24** (1977), 153–165.
10. Sums and differences of additive arithmetic functions in mean square.
    J. reine und angewandte Mathematik **309** (1979), 21–54.
11. *Probabilistic Number Theory*, two volumes.
    Grundlehren, Vols. 239, 240. Springer-Verlag, New York, Berlin, Heidelberg, 1979/80.
12. On sums of an additive arithmetic function with shifted arguments.
    J. London Math. Soc. (2) **22** (1980), 25–38.
13. A remark on the Dirichlet values of a completely reducible polynomial (mod p).
    J. Number Theory **13** (1981) (1), 12–17.
14. Representing integers as products of integers of a specified type.
    Séminaire de Théorie des Nombres, Université de Bordeaux I, 1981–82, exposé no. **38**, 11 juin 1982.
15. The exponential function characterized by an approximate functional equation.
    Illinois J. Math. **26** (1982), 503–518.

16. On representing integers as products of integers of a prescribed type.
    J. Australian Math. Soc. (Series A) **35** (1983), 143–161.
17. Cauchy's functional equation in the mean.
    Adv. Math. **51** (3) (1984), 253–257.
18. Subsequences of primes in arithmetic progressions with large moduli.
    Turán Memorial Volume of the J. Bolyai Math. Soc., Hungary.
19. High power analogues of the Turán–Kubilius inequality and an application to number theory.
    Canadian J. Math. **32** (1980), 893–907.
20. Mean value theorems for multiplicative functions bounded in mean α-power, $\alpha > 1$.
    J. Australian Math. Soc. (Series A), **29** (1980), 177–205.
21. Multiplicative functions and Ramanujan's τ-function.
    J. Australian Math. Soc. (Series A) **30** (1981), 461–468.
22. The simultaneous representation of integers by products of certain rational functions.
    J. Australian Math. Soc. (Series A) **35** (1983), 404–420.
23. On the distribution of the roots of certain congruences and a problem for additive functions.
    J. Number Theory **16** (1983), 267–282.
24. On additive arithmetic functions $f(n)$ for which $f(an+b) - f(cn+d)$ is bounded.
    J. Number Theory **16** (1983), 285–310.
25. On the position of a theorem of Kubilius in probabilistic number theory.
    Math. Ann. **209** (1974), 201–209.
26. On representing integers as products of the $p+1$.
    Monatshefte für Math. **97** (2) (1984), 85–97.

Elliott, P. D. T. A., Moreno, C. J., Shahidi, F.
1. On the absolute value of Ramanujan's τ-function.
   Math. Ann. **266** (1984), 507–511.

Elliott, P. D. T. A., Ryavec, C.
1. The distribution of the values of additive arithmetical functions.
   Acta math. **126** (1971), 143–164.

Erdös, P.
1. On the density of some sequences of numbers, III.
   J. London Math. Soc. **13** (1938), 119–127.
2. On the distribution function of additive functions.
   Ann. Math. **47** (1946), 1–20.

Erdös, P., Ryavec, C.
1. A characterization of finitely monotonic additive functions.
   J. London Math. Soc. (2) **5** (1972), 362–367.

Erdös, P., Wintner, A.
1. Additive arithmetical functions and statistical independence.
   Amer. J. Math. **61** (1939), 713–721.

Faddeev, D. K.
1. On the concept of entropy of a finite probability scheme.
   Uspekhi Mat. Nauk. (N.S.), **11** (1956), No. 1 (67), 227–231.

Gallagher, P. X.
1. The large sieve.
   Mathematika **14** (1967), 14–20.

Gnedenko, B. V., Kolmogorov, A. N.
1. *Limit Distributions for Sums of Independent Random Variables.*
   Translated from the Russian and annotated by K. L. Chung. Addison-Wesley, Reading, Mass., London, 1968.

Halász, G.
1. Über die Mittelwerte multiplikativer zahlentheoretischer Funktionen.
   Acta Math. Acad. Sci. Hung. **19** (1968), 365–403.
2. On the distribution additive arithmetic functions.
   Acta Arith. **27** (1975), 143–152.

Halberstam, H., Davenport, H.
1. The values of a trigonometrical polynomial at well-spaced points.
   Mathematika **13** (1966), 91–96. See also, Corrigendum and addendum. Mathematika **14** (1967), 229–232.

Halberstam, H., Richert, H.-E.
1. Sieve Methods.
   Academic Press, London, New York, 1974.

Halberstam, H., Roth, K. F.
1. Sequences, I.
   Clarendon Press, Oxford, 1966.

Hardy, G.H., Wright, E.M.
1. An Introduction to the Theory of Numbers, fourth edition.
   Clarendon Press, Oxford, 1960.

Heath-Brown, D. R.
1. Review of Probabilistic Number Theory.
   Bull. London Math. Soc. **13** (1981), 367–372.

Hildebrand, A.
1. Sur les moments d'une fonction additive.
   Ann. Inst. Fourier (à paraître) Zbl. **486** (1983), 10043.
2. An asymptotic formula for the variance of an additive function.
   Math. Zeit. **183** (1983) (2), 145–170.

Hildebrand, A., Spilker, J.
1. Charakterisierung der additiven, fast-geraden Funktionen.
   Manuscripta Math. **32** (1980), 213–230.

Kac, M.
1. Statistical Independence in Probability, Analysis and Number Theory.
   Amer. Math. Soc. Carus Monographs, Vol. 12. Wiley, New York, 1959.

Kantorovich, L. V., Akilov, G. P.
1. Functional Analysis in Normed Spaces.
   Translated from the Russian. Pergamon, Macmillan, New York, 1964.

Kátai, I.
1. On sets characterizing number-theoretical functions.
   Acta Arith. **13** (1968), 315–320.
2. On sets characterizing number-theoretical functions, II. (The set of "prime plus one"'s is a set of quasi-uniqueness.)
   Acta Arith. **16** (1968), 1–4.
3. Some results and problems in the theory of additive functions.
   Acta Sc. Math. Szeged. **30** (1969), 305–311.
4. On number-theoretical functions.
   Colloquia Mathematica Societas Janos Bolyai, Vol. 2. North-Holland, Amsterdam 1970, 133–136.
5. On a problem of P. Erdös.
   J. Number Theory **2** (1970), 1–6.

Kolmogorov, A. N., Gnedenko, B. V.
1. Limit Distributions for Sums of Independent Random Variables.
   Translated from the Russian and annotated by K. L. Chung. Addison-Wesley, Reading, Mass., London, 1968.

Kubilius, J. P.
1. Probability methods in number theory.
   Vestnik Leningrad Univ. **10** (1955), no. 11, 59–60.
2. Probabilistic methods in the theory of numbers.
   Uspekhi Mat. Nauk (N.S.) **11** (1956), 2 (68), 31–66 = *Amer. Math. Soc. Transl.* **19** (1962), 47–85.
3. *Probabilistic Methods in the Theory of Numbers.*
   American Math. Soc. Translations of Math. Monographs, Vol. 11, Providence, Rhode Island, 1964.
4. On the estimation of the second central moment for strongly additive arithmetic functions.
   Liet. Mat. Rinkinys = *Litovsk Mat. Sbornik* **23** (1983), (1) 122–123.
5. On the estimate of the second central moment for arbitrary additive arithmetic functions.
   Liet. Mat. Rinkinys = *Litovsk Mat. Sbornik* **23** (1983) (2), 110–117.

Lambek, J., Moser, L.
1. On monotone multiplicative functions.
   Proc. Amer. Math. Soc. **4** (1953), 544–545.

Lang, S.
1. *Algebra.*
   Addison-Wesley, Reading, Mass., 1965.

LeVeque, W. J.
1. On the size of certain number-theoretic functions.
   Trans. Amer. Math. Soc. **66** (1949), 440–463.

Linnik, Yu. V.
1. *The Dispersion Method in Binary Additive Problems.*
   Amer. Math. Soc. Translations of Math. Monographs, Vol. 4, Providence, Rhode Island, 1963.

Manstavičius, E.
1. The degenerate distribution for additive arithmetic functions.
   Liet. Mat. Rinkinys = *Litovsk Mat. Sbornik* **16** (1976), 194–195.

Mauclaire, J.-L.
1. Sur la régularité des fonctions additives.
   Séminaire Delange–Pisot–Poitou, Théorie des Nombres, Paris **15** (1973/74), no. 23.

Meyer, J.
1. Ensembles d'unicité pour les fonctions additives. Étude analogue dans le cas des fonctions multiplicatives.
   Journées de Théorie Analytique et Élémentaire des Nombres, Orsay, 2 et 3 Juin, 1980.
   Publications Mathématiques d'Orsay, 50–66.
2. Réprésentation multiplicative des entiers et fonctions multiplicatives.
   Séminaire de Théorie des Nombres, Université de Bordeaux 1, 1981–82, exposé no. 27, 9 Avril.
3. Représentation multiplicative des entiers à l'aide de l'ensemble $P+1$, II.
   Astérique **94** (1982), 133–142.

Miech, R. J.
1. On the equation $n = p + x^2$.
   Trans. Amer. Math. Soc. **130** (1968), 494–512.

Mirsky, L.
1. *An Introduction to Linear Algebra.*
   Clarendon Press, Oxford, 1955.

Misevičius, G.
1. Application of the method of moments in the probabilistic theory of numbers.
   Liet. Mat. Rinkinys = *Litovsk Mat. Sbornik* **5** (1965), 275–289.

Montgomery, H. L.
1. *Topics in Multiplicative Number Theory.*
   Lecture Notes in Mathematics, Vol. 227, Springer-Verlag, New York, Berlin, Heidelberg, 1971, Corollary 3.2, p. 25.
2. Maximal variants of the large sieve.
   J. Fac. Sci. Univ. Tokyo (1981), Sect. I A **28**, 805–812.

Montgomery, H. L., Vaughan, R. C.
1. On the large sieve.
   Mathematika **20** (1973), 119–134.
2. Hilbert's inequality.
   J. London Math. Soc. (2) **8** (1974), 73–82.

Mordell, L. J.
1. On Mr. Ramanujan's empirical expansions of modular functions.
   Math. Proc. Camb. Philos. Soc. **19** (1971), 117–124.

Moreno, C. J., Shahidi, F., Elliott, P. D. T. A.
1. On the absolute value of Ramanujan's τ-function.
   Math. Ann. **266** (1984), 507–511.

Prachar, K.
1. *Primzahlverteilung.*
   Grundlehren, Vol. 91, Springer-Verlag, Berlin, 1957.

Raikov, D. A.
1. On the addition of point-sets in the sense of Schnirelmann.
   Mat. Sbornik (N.S.) **5** (1939), 425–440.

Rainville, Earl D.
1. *Special Functions.*
   Macmillan, New York, 1960.

Ram Murty, M.
1. Oscillations of Fourier coefficients of modular forms.
   Math. Ann. **262** (1983), 431–446.

Rankin, R. A.
1. Contributions to the theory of Ramanujan's function $\tau(n)$ and similar arithmetical functions.
   Math. Proc. Camb. Philos. Soc. **35** (1934) 357–372.
2. Sums of powers of cusp form coefficients.
   Math. Ann. **263** (1983), 227–236.

Rényi, A.
1. On a theorem of P. Erdös and its application in information theory.
   Mathematica (Cluj), **1** (1959), 341–344.

Richert, H.-E., Halberstam, H.
1. *Sieve Methods.*
   Academic Press, London, New York, 1974.

Roth, K. F., Halberstam, H.
1. *Sequences I.*
   Clarendon, Oxford 1966.

Ruzsa, I. Z.
1. General multiplicative functions.
   Acta Arith. **32** (1977), 313–347.
2. Additive functions with bounded difference.
   Periodica Math. Hung. **10** (1979), 67–70.
3. On the variance of additive functions.
   To appear in the *Turán Memorial Volume* of the J. Bolyai Math. Soc., Hungary. 16 pp.

4. Review of *Probabilistic Number Theory*.
   Zentralblatt **431** (1981) 10029, 10030.
5. The law of large numbers for additive functions.
   Studia Sci. Math. Hungar. **14** (1979), nos. 1–3, 247–253 (1982).
6. Generalized moments of additive functions.
   J. Number Theory, **18** (1984), 27–33.
7. Effective Results in Probabilistic Number Theory.
   Preprint; *Budapest*, September 1982, 24 pp.

Ryavec, C., Elliott, P. D. T. A.
1. The distribution of the values of additive arithmetical functions.
   Acta math. **126** (1971), 143–164.

Ryavec, C., Erdös, P.
1. A characterization of finitely monotonic additive functions.
   J. London Math. Soc. (2) **5** (1972), 362–367.

Salié, H.
1. Über die Kloostermanschen Summen $S(u, v; q)$.
   Math. Zeit. **34** (1931), 91–109.

Samuel, P., Zariski, O.
1. *Commutative Algebra*.
   Van Nostrand, Princeton, New Jersey, 1958.

Schmidt, W. M.
1. *Equations Over Finite Fields. An Elementary Approach*.
   Lecture Notes in Mathematics, Vol. 536, Springer-Verlag, New York, Berlin, Heidelberg, 1976.

Schönberg, I. J.
1. On two theorems of P. Erdös and A. Rényi.
   Illinois J. Math. **6** (1962), 53–58.

Schwarz, W.
1. *Fourier-Ramanujan Entwicklungen zahlentheoretischer Funktionen und Anwendungen*.
   Festschrift der wissenschaftlichen Gesellschaft an der Johann Wolfgang Goethe-Universität, Frankfurt am Main.

Schwarz, W., Spilker, J.
1. Remarks on Elliott's Theorem on Mean-Value of Multiplicative Functions.
   Recent Progress in Analytic Number Theory. Proceedings of the Durham Conference on Number Theory, 22, pp. 235–239. Academic Press, London, 1981.

Shahidi, F., Elliott, P. D. T. A., Moreno, C. J.
1. On the absolute value of Ramanujan's τ-function.
   Math. Ann. **266** (1984), 507–511.

Shapiro, H. N.
1. Distribution functions of additive arithmetic functions.
   Proc. Nat. Acad. Sci. U.S.A. **42** (1956), 426–430.

Sós, V. T.
1. Solution of problem 28.
   Matematikai Lapok. **3** (1952), 91.

Spilker, J., Hildebrand A.
1. Charakterisierung der additiven, fast-geraden Funktionen.
   Manuscripta Math. **32** (1980), 213–230.

Spilker, J., Schwarz, W.
1. Remarks on Elliott's Theorem on Mean-Value of Multiplicative Functions.
   Recent Progress in Analytic Number Theory. Proceedings of the Durham Conference on Number Theory, 22, pp. 235–239. Academic Press, London 1981.

Stephens, N. M.
 1. On a conjecture of Chowla and Chowla.
    J. Number Theory **8** (1977), 276–277.

Stilwell, J.
 1. The word problem and the isomorphism problem for groups.
    Bull. Amer. Math. Soc. **6** (1982) (1), 33–56.

Suck, R.
 1. Zur Charakterisierung der zahlentheoretischer Funktionen log n.
    Dissertation, Universität Ulm, 1976.

Szüsz, P.
 1. Remark to a theorem of P. Erdös.
    Acta Arith. **26** (1974), 97–100.

Timofeev, N. M.
 1. Convergence to Discontinuous Distributions.
    Liet. Mat. Rinkinys = Litovsk Mat. Sbornik **22** (1982) (1), 159–171.

Titchmarsh, E. C.
 1. *The Theory of the Riemann Zeta-Function.*
    Clarendon Press, Oxford, 1951.

Turán, P.
 1. On a theorem of Hardy and Ramanujan.
    J. London Math. Soc. **9** (1934), 274–276.

Vaughan, R. C.
 1. Sommes trigonométriques sur les nombres premiers.
    C.R. Acad. Sci. Paris, Sér. A **258** (1977), 981–983.
 2. *The Hardy–Littlewood Method.*
    Cambridge University Press, London, 1981.

Vaughan, R. C., Montgomery, H. L.
 1. On the large sieve.
    Mathematika **20** (1973), 119–134.
 2. Hilbert's inequality.
    J. London Math. Soc. (2) **8** (1974), 73–82.

Vinogradov, I. M.
 1. *The Method of Trigonometrical Sums in the Theory of Numbers.*
    Translated from the Russian, revised and annotated, by Anne Davenport and K. F. Roth., Chap. IX. Interscience, London, New York.
 2. *An Introduction to the Theory of Numbers.*
    Pergamon, London and New York, 1955.

Volkmann, B., Dress, F.
 1. Ensembles d'unicité pour les fonctions arithmétiques additives ou multiplicatives.
    C.R. Acad. Sci. Paris, Sér A **287** (1978), 43–46.

van der Waerden, B. L.
 1. *Modern Algebra.*
    1931, translated from the second revised German edition by Fred Blum, with revisions and additions by the author. Ungar, New York, 1949.

Weyl, H.
 1. Über die Gleichverteilung von Zahlen mod Eins.
    Math. Ann. **77** (1916), 313–352.

Wirsing, E.
 1. A characterization of log n as an additive arithmetic function.
    Symposia mathematica IV, 45–57, Academic Press, London and New York, 1970.

2. Characterization of the logarithm as an additive function.
     Amer. Math. Soc. Proc. of Symp. in Pure Math. **XX** (1971), 375–381.
  3. Additive and Completely Additive Functions with Restricted Growth.
     Recent Progress in Analytic Number Theory. Proc. of Symposium, Durham 1979, Volume 2, 231–280, Academic Press, London, 1981.
  4. Additive functions with restricted growth on the numbers of the form $p+1$.
     Acta Arith. **37** (1981), 345–357.

Wolke, D.
  1. Farey-Brüche mit primen Nenner und das große Sieb.
     Math. Zeit. **114** (1970), 145–158.
  2. Über die mittlere Verteilung der Werte zahlentheoretischer Funktionen auf Restklassen I.
     Math. Ann. **202** (1973), 1–25.
  3. Bemerkungen über Eindeutigkeitsmengen additiver Funktionen.
     Elem. der Math. **33** (1978), 14–16.

Wright, E. M., Hardy, G. H.
  1. *An Introduction to the Theory of Numbers*, fourth edition.
     Clarendon Press, Oxford, 1960.

Yosida, K.
  1. *Functional Analysis*,
     Grundlehren, third edition. Vol. 123. Springer-Verlag, New York, Berlin, Heidelberg, 1971.

Zariski, O., Samuel, P.
  1. *Commutative Algebra.*
     Van Nostrand, Princeton, New Jersey, 1958.

Subject Index

Additive function 1-3, 7, 9, 12-16, 19, 53-56, 82, 99, 100, 153, 180, 244, 247, 249, 259, 278, 282, 286, 292, 300, 356-357, 360, 372-373, 375, 381, 395-400, 404-405, 417-418, 420-421, 423, 430, 433, 436, 442, 447, 448
  completely 1, 9, 19, 173, 249, 270, 282, 313-314, 320, 325, 329, 331, 335, 349
  in residue classes 121, 241
  strongly 97, 413-414, 442
adjoint map 84, 94
algebraicanalytic inequalities 149
algebraic structure of coordinates 153-154
algorithms 3, 77, 287, 394, 416-417
analytic continuation 99
approximate functional equation 14, 100, 175, 177, 180, 183, 210, 241, 243, 249, 367, 435
arithmetic function 1, 3-6, 17, 129, 153, 177, 299, 393, 441
arithmic groups 392-393

Basis 85, 95, 431
Bombieri-Vinogradov theorem 150, 250, 257, 411
bounded order 277
Brun-Titchmarsh 133, 150

Cauchy-Davenport theorem 423
Cauchy's functional equation 177-180, 404
Central Limit Theorem 356
character sums, Dirichlet 36, 106, 116, 135, 410
characteristic function 11, 15, 356, 369-370
characters on Q^* 2, 7, 277
concentration function 448
continuity, rôle and analogue 17
convergence, weak 9-12, 15, 356-357, 369, 371, 415, 444-447
convolutions, Dirichlet 129
correlation 15, 16
correspondence, metamathematical 8
critical strip 99, 109

Density
  asymptotic 215, 372-373, 388-389, 391, 402, 442
  lower 375-376, 401, 409, 415
  multiplicative 405
  Schnirelmann 375
  upper 372, 381, 387, 401-403, 406, 414
Desargues' theorem 8
differential equation, approximate 178, 196
Dirichlet character 4, 7, 327, 334, 336-340, 398-399, 418
Dirichlet character sums 36, 106, 116, 135, 410
Dirichlet convolution 129
Dirichlet L-series 109, 119, 141, 448
Dirichlet series 7, 13, 95, 98, 243
Dirichlet's divisor function 1, 26-27
Dirichlet's theorem on primes 7
distribution function, improper 180, 447
divisible groups 278, 280, 304, 379, 383, 385
divisible modules 278
dual argument 8, 11
dual group 2, 7, 277-278, 384, 389
dual groups, measures on 380, 384
dual map 8, 93

dual space 84. 93, 95
duality
  in finite spaces 7, 11, 81, 131, 137, 318, 373, 437
  in general 7, 93
  in Hilbert space 92-93, 95, 102

Eigenfunction 425, 428, 433
eigenvalue 85, 95, 139, 424, 426-427, 429-431, 433-434
eigenvector 85, 95, 139, 425, 431, 433
elliptic functions 102
elliptic power sums 5, 318, 323, 328
entropy, Shannon's 16, 345-346, 352
Eratosthenes 10
Euler's function 1, 38, 59
exceptional primes 291-292, 372, 382, 384, 386

Fejér kernel 413
finitely distributed 244
fixed-divisor pairs 54, 205, 250, 261, 265
forms
  non-negative definite 83, 91
  positive definite 83
free subgroup of $Q^*/\Gamma(k)$ 291
freedom of Q^* 9, 286
frequency function 7, 9, 356
functional equations, approximate 14, 100, 175, 177, 180, 183, 210, 241, 243, 249, 367, 435
Fundamental Lemma of Kubilius 33

Geometry, projective 8
groups
  arithmic 392-393
  denumerable abelian 3, 393
  divisible 286, 379
  dual 2, 7, 277-278, 380-390, 393
  extra-divisible 286
  of bounded order 3, 277, 282-283, 306, 379, 383, 385, 393
  $G = Q^*/\Gamma$ 3, 20, 78, 277, 289, 291, 310, 325, 328, 331-332, 341-342, 381, 390-391, 398, 419

Haar measure 181, 380, 384, 393
Halász' method 16, 240-241, 243

hermitian form 83, 91, 102
hermitian map 83
hermitian matrix 84, 91, 430-431

Independence and divisibility 9
infinitely divisible distribution 357, 370
infinitesimal random variables 360
information 15, 276, 343-345
  as algebraic object 353
information equation 344, 355
inner product 83-85, 430, 433
isometry 436

Kakutani/Syracuse map 20, 262
Khinchine-Lévy representation 357
Kloosterman sum 13, 34, 40, 51
Kolmogorov representation 357
Kolmogorov's inequality 435
Kubilius class H 447

Large sieve 95, 101, 109, 153
  maximal 436
L-class laws 357
Legendre symbol 397
Lévy distance 10, 358-359, 361, 365
Lévy-Khinchine representation 357
limiting distribution 9, 15, 356, 415-416, 437
linear recurrence 5, 310, 320-321, 325-326, 336
L^α norm, $\alpha > 1$ 12, 435
L^2 norm 11-12, 14, 95, 152
local argument 14
logarithms of algebraic integers 328
loop, the 14, 155, 174-175, 183, 223
L^2 operator 12
  dual of 12
L-series 448
L^2 space 20, 95, 154, 180-181, 250, 425
L^τ space 177, 179, 441-443

Mann's theorem 423
mean of frequency 232, 237-239, 265
mean-values 374, 385, 387, 393, 417, 441-443
measures on dual groups 380
Mellin transform 100, 112, 125, 141

Subject Index   461

metamathematical notion 8
modules
  divisible 3, 278
  extra-divisible 3, 284
moments of additive functions 435-436
Motives 14, 19, 93, 97, 100, 177
multiplicative function 1, 11, 15-16, 23, 150-151, 285, 376, 417, 420, 441-443
  completely 2, 285, 393

Norm 8, 11-12, 14, 17, 93, 95, 152-153, 435-436, 442

Operator 8, 11-12, 17, 93
  dual of 8, 11-12, 93
  norm of 8, 11, 17, 93
  self-adjoint 429
orthonormal basis 85, 431

P-adic information function 355
persistence of form 412
PNT 17, 268, 269, 423, 433, 435, 437, 440-442, 444-445, 447-448
Pólya-Vinogradov inequality 36
power sums, elliptic 318-320, 323-324
pre-valuation 311
prime number sums 106-111, 116
prime number theorem 13, 99, 172, 184, 234, 236, 275, 374
prime pairs 2
primes in short arithmetic progressions 134
probabilistic number theory 8, 15-16, 178, 180, 278, 356, 358, 420, 423, 428
product representation by integers 2-3, 15-16, 277, 394-398, 401-403, 405-406, 409-411, 413, 417-420
  simultaneous 5-6, 286, 309, 329, 333, 391-392, 395, 401, 418-419
projective geometry 8
$\pi(x)$. See Notation

Q_2 309
Q^* the group of positive rationals 1-3, 9, 20, 304, 309, 354, 392
quadratic non-residues, least pair 414
quasiprimes 38, 67, 70

Ramanujan's τ-function 1, 443-444
rational function $R(x)$ 4-5, 297, 309, 400-401, 418
rational probability space 16, 343
recurrences, linear 310-313, 315, 320-321, 325-326, 336
relations on a group 3, 5, 416-417
ring of operators 297, 336

Selberg sieve 33, 41, 46-49, 52, 68-70, 251-258, 314
self-adjoint map 17, 83-85, 94, 424, 429, 433
self-adjoint matrix 85
sets of uniqueness 281
Shannon's entropy function 16, 345-346, 352
Shapiro's conjecture 447
Siegel-Walfisz theorem 13, 139
sieve of Eratosthenes 10
spectral theorem 85, 431
submultiplicative arithmetic functions 23
surrealistic continuity theorem 381

Tensor product 347-348
torsion subgroup of $Q^*/\Gamma(k)$ 291, 294, 296, 381
trace of map 88, 427-428, 431
trace of matrix 88, 92, 427-428
transcendence measure 328
Turán-Kubilius inequality (including analogues and variants) 9-10, 17, 27, 30, 70, 74, 83, 153, 213, 225, 240, 352-353, 414, 423, 433, 435, 437, 443, 445, 448
  dual of 11, 83, 241, 414
Turán's method 30, 33, 240

Variance of frequency 356-357, 444-447
Vinogradov's method in Vaughan's form 114

Weak convergence 9-12, 15, 356-357, 369, 371, 415, 444-447
well distribution (mod 1) 440
Weyl's criterion 7
Weyl's inequality 318
word problem for groups 416-417

Grundlehren der mathematischen Wissenschaften

Continued from page ii

225. Schütte: Proof Theory
226. Karoubi: K-Theory, An Introduction
227. Grauert/Remmert: Theorie der Steinschen Räume
228. Segal/Kunze: Integrals and Operators
229. Hasse: Number Theory
230. Klingenberg: Lectures on Closed Geodesics
231. Lang: Elliptic Curves: Diophantine Analysis
232. Gihman/Skorohod: The Theory of Stochastic Processes III
233. Stroock/Varadhan: Multi-dimensional Diffusion Processes
234. Aigner: Combinatorial Theory
235. Dynkin/Yushkevich: Markov Control Processes and Their Applications
236. Grauert/Remmert: Theory of Stein Spaces
237. Köthe: Topological Vector-Spaces II
238. Graham/McGehee: Essays in Commutative Harmonic Analysis
239. Elliott: Probabilistic Number Theory I
240. Elliott: Probabilistic Number Theory II
241. Rudin: Function Theory in the Unit Ball of C^n
242. Blackburn/Huppert: Finite Groups I
243. Blackburn/Huppert: Finite Groups II
244. Kubert/Lang: Modular Units
245. Cornfeld/Fomin/Sinai: Ergodic Theory
246. Naimark: Theory of Group Representations
247. Suzuki: Group Theory I
248. Suzuki: Group Theory II
249. Chung: Lectures from Markov Processes to Brownian Motion
250. Arnold: Geometrical Methods in the Theory of Ordinary Differential Equations
251. Chow/Hale: Methods of Bifurcation Theory
252. Aubin: Nonlinear Analysis on Manifolds, Monge—Ampère Equations
253. Dwork: Lectures on p-adic Differential Equations
254. Freitag: Siegelsche Modulfunktionen
255. Lang: Complex Multiplication
256. Hormander: The Analysis of Linear Partial Differential Operators I
257. Hormander: The Analysis of Linear Partial Differential Operators II
258. Smoller: Shock Waves and Reaction-Diffusion Equations
259. Duren: Univalent Functions
260. Freidlin/Wentzell: Random Perturbations of Dynamical Systems
261. Remmert/Bosch/Güntzer: Non Archemedian Analysis—A Systematic Approach to Rigid Analytic Geometry
262. Doob: Classical Potential Theory & Its Probabilistic Counterpart
263. Krasnoselškii/Zabreĭko: Geometrical Methods of Nonlinear Analysis
264. Aubin/Cellina: Differential Inclusions
265. Grauert/Remmert: Coherent Analytic Sheaves
266. de Rham: Differentiable Manifolds
267. Arbarello/Cornalba/Griffiths: Geometry of Algebraic Curves, Vol. I